Discrete Mathematical Structures

Discrete Mathematical Structures and Their Applications

HAROLD S. STONE, Stanford University

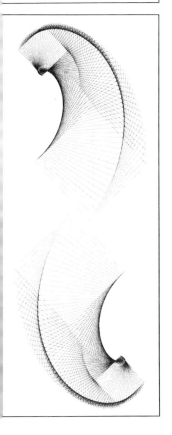

SRA

SCIENCE RESEARCH ASSOCIATES, INC.
Chicago, Palo Alto, Toronto, Henley-on-Thames, Sydney

A Subsidiary of IBM

The SRA Computer Science Series:

Gear: INTRODUCTION TO COMPUTER SCIENCE
Elson: CONCEPTS OF PROGRAMMING LANGUAGES
Stone: DISCRETE MATHEMATICAL STRUCTURES

Printed in the United States of America
Library of Congress Catalog Card Number 72–96797

Contents

Preface   *vii*

Introduction   *x*

One   Foundations of Discrete Mathematics   1
    1.1   Elementary Logic   2
    1.2   Sets   11
    1.3   Graphs   20
    1.4   Relations   23
    1.5   Functions   30
    1.6   Composition of Functions   35
    1.7   Binary Operations   39
    1.8   Algebraic Structures and Structure-
    Preserving Maps   45

Two   Groups   53
    2.1   Group Axioms   55
    2.2   Generators and Group Graphs   64
    2.3   Permutation Groups   74
    2.4   Subgroups, Cosets, and Group
    Homomorphisms   82
    2.5   Special Families of Groups   95
    2.6   Symmetry Groups   105

Three   The Pólya Theory of Enumeration   113
    3.1   Counting Equivalence Classes à la
    Burnside   115
    3.2   Inventories of Functions   121
    3.3   Cycle Index Polynomials   126
    3.4   The Redfield-Pólya Theorem   130
    3.5   Historical Background and Applications   134

Four | Applications of Group Theory to Computer Design | 137
 | 4.1 How Fast Can We Add? | 138
 | 4.2 The Residue Number System | 147
 | 4.3 Permutation Interconnections for Dynamic Memories | 151

Five | Group Codes | 163
 | 5.1 An Error Model for Computer Systems | 164
 | 5.2 Parity-Check Codes for Independent Errors | 166
 | 5.3 Arithmetic Codes | 179

Six | Semigroups | 185
 | 6.1 Semigroups and Monoids | 187
 | 6.2 Subsemigroups and Submonoids | 196

Seven | Finite-State Machines | 203
 | 7.1 Semigroups and Finite-State Machines | 205
 | 7.2 State Reduction and State Equivalence | 214
 | 7.3 Machine Homorphisms and Machine Simulation | 223
 | 7.4 Sequential Machines as Sequence Recognizers | 238

Eight | Rings and Fields | 259
 | 8.1 Algebraic Structures with Two Operations | 261
 | 8.2 Finite Fields | 267
 | 8.3 The Structure of Finite Fields | 276
 | 8.4 The Representation of Finite Fields | 290

Nine | Linear Finite-State Machines | 295
 | 9.1 Linear Machines | 296
 | 9.2 Autonomous Linear Machines | 302
 | 9.3 The Cycle Structure of Linear Feedback Shift-Registers | 313
 | 9.4 A Reprise: Pólya Theory | 322
 | 9.5 Primitive Shift-Registers and Maximum-Length Sequences | 335
 | 9.6 Shift-Register Decoders and Encoders | 350

Ten | Boolean Algebra with Applications to Computer Design | 363
 | 10.1 Lattices | 364
 | 10.2 Boolean Algebras and Switching Functions | 375

Bibliography | 388

Index | 393

Index to Notation | 400

Preface (to the Instructor)

This text is intended for use in a first course in discrete mathematics for computer-science and computer-engineering students. It is written for undergraduates who have had some experience programming computers, preferably both in high-level compiler languages and in assembly language. The programming experience is necessary, not so much to grasp the material presented here, but to provide the requisite context for discussions of computer arithmetic, sequential machines, compilers, and similar topics.

The text includes both a condensed survey of algebraic structures and a discussion of their applications to computers. The mathematical material in itself constitutes a significant portion of a modern algebra text, with some notable omissions of topics that have little relevance to computers. Interspersed with the mathematical treatments are self-contained sections on applications to computers. Although these applications tend to be oriented largely toward computer design, they are of interest to all computer specialists, whether programmers or designers.

The scope of the text is close in spirit to the scope of the discrete structures course (Course B3) of the 1968 ACM curriculum. This text differs from that course primarily in that it contains much less graph theory and substantially more material on modern algebra. Moreover, the applications covered here are far more extensive than those suggested in the ACM recommendations. Course prerequisites for this text should include a first course in computer programming (Course B1 in the ACM curriculum, or its equivalent) and an introduction to the internal structure of computers (Course B2 in the ACM curriculum).

College juniors and seniors should have sufficient background to use this text effectively. At Stanford, the material has been used in a course given to first-year graduates and undergraduate seniors. When teaching this course in the late 1960s, the author found that most of the graduate students had not been exposed previously to even the most elementary aspects of discrete mathematics. Over the past few years, however, the situation has gradually changed to the point that a significant fraction of the graduate students have mastered most of the material in Chapter

One before taking this course and are well prepared for the material in the subsequent chapters. This change is due to better preparation in undergraduate studies and in high school. Indeed, the material in Chapter One is covered in its entirety by one of the mathematics texts in statewide high-school use in California today. Undoubtedly this trend to teach discrete mathematics at lower grade levels will continue. Therefore, at some point in time and at some schools, freshmen and sophomores may have adequate preparation to use this text successfully.

The material in the text is more than adequate for a one-semester course in discrete mathematics, and could easily be sufficient for a two-quarter course sequence. Some of the self-contained material can be eliminated without marring the development of the major topics, so that the text is suitable for use in one-quarter courses as well. Fig. P.1 shows a chart of prerequisites for the chapters. Material on the algebraic structures is presented in Chapters One, Two, Six, Eight, and Section 10.1. Note that the three advanced topics—semigroups in Chapter Six, rings and fields in Chapter Eight, and lattices in Section 10.1—are independent of one another. Thus the instructor can choose among these topics if he cannot treat them all.

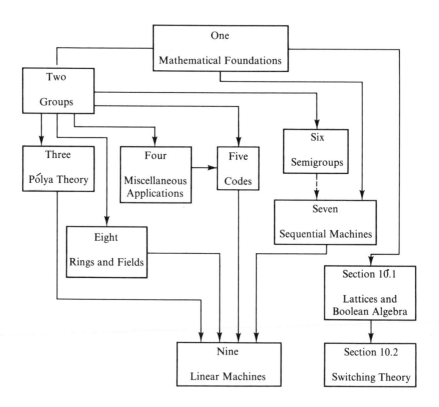

FIG. P.1   The Prerequisite Structure of This Text

Fig. P.1 shows that the applications chapters are dependent on the immediately-preceding mathematical chapters, with occasional dependencies on earlier applications chapters. Chapter Six (semigroups) is shown as a quasi-prerequisite for Chapter Seven (sequential machines) because the applications in Chapter Seven are related only marginally to the theory of semigroups, although other aspects of sequential machines do have close ties with the theory.

While teaching this course, the author experimented with a number of strategies. The students in the courses tended to be most receptive to applications and least receptive to the pure abstract mathematics. One good way to provide motivation for the mathematical material is to introduce it by posing a practical problem whose solution is not immediately evident, but whose solution is a direct consequence of the mathematical theory. Students are invited to solve the problem both before and after studying the mathematical material. A good example of this approach is the use of counting problems and error-correcting coding problems to demonstrate the usefulness of group theory.

The author recommends the use of take-home rather than in-class examinations. In-class examinations tend to test shallow knowledge rather than deep insight. On a take-home examination, the student has some time to organize his thoughts and to demonstrate his understanding of major principles. Many questions from past take-home exams appear among the exercises.

Harold S. Stone
Stanford University
January 1973

Introduction (to the Reader)

Is there a number n with the property that $1 \cdot n \neq n$? If the operation is multiplication in the usual sense, we know that there is no such number among the real or complex numbers. Thus we are tempted to give a definite negative reply to the question and lay the matter to rest. It is rather surprising to find that there does exist a number system with a reasonably satisfactory definition of multiplication in which $1 \cdot n$ is not always equal to n. It is even more surprising to find that this number system is not just a mathematical curiosity, but is in daily use by millions of people. The number system is the floating-point arithmetic system of a widely-used family of computers.

Although computer-arithmetic systems need not exhibit this particular anomalous behavior, they inherently must be different in some way from conventional arithmetic systems. A digital computer can deal with only a finite number of different numbers; therefore computer arithmetic at best is a close imitation of conventional arithmetic, but it cannot be an exact duplicate.

If the computer arithmetic is not identical to conventional arithmetic, then how can we do computation with computer arithmetic? We do arithmetic on computers just as we would do ordinary arithmetic with infinite precision. In most cases, computer arithmetic is indistinguishable from ordinary arithmetic to within the useful accuracy of the results. In some cases, we must take a few precautions to be sure that the discrepancies are not too large between the computed results and the true answers, but the majority of computer users never meet such cases.

The example of computer arithmetic illustrates several themes of this text. One of the major themes is the study of various mathematical systems and their applications to computer science. The bulk of the practical problems and applications in the text relate to computer arithmetic, computer design, and switching theory.

Secondly, the book is oriented toward discrete mathematics as opposed to continuous mathematics. Computers are inherently finite. Inherently discrete mathematical problems arise in their design, construction, and utilization.

Another major goal of the text is to show that many properties of mathematical systems can be derived from a few defining properties.

Every mathematical system that has a particular set of defining properties must have all of the consequent characteristics. Thus, when we explore mathematical systems in which equations of the form $a \cdot x = b$ are always uniquely solvable for x, we find that we can apply the results both to arithmetic systems and to sets of permutations, because both satisfy the solvability criteria.

One goal in approaching the subject from this direction is to show that particular problems may arise in many different disguises. For example, the alternating group of degree n always has half as many elements as the symmetric group of degree n, and a particular puzzle game can be placed in only half of the $n!$ different positions available. How is the puzzle related to the alternating group? If we can find a correspondence, we can prove that only half of the puzzle's states are attainable. As another example, we discover that a combinatorial derivation of properties of linear sequential machines can be applied to the vastly different problem of counting equivalence classes of switching functions. In these examples and in many others, a few crucial properties are common to two different problems, but these properties alone determine the observable features of the problems. Obviously, a solution to one problem can be used to solve an analogous problem, so that knowledge of discrete mathematics is widely applicable.

Another aspect of the text that is illustrated by the example of computer arithmetic is the notion of simulation. In this instance, we wish to perform ordinary arithmetic, but we are forced to simulate that arithmetic by computer arithmetic. The simulation is inexact but satisfactory for most purposes. Computers can simulate some mathematical systems exactly, however. These systems, of course, are inherently finite. In several chapters we explore the conditions under which one mathematical system can simulate another. The key notion here is that of homomorphism. In the context of computer design, we often wish to have the smallest possible catalog of spare parts. We can construct small catalogs if we can have some parts do double duty. If part A can simulate part B, then we need not stock both A and B. Part A alone will suffice. The practical application of this notion extends beyond hardware into computer software, because we can treat subroutines in program libraries as catalogued parts, and some subroutines can simulate others in particular environments.

Many people have contributed to the development of this text. E. J. McCluskey was the most instrumental and influential of these, for it is he who saw the need for the course and took steps to implement it. Edward Davidson contributed ideas along the way and shared the teaching of the course. Bernard Elspas and William Kautz helped immeasurably with their well-known and beautifully written contributions to several topics covered here. Martha Evans Sloan made extensive contributions to the latter stages of preparation, and she has lent her expertise to place the text on a firm pedagogical footing. Others whose contributions appear in one form or another include John Wakerly, Tomas Lang, and Alan Usas. Gillian Pickles proved that people can type manuscripts perfectly, unlike the computers I know. Stephen Mitchell, through his encouragement and critiques helped bring the manuscript into its final form. Finally, George Forsythe deserves much credit for his leadership as an educator and computer scientist, as well as for his specific comments and encouragement. We each lost some of ourselves in his passing.

Harold S. Stone
Stanford University
January 1973

To the memory of George Forsythe

The use of things is all, and not the store.

—Jonson

If we have a correct theory but merely prate about it,
pigeonhole it and do not put it into practice,
then that theory, however good, is of no significance.

—Mao

ONE Foundations of Discrete Mathematics

ONE | Foundations of Discrete Mathematics

If we could observe the voltages on the electronic circuits at various places within a digital computer, we would quickly discover that each voltage has one of two values. That is, each signal line to each circuit is either at a particular low voltage value or at a particular high value; no other values are found. Although the voltages on the signal lines may change rapidly—as often as 10^9 times per second—they always alternate between the two fixed values. Of course, the physical voltages cannot change from one value to another instantaneously, but the changes are so rapid that we can properly model them as instantaneous.

In a historical perspective, the discrete character of computer signals is somewhat unusual. Prior to the development of digital computers, the vast majority of electronic circuits had continuously variable signal voltages. Continuous mathematics is an appropriate tool for the design and study of such circuits. We shall find, however, that discrete mathematics is the appropriate mathematical tool for computer design and applications.

This chapter introduces several elementary notions of discrete mathematics. The first two sections briefly examine mathematical logic and sets, establishing the notation and mathematical foundations for the remainder of the chapter. Graphs, introduced in the third section, provide a means of illustrating discrete mathematical concepts and are excellent aids to understanding and intuition. Following sections treat relations, functions, binary operations, and homomorphisms from a set-theoretical point of view. Each of these topics appears throughout the text, both in theoretical and applied material.

1.1 Elementary Logic

Logic focuses on the study of statements that are either true or false. "It is raining now" and "Ten is a prime number" are typical examples of such statements. The formal study of logic, including the development of many of the notions set forth in this text, can be traced back to the works of Aristotle. In addition to its philosophical role, logic has proved central to the development of discrete mathematics, set theory, and other mathematical disciplines. Only in relatively modern times has logic also acquired importance in computer design. C. E. Shannon (1938) in America and V. I. Shestakoff (1941) in Russia independently saw the analogy between logic and those electronic circuits whose signals can assume only one of two possible values. The values of "true" and "false" for statements correspond to the high and low voltages in the electronic circuits of computers. Shannon and Shestakoff essentially showed how to design two-valued circuits through the use of the logical ideas described in this section. Thanks primarily to their work, the early pioneers in computer development were able (in the late 1940s) to make use of systematic design procedures in the construction of early prototype computers. This section describes the elementary aspects of logic that apply to our study of discrete mathematics.

We begin the study by noting that some English sentences are statements of fact. Some statements are true: "A square has four sides"; "George Washington was a president of the United States." Other purported statements of fact are clearly incorrect: "A square has five sides"; "Aristotle was a president of the United States."

In each case, the statement must be either true or false and, from a knowledge of geometry or history, it is possible to determine the truth or falsity of each statement. Other statements—such as "Francis Bacon wrote all of the plays commonly attributed to Shakespeare"—are similar in that they must be true or false, but we cannot tell which is the case. Most people believe that the statement about Bacon is false, but a few people believe that it is true. The evidence presently at hand is insufficient to establish with certainty the correctness or incorrectness of either belief. Statements about what will or will not happen in the future inherently possess a similar uncertainty. For example, the statement "It will rain here tomorrow" must be either true or false, but its truth or falsity cannot be established with certainty today.

The class of statements we have described plays a major role in discrete mathematics. The principal attribute of these statements is that they must have only one of two possible discrete values; they must be either true or false. For convenience, the word *statement* is used (by definition) only to describe a sentence that must be either true or false. By assumption, a statement cannot be both true and false—although there are cases where it is impossible to determine which value should be assigned to a given statement.

An important characteristic of statements is that we can construct compound statements from two or more individual ones. We normally do so by using the logical connectives "and," "or," "not," "implies," and "if and only if." For example, we can join the two statements "All cows eat grass" and "Columbus discovered America" to form the compound statement "All cows eat grass and Columbus discovered America." Because we used "and" to connect the statements, the compound statement is true if both of its constituent statements are true; otherwise it is false.

A notation similar to that of algebra is useful in the study of compound statements. If we let p denote the statement "All cows eat grass" and let q denote the statement "Columbus discovered America," then we can write the compound statement as "p and q." Logicians frequently use the symbol $\wedge$ to denote "and," so that the compound statement usually is written as $p \wedge q$. Table 1.1.1 shows each of the logical connectives and tabulates the relation between the truth value of the compound statement and the values of its constituent statements.

The table reveals some interesting aspects of the connectives. Only "not" operates on a single statement; each of the other connectives operates on two statements. The meanings of "and" and "or" are obvious and require no further explanation. The logical meaning of "implies" is a bit trickier; logicians use "p implies q" to mean that q is true whenever p is true. This definition does not correspond completely to the common usage of the word *implies*. Note that $p \Rightarrow q$ is true when p is false, regardless of the truth value of q. Conversely, $p \Rightarrow q$ is true if q is true, regardless of the truth value of p. The statement $p \Rightarrow q$ is false only when p is true and q is false. The last connective in the table, "if and only if," is defined such that $p \Leftrightarrow q$ is true when its constituent statements have identical truth values.

To construct examples of compound statements, let p denote the statement "Henry VIII had six wives," and let q denote the statement "21 is a prime number." In this case, p is true and q is false. The statement $\neg p$ is equivalent to the statement "Not Henry VIII had six wives," or (in a more proper grammatical form) "Henry VIII did not have six wives." Because p is true, $\neg p$ is false. The statement $p \vee q$ ("Henry VIII had six wives or 21 is a prime number") is true, because at least one of its constituent statements is true. The statement $p \Rightarrow q$ ("Henry VIII had six wives implies 21 is a prime number") is false. However, note that $q \Rightarrow p$ ("21 is a prime number implies Henry VIII had six wives") is true, even though 21 is not a prime number. In fact, because Henry VIII did have six wives, we can substitute any statement whatsoever for q and obtain the true statement $q \Rightarrow p$. Similarly, the statement $p \Leftrightarrow \neg q$ ("Henry VIII had six wives if and only if 21 is not a prime number") is true, although it seems strange to suggest that the divisibility of 21 has anything to do with Henry's marital status. The connectives "implies" and "if and only if" in logic do not indicate causal relations, but we generally do use them to connect statements that have some relation to each other. For example, the following state-

TABLE 1.1.1   Truth Values of Compound Statements*

p	q	$\neg p$ ("not p")	$p \wedge q$ ("p and q")	$p \vee q$ ("p or q")	$p \Rightarrow q$ ("p implies q")	$p \Leftrightarrow q$ ("p if and only if q")
F	F	T	F	F	T	T
F	T	T	F	T	T	F
T	F	F	F	T	F	F
T	T	F	T	T	T	T

*T and F denote "true" and "false," respectively.

ments are true and capture the intent of the use of these connectives: "21 is divisible by 3 implies 21 is not a prime number"; "21 is an even number if and only if 21 is divisible by 2."

We can construct very complex statements by using connectives to nest constituent statements to any level. For example, if p, q, r, and s are statements, then "p implies q" is a statement, as is "r if and only if s." We can connect the latter two statements by "or" to form the statement "p implies q or r if and only if s." To eliminate ambiguity, we normally parenthesize the terms. Thus we write the final statement in parenthesized logical notation as $(p \Rightarrow q) \lor (r \Leftrightarrow s)$. To compute the truth value of this statement, we substitute the individual truth values of p, q, r, and s, and then use Table 1.1.1 to evaluate the compound expressions. For example, suppose that p and r are false and that q and s are true. In this case, $p \Rightarrow q$ is true and $r \Leftrightarrow s$ is false. The complete statement is true, because one constituent of the "or" connective is true. Table 1.1.2 gives truth values of the complete statement for each of the sixteen possible combinations of truth values of the statements p, q, r, and s. Approached in this fashion, the manipulation and evaluation of compound statements—even very complex ones—is relatively straightforward.

Some sentences are not statements because they contain unspecified variables. For example, we cannot assign a truth value to the sentence "He was a president of the United States" until a proper name is substituted for the pronoun "he." We call a sentence that contains un-

TABLE 1.1.2   Truth Values for the Compound Statement $(p \Rightarrow q) \lor (r \Leftrightarrow s)$

p	q	r	s	$p \Rightarrow q$	$r \Leftrightarrow s$	$(p \Rightarrow q) \lor (r \Leftrightarrow s)$
F	F	F	F	T	T	T
F	F	F	T	T	F	T
F	F	T	F	T	F	T
F	F	T	T	T	T	T
F	T	F	F	T	T	T
F	T	F	T	T	F	T
F	T	T	F	T	F	T
F	T	T	T	T	T	T
T	F	F	F	F	T	T
T	F	F	T	F	F	F
T	F	T	F	F	F	F
T	F	T	T	F	T	T
T	T	F	F	T	T	T
T	T	F	T	T	F	T
T	T	T	F	T	F	T
T	T	T	T	T	T	T

specified variables a *predicate*. For example, "*S* is green" and "*X* discovered America" are predicates; *S* and *X* are unspecified variables that may be replaced by various nouns. When we replace *X* by "George Washington," the resulting statement is false. However, the predicate itself is neither true nor false; its truth value depends upon the name that replaces *X*. In these examples, we call "he," *S*, and *X free variables* in their respective predicates.

In Section 1.2 we make use of predicates to define sets, but predicates also have a more practical application. The circuits that produce two-valued signals in computer systems are analogous to predicates; the inputs to these circuits are analogous to free variables. When we impress signal voltages on all of the inputs to a circuit, the circuit produces one of the two possible output voltages. Some computer engineers carry the analogy with logic farther by calling the two signal values "true" and "false," although the names 1 and 0 are used more commonly.

To close this section, we discuss briefly and informally the use of logic in mathematical proofs in this text. Theorems to be proved generally have the form "If *p* is true, then *q* is true," or equivalently $p \Rightarrow q$. For example, a typical theorem is "If *S* is a square, then *S* is a rectangle." The theorem does not assert that *p* is true or false (in the example, that *S* is or is not a square). Rather, it asserts something about the relation between the truth values of *p* and *q*. We prove the theorem by demonstrating that *q* is true whenever *p* is true; we need not demonstrate anything about the truth value of *q* when *p* is false.

In this text, we make use of several different methods to prove theorems of the form $p \Rightarrow q$. One way involves finding some other statement *r* with the property that "*p* implies *r*" is true and "*r* implies *q*" is true. If *p* is true, the chain of implications shows that *q* is true; thus "*p* implies *q*" also is true. We can demonstrate the validity of this inference rigorously by showing that

$$((p \Rightarrow r) \wedge (r \Rightarrow q)) \Rightarrow (p \Rightarrow q)$$

is always a true statement. With a table similar to Table 1.1.2, we can demonstrate that this compound statement is true for all eight possible combinations of truth values for *p*, *q*, and *r*. We leave this demonstration as an exercise for the reader (Ex. 1.1.4).

We can use the method just described to prove the assertion "If *S* is a square, then *S* is a rectangle." The statement can be written as $p \Rightarrow q$, where *p* denotes "*S* is a square," and *q* denotes "*S* is a rectangle." To construct the proof, let *r* denote the following statement: "*S* is a polygon with four sides and four right angles." Both $p \Rightarrow r$ and $r \Rightarrow q$ are true, because these statements follow from the definitions of square and rectangle. Hence, $p \Rightarrow q$ is true, and the assertion is proved.

Another method of proof is particularly useful in many cases because it does not require an intermediate statement *r*. In this method, we prove the assertion $p \Rightarrow q$ by proving the assertion $(\neg q) \Rightarrow (\neg p)$. We can easily give a nonrigorous demonstration of the validity of this method. If p is false whenever q is false, then it is impossible for p to be true when q is false. Hence, q is true whenever p is true, or "p implies q." More formally, we can show that the statement

$$((\neg q) \Rightarrow (\neg p)) \Rightarrow (p \Rightarrow q)$$

is always true, regardless of the truth values of p and q (Ex. 1.1.5). In other words, to prove that $p \Rightarrow q$ is true, it is sufficient to prove that $(\neg q) \Rightarrow (\neg p)$. In this text, we make frequent use of such proofs by contradiction, or contrapositive proofs.

We can use this method also to prove the assertion "If S is a square, then S is a rectangle." In this case, a contrapositive proof involves a demonstration of the truth of the statement "If S is not a rectangle, then S is not a square." It is easy to show from geometrical definitions that if S fails to satisfy the definition of a rectangle, then it must also fail to satisfy the definition of a square. Such a demonstration implies the validity of the assertion to be proved.

Some theorems in this text deviate slightly from the form "If p, then q." Most of these other theorems take the form "p if and only if q." This assertion, however, is equivalent to the statement $(p \Rightarrow q) \wedge (q \Rightarrow p)$. (Two statements are said to be equivalent if they have the same truth tables.) We can demonstrate this equivalence informally by noting that if both $p \Rightarrow q$ and $q \Rightarrow p$ are true, then either both p and q are true or both are false; in other words, q is true only when p is true and p is true only when q is true. But the latter statement is equivalent to $p \Leftrightarrow q$. We can give a more formal demonstration by showing that the statement

$$((p \Rightarrow q) \wedge (q \Rightarrow p)) \Rightarrow (p \Leftrightarrow q)$$

is always true (Ex. 1.1.6).

We use a two-step process to prove theorems of the form "p if and only if q." First, we show that $p \Rightarrow q$ is true; then we show that $q \Rightarrow p$ is true. Sometimes we prove one of the implications through a contrapositive assertion. For example, a satisfactory two-step proof of $p \Leftrightarrow q$ is a proof of $p \Rightarrow q$ followed by a proof of $(\neg p) \Rightarrow (\neg q)$.

One type of theorem requires a more powerful technique of proof than those mentioned thus far. Theorems of this type generally assert that some property is true for all positive integral values of a variable n. In essence, such a theorem states that a predicate containing the free variable n is true when $n = 1, 2, \ldots$, for all positive integral values of n. Obviously, it is impractical to construct a separate proof for each possible value of n. A suitable method of proof is the method of *mathematical induction*:

Let $p(n)$ be a predicate that includes the positive integer n, and let $p(1)$ be the statement obtained by substituting the value 1 for n.

If we show that
   (i)   [*basis step*] $p(1)$ is true, and that
   (ii) [*induction step*] $p(n) \Rightarrow p(n + 1)$ for $n \geq 1$,
then $p(n)$ must be true for all positive integral values of n.

We cannot prove the principle of mathematical induction, but we can give an intuitive justification for its use. Suppose that conditions (*i*) and (*ii*) are satisfied by a predicate $p(n)$. Then $p(1)$ is true because of the basis step, and $p(2)$ is true from letting $n = 1$ in the induction step. Because $p(2)$ is true, the induction step guarantees that $p(3)$ is true, and so forth as far as we might wish to go through the positive integers. This demonstration is not rigorous because it does not include a formal proof that an infinite number of repetitions of the step would include all positive integral values of n.

As an example of mathematical induction, consider a proof for the well-known formula for the sum of an arithmetic series:

$$\sum_{i=1}^{n} i = \frac{n(n + 1)}{2}, \quad \text{for } n \geq 1.$$

An inductive proof of this formula involves **(i)** the basis step ($n = 1$),

$$\sum_{i=1}^{1} i = 1 = \frac{1(1 + 1)}{2},$$

and **(ii)** the induction step,

$$\text{assuming } \sum_{i=1}^{n} i = \frac{n(n + 1)}{2}, \quad \text{prove } \sum_{i=1}^{n+1} i = \frac{(n + 1)(n + 2)}{2}:$$

$$\sum_{i=1}^{n+1} i = n + 1 + \sum_{i=1}^{n} i$$

$$= n + 1 + \frac{n(n + 1)}{2} \quad \text{[by induction hypothesis]}$$

$$= \frac{n^2 + 3n + 2}{2} = \frac{(n + 1)(n + 2)}{2}. \qquad \square$$

This completes the inductive proof, as indicated by the end-of-proof symbol $\square$.

The reader interested in pursuing formal logic in more detail will find many elementary texts on the subject (see Suppes 1957). The relation between computer design and logic is explored in greater depth in following chapters of this text.

Exercises

1.1.1 Tabulate the truth values of the following compound statements in the form of Table 1.1.1:

    (a) $(\neg p) \vee q$;

    (b) $\neg(p \vee q)$;

    (c) $(\neg p) \wedge (\neg q)$;

    (d) $(p \Rightarrow q) \wedge (q \Rightarrow p)$;

    (e) $\neg(p \Leftrightarrow q)$;

    (f) $(\neg p) \Leftrightarrow q$; and

    (g) $(\neg p) \Leftrightarrow (\neg q)$.

1.1.2 How many distinct compound statements can be constructed from the statements p and q? NOTE: r and s are distinct statements if their truth tables are different—in other words, if they are not equivalent.

1.1.3 Prove that the compound statement $\neg(p \wedge q)$ is equivalent to the compound statement $(\neg p) \vee (\neg q)$.

1.1.4 Show that the statement $((p \Rightarrow r) \wedge (r \Rightarrow q)) \Rightarrow (p \Rightarrow q)$ is true for all possible combinations of p, q, and r. NOTE: A statement that is always true is called a *tautology*.

1.1.5 Show that $((\neg q) \Rightarrow (\neg p)) \Rightarrow (p \Rightarrow q)$ is a tautology.

1.1.6 Show that $((p \Rightarrow q) \wedge (q \Rightarrow p)) \Rightarrow (p \Leftrightarrow q)$ is a tautology.

1.1.7 Show that $(p \Rightarrow q) \wedge (q \Rightarrow p)$ is equivalent to $p \Leftrightarrow q$.

1.1.8 Show that $p \Rightarrow q$ is equivalent to $(\neg p) \vee q$.

1.1.9 Prove that $(p \Leftrightarrow q) \Rightarrow (p \wedge q)$ is not a tautology.

1.1.10 A statement that is always false is called an *absurdity*. A statement that may be true or false, depending upon the values of its constituent statements, is called a *contingency*. Determine whether each of the following statements is a tautology, an absurdity, or a contingency:

    (a) $(\neg p) \vee p$;

    (b) $(\neg p) \wedge p$;

    (c) $(\neg(p \Rightarrow q)) \Rightarrow (q \Rightarrow (\neg p))$;

    (d) $(((\neg p) \wedge (\neg q)) \vee (p \wedge q)) \Leftrightarrow (p \Leftrightarrow q)$;

    (e) $(p \Leftrightarrow q) \Rightarrow (p \vee q)$; and

    (f) $(p \Leftrightarrow q) \Rightarrow ((\neg p) \Leftrightarrow (\neg q))$.

1.1.11 Show the logical connectives in the following argument, and determine if the argument is valid:

    (A) "To be great is to be misunderstood." [Emerson]

    (B) He is misunderstood.

    (C) Therefore, he is great.

    NOTE: The argument is valid if and only if $(A \wedge B) \Rightarrow C$. That is, the conclusion must be implied by the premises. Thus, an argument may be valid even though its premises or its conclusion be false. For example, any conclusion follows validly from hypotheses that are absurdities (although such an argument sometimes is said to have "hollow validity"). Similarly, an argument may be invalid even though its premises and conclusion are true.

1.1.12 Is the following argument valid?
(*A*) Blue birds sing sweetly.
(*B*) All birds that sing sweetly build nests in tall trees.
(*C*) Robins nest in low trees.
(*D*) Therefore, robins are not blue birds.

1.1.13 Determine if the following argument is valid:
(*A*) Lawyers make good judges.
(*B*) People without a legal education do not make good judges.
(*C*) Therefore, a person can be a good judge if and only if he has a legal education.

1.1.14 Show the logical connectives in the following argument, and determine if the argument is valid:
(*A*) A computer designer who knows discrete mathematics produces only good computer designs.
(*B*) A good computer design is economical.
(*C*) Therefore, if a computer designer produces an uneconomical design, then he does not know discrete mathematics.

1.1.15 Determine if the following argument is valid:
(*A*) A computer designer who knows discrete mathematics produces only good computer designs.
(*B*) An economical design is a good design.
(*C*) Therefore, if a computer designer produces an uneconomical design, then he does not know discrete mathematics.

1.1.16 Given the infinite sequence $a_1, a_2, \ldots, a_i, \ldots$, let $x_1 = a_1$, and let $x_i = a_i + x_{i-1}$ for $i \geq 2$. Prove by induction that

$$x_n = \sum_{i=1}^{n} a_i \quad \text{for } n \geq 1.$$

1.1.17 Given the infinite sequence $a_1, a_2, \ldots, a_i, \ldots$, let $u_1 = a_1$, and let $u_i = a_i + 1/u_{i-1}$ for $i \geq 2$. The sequence $u_1, u_2, \ldots, u_i, \ldots$ is said to be *well defined* if the a sequence is such that no u_i is zero. Let $q_1, q_2, \ldots, q_i, \ldots$ be the infinite sequence defined by $q_1 = a_1$, $q_2 = a_2 a_1 + 1$, and $q_i = a_i q_{i-1} + q_{i-2}$ for $i \geq 3$. Prove by induction that $u_i = q_i/q_{i-1}$ for $i \geq 2$ if the u sequence is well defined.

1.1.18 For the sequences defined in Ex. 1.1.17, prove by induction that

$$q_i = \prod_{j=1}^{i} u_j \quad \text{for } i \geq 1$$

if the u sequence is well defined.

1.1.19 For $n \geq 1$, $n!$ is defined as $n \cdot (n - 1) \cdot (n - 2) \cdot \ldots \cdot 1$. Also by definition, $0! = 1$, and

$$\binom{n}{i} = \frac{n!}{i!(n-i)!} \quad \text{for } 0 \leq i \leq n.$$

Prove by induction that

$$\binom{n}{i} = \binom{n-1}{i} + \binom{n-1}{i-1} \quad \text{for } 1 \leq i \leq n - 1.$$

1.1.20 Use induction to prove that

$$\sum_{i=0}^{m} \binom{n+i}{i} = \binom{n+m+1}{m} \quad \text{for } m \geq 0.$$

1.2 Sets

Throughout the study of computer science, we encounter collections of related things. One example is the collection of representable numbers for any given computer. Another is the collection of legal FORTRAN identifiers. Yet another is the set of character sequences that constitute syntactically correct FORTRAN programs. In each of these examples, we can speak of the collection as an entity in itself, distinct from the items in the collection. The collection of representable numbers, for example, has some special meaning quite apart from the meaning that is attached to any number in this collection. In some sense, we have categorized the numbers that we can manipulate on our computer by saying that they form the collection of representable numbers. Formation of collections of related objects is the essential idea of *set theory*. It is this idea that enables us to study relationships among objects.

To begin the study of sets, we use the informal definition that a *set* is a collection of objects. The objects are called the *elements*, or *members*, of the set to which they belong.

We usually describe a particular set by listing its elements between braces, as for the set $S = \{true, false\}$. To denote that s is an element of S, we use the notation $s \in S$. Thus, $true \in S$, but 0 is not an element of S (this can be denoted by $0 \notin S$).

Our development of set theory in this section is rather informal. Set theoreticians normally work with a system of axioms that they assume to hold for all sets; they derive all of the characteristics of sets from these axioms. To be mathematically precise, our definition of "set" should specify the axioms that a set must satisfy. However, we shall simply state the important properties of sets, without identifying those properties that are directly or indirectly derivable from the axioms.

The most primitive operation on sets is that of comparing two sets. Our first definition permits us to determine if two sets are equal.

Definition   Two sets S and T are *equal* if and only if every element of S is an element of T and every element of T is an element of S.

For example, $\{1, 2, 3\} = \{3, 1, 2\}$, but $\{1, 2, 3\} \neq \{1, 2\}$. In particular, note that the order in which elements appear between braces is unimportant. How shall we interpret the definition of set equality when sets have repetition? If $S = \{1, 1, 2, 2, 3\}$ and $T = \{1, 2, 3\}$, then does $S = T$? The definition of set equality suggests that we can determine set equality in this case by using only the predicates $x \in S$ and $x \in T$. These predicates are true when we substitute 1, 2, or 3 for x; otherwise they are false. Therefore, every element of S is an element of T, every element of T is an element of S, and the sets are equal. Implicit in the definition of set equality is the notion that repetition of elements in a set has no significance. In any set, we can delete all but one occurrence of any repeated element without altering the set.

For some applications (notably sorting), it is convenient to deal with collections in which the multiplicity of elements does have significance. Such collections have been called *multisets* and *bags* in the literature. The multiset $S = \{1, 1, 2, 2, 3\}$ is not equal to the multiset $T = \{1, 2, 3\}$ because S has two 1's and two 2's, whereas T has just one of each. The properties of sets are sufficient for our purposes, however, and we give no further discussion of multisets in this text.

In computer applications, sets arise in numerous ways. We state early in this chapter that signals generated internally in computers must have one of only two possible values. An equivalent statement in terms of set theory is that there is a set of signal values—a set that has only two elements. Among many other useful sets in the study of computers are the set of representable numbers, the set of states of a computer, the set of valid subroutine parameter values, and the set of names in a directory. In following chapters we use set theory for far more than just a useful terminology for computer applications; we often use the theory itself.

Thus far, we have assumed that the elements of sets are concrete objects, but the elements of sets may themselves be sets. Indeed, we can develop all of set theory in terms of sets that contain only sets as their elements. We can construct new sets from any two sets S and T, and we can iterate the constructions indefinitely so as to construct an infinite number of different sets. Before discussing the constructions, we introduce the following definitions to aid us in comparing sets.

Definition A set T is said to be a *subset* of a set S (denoted by $T \subseteq S$) if every element of T is also an element of S. If $T \subseteq S$ and $T \neq S$, T is said to be a *proper subset* of S (denoted by $T \subset S$).

Definition The *cardinality* of S, written $|S|$, is the number of elements in S.

This definition of the cardinality of a set is unambiguous for a finite set, but it leads to some problems for sets with infinite numbers of elements. We avoid these problems by limiting the use of cardinality to finite sets.

We can construct new sets from the sets S and T in many different ways, including the following:

(1) $\{S, T\}$, the *unordered pair* of S and T, which is the set containing as elements just the sets S and T;

(2) $S \cup T$, the *union* of S and T, which is the set containing all those elements that are elements of either S or T;

(3) $S \cap T$, the *intersection* of S and T, which is the set containing just those elements that appear in both S and T;

(4) $S - T$, the *set difference* of S and T, which is the set of elements of S that are not also elements of T;

(5) $\bar{T}$ in S, the *relative complement* of T in S, which is defined only if $T \subseteq S$ and then is equal to $S - T$; and

(6) $\mathscr{P}(S)$, the *power set* of S, which is the set of all subsets of S.

Fig. 1.2.1 depicts the union, intersection, set difference, and relative complements of S and T. In this figure, S and T are represented as disks, and the shaded areas indicate the constructed sets. This type of figure is called a *Venn diagram*.

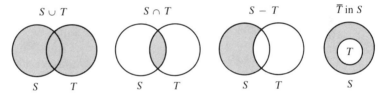

FIG. 1.2.1   Venn Diagrams of Constructed Sets

We illustrate these set constructions with the three sets

$$X = \{1, 2, 3, 4, 5\}, \quad Y = \{1, 4\}, \quad \text{and} \quad Z = \{2, 4, 5\}.$$

Among the sets that can be constructed from these three sets are the following:

unordered pair:	$\{X, Y\} = \{\{1, 2, 3, 4, 5\}, \{1, 4\}\},$
	$\{Y, Z\} = \{\{1, 4\}, \{2, 4, 5\}\};$
union:	$X \cup Y = \{1, 2, 3, 4, 5\},$
	$Y \cup Z = \{1, 2, 4, 5\};$
intersection:	$X \cap Y = \{1, 4\},$
	$Y \cap Z = \{4\};$
set difference:	$X - Y = \{2, 3, 5\},$
	$Y - Z = \{1\},$
	$Z - Y = \{2, 5\};$
relative complement:	$\bar{Y} \text{ (in } X) = \{2, 3, 5\},$
	$\bar{Z} \text{ (in } X) = \{1, 3\},$
	$\bar{Y} \text{ (in } Z) = \text{undefined};$
power set:	$\mathscr{P}(Y) = \{\{\ \}, \{1\}, \{4\}, \{1, 4\}\}.$

Some aspects of these constructions require discussion. First, note that $|\{S, T\}| = 2$, regardless of the numbers of elements in S and T. The example, $\{X, Y\} = \{\{1, 2, 3, 4, 5\}, \{1, 4\}\}$, is a set of cardinality 2, *not* a set of cardinality 5 or 7. The cardinality of $\{X, Y\}$ is not the sum of the cardinalities of X and Y, nor is it the cardinality of $X \cup Y$. A useful analogy here is to think of a set as a box, with the cardinality of the set representing the number of things in that box. Some of the things in the box may themselves be boxes. The unordered pair $\{S, T\}$ is analogous to a box that contains two boxes.

We also call attention to the power-set example. In this example, $|Y| = 2$ and $|\mathscr{P}(Y)| = 4$. Note that one of the subsets of Y is $\{\ \}$, a set with no elements. This set is of particular significance, as indicated by the following definition.

Definition The *empty set* (sometimes called the *null set*) is the set with no elements. We denote this set as $\varnothing$.

Because $\varnothing$ is defined as the set with no elements, the predicate $x \in \varnothing$ is false for all x, and thus the predicate $(x \in \varnothing) \Rightarrow (x \in S)$ is true for all x and for all sets S. This is simply a formal way of saying $\varnothing \subseteq S$ for all sets S. Therefore, $\varnothing$ is an element of the power set of any set.

Note that $|\mathscr{P}(S)| = 2^{|S|}$ when $|S|$ is finite. We can derive this formula by noting that each subset of S can be constructed by deciding separately for each possible element whether or not it is in the subset. Because there are $|S|$ elements in S, there are $2^{|S|}$ different ways of making the $|S|$ decisions, and each selection of $|S|$ decisions yields a distinct subset. When $|S|$ is infinite, $2^{|S|}$ is infinite as well; the meaning of $2^{|S|}$ in this case is a deep subject of philosophy. In this text we restrict our attention to power sets of finite sets, so that $\mathscr{P}(S)$ is always a finite set.

Infinite sets are of some interest, however. For example, consider the sequence of sets that set theorists call the nonnegative integers. To construct this sequence, we let n be a set and define the *successor* of n (denoted n^+) to be the set $n \cup \{n\}$. We can relate the sets to the integers by letting each n be an integer such that the successor of n is the integer one greater than n. We begin by letting the null set be the integer 0:

$$0 = \varnothing = \{\ \};$$
$$1 = \varnothing \cup \{\varnothing\} = \{\varnothing\} = \{0\};$$
$$2 = 1 \cup \{1\} = \{\varnothing\} \cup \{\{\varnothing\}\} = \{\varnothing, \{\varnothing\}\} = \{0, 1\};$$
$$3 = 2 \cup \{2\} = \{0, 1\} \cup \{2\} = \{0, 1, 2\}; \ldots.$$

Note carefully that we distinguish between $\varnothing$ and $\{\varnothing\}$. The null set has no elements, but the set containing the null set has one element. Using our box analogy, we recall that the null set is analogous to an empty box, but $\{\varnothing\}$ is analogous to a box that contains an empty box within it. Therefore $\{\varnothing\}$ is not an empty set. It is clear that we can construct the set named "n" for any finite integer n by repeating this construction n times. One of the axioms of set theory assures us that we can construct infinite sets by repeating this construction an infinite number of times.

The constructions discussed thus far enable us to obtain new sets from old ones. We also can construct new sets from predicates.

Definition For any predicate $p(x)$ and any set S, the set $T = \{x | x \in S \wedge p(x)\}$ is the set containing all x such that $x \in S$ and $p(x)$ are both true statements. We often use the simpler notation $T = \{x | p(x)\}$, with the set S being identified by the context of the statement.

The following are some examples of specification of sets by predicates:

$\{x | (x \in \{\text{the integers}\})$
$\quad \wedge$ (there exists an integer y for which $x = 2y)\}$
$$= \{\text{the even integers}\};$$

$\{x | (x \text{ is real}) \wedge (0 \leq x) \wedge (x < 100)\}$
$$= \{\text{the real numbers in the interval } 0 \leq x < 100\};$$

$\{x | (x \text{ is a word with six or fewer alphanumeric characters})$
$\quad \wedge$ (the first character of x is alphabetic)$\}$
$$= \{\text{the valid FORTRAN identifiers}\};$$

$\{x | (x \text{ is the text of a computer program}) \wedge (\text{a correct FORTRAN}$
$\quad$ compiler that processes x will find no syntax errors)$\}$
$$= \{\text{the syntactically correct FORTRAN programs}\};$$

$\{x | (x \text{ is a positive integer}) \wedge (\text{for all positive integers } y,$
$\quad (y \text{ is a divisor of } x) \Rightarrow ((y = 1) \vee (y = x)))\}$
$$= \{\text{the prime numbers}\}.$$

These examples bear deeper examination. First, note that we use predicates that begin "there exists an integer y ..." and "for all positive integers y ...." Both of these predicates include the variable y, which is not a free variable. In each case, specification of the free variable x converts the predicate into a statement that can be evaluated. The variable y can be any element of a particular set, but it is not a parameter of the predicate. In one example y must be in the set of integers; in the other example y is restricted to the set of positive integers. We say that these two predicates are *quantified over y*; we may possibly have to evaluate the statement for every possible y in order to find out if it is true. One predicate begins "there exists an ...."; the resulting statement is true if we can find at least one y that satisfies the predicate for the x we have chosen. The other predicate begins "for all ...," and the resulting statement must be true for any y that we select. These observations lead to the next definition.

Definition Let S be an arbitrary set. We say that predicates of the form "for all s in S, $p(x)$" are *universally quantified*. We can write such a predicate in the form $\forall\, s \in S\, p(x)$, where $\forall$ is the *universal quantifier*, which can be read as "for all." We say that predicates of the form "there exists an *s* in *S* such that $p(x)$" are *existentially quantified*. We can write such a predicate in the form $\exists\, s \in S\, p(x)$, where $\exists$ is the *existential quantifier* and can be read as "there exists an."

We may have to do an extensive computation in order to determine if substitution of a particular *x* leads to a true statement. The example of the set of syntactically correct FORTRAN programs illustrates a case in which the computation may be even more complex than that required for a quantified predicate. Quantification, after all, requires only a search of the elements of a set, which is a simple task in principle although the actual enumeration of the set elements may be a tedious or lengthy task. In the example of FORTRAN programs, we determine if a particular *x* is in a set by applying a complex algorithm to *x*. Thus we see that the predicates used to specify sets may range from easily evaluated predicates through quantified predicates (which involve an enumeration of the elements of a set) to complex algorithms that have two possible outcomes.

Fig. 1.2.2 illustrates a conceptual view of a network for evaluating predicates in which the only logical connectives are $\land$, $\lor$, and $\lnot$, and in which the only elementary predicate is set membership. The network in the figure evaluates the predicate $x \in (X \cup Y) - (X \cap Y)$. This predicate is true only if *x* is in *X* or in *Y* but not in both *X* and *Y*. Hence, the network computes the logical function

$$((x \in X) \land \lnot(x \in Y)) \lor ((x \in Y) \land \lnot(x \in X)).$$

The modules that compute the predicates $x \in X$ and $x \in Y$ accept an input symbol that represents *x* and place "true" or "false" on their outputs, depending on the set membership of *x*. The $\land$, $\lor$, and $\lnot$ modules produce "true" or "false" on their outputs, as determined by

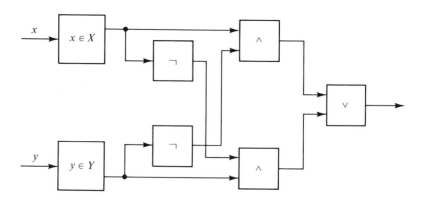

FIG. 1.2.2   A Network for Evaluating the Predicate $x \in (X \cup Y) - (X \cap Y)$

their input values and by their respective definitions. Thus, we can place a particular x on the network input, and the network reports "true" or "false" at its output, depending on whether or not x is in $(X \cup Y) - (X \cap Y)$.

In Fig. 1.2.2, the set of possible network inputs may be many-valued, but all intermediate network signals are two-valued. As such, the network is suggestive of actual networks within computers. We explore the design of computer networks more closely in other chapters of this text. Although the predicates of Fig. 1.2.2 are unquantified, we need not limit networks to such predicates. In order to evaluate quantified predicates, we must be able to store or to generate all of the elements of the quantified set. The sequential generation of elements of a set implies that the network must behave sequentially, and we have not modeled such behavior in Fig. 1.2.2, although computer networks do exhibit such behavior. Perhaps a better model for generating the elements of a set and evaluating quantified predicates is the execution of a computer program, because such operations more commonly are done with programs rather than with hardware.

Throughout this text we make frequent reference to several well-known sets, which are defined as follows:

$$\mathbf{N} = \{0, 1, 2, 3, \ldots\} = \{\text{the natural numbers}\};$$
$$\mathbf{Z} = \{0, \pm 1, \pm 2, \pm 3, \ldots\} = \{\text{the integers}\};$$
$$\mathbf{P} = \{1, 2, 3, \ldots\} = \{\text{the positive integers}\};$$
$$\mathbf{Q} = \{x \text{ such that } x = p/q \text{ and } p \in \mathbf{Z} \text{ and } q \in \mathbf{P}\}$$
$$\quad = \{\text{the rational numbers}\};$$
$$\mathbf{R} = \{\text{the real numbers}\};$$
$$\mathbf{C} = \{\text{the complex numbers}\};$$
$$\mathbf{Z}_n = \{0, 1, 2, \ldots, (n - 1), \text{ where } n \geq 2\}$$
$$\quad = \{\text{the integers modulo } n\}.$$

Of these sets, only $\mathbf{Z}_n$ is finite. We use the others in examples primarily because we assume that the reader is somewhat familiar with them. The set $\mathbf{Z}_n$ appears because the set of manipulable integers for any computer is $\mathbf{Z}_n$, where the n varies from computer to computer.

In defining sets, we remarked that there is no significance to the order in which the elements are written. However, there are collections of items in which order has significance. In dealing with such collections, we have occasion to speak of the first, second, last, or ith element for any particular value of i. This need motivates the next definition.

Definition   Let n be a positive integer. An *n-tuple* V is an ordered collection of n items. The elements of V are $v_1, v_2, \ldots, v_n$. We usually write V with angle brackets as $V = \langle v_1, v_2, \ldots, v_n \rangle$. Because order is important in n-tuples, two n-tuples $U = \langle u_1, u_2, \ldots, u_n \rangle$ and $V = \langle v_1, v_2, \ldots, v_n \rangle$ are equal if and only if $u_i = v_i$ for $1 \leq i \leq n$.

As one example of n-tuples, we can treat complex numbers as 2-tuples (often called *ordered pairs*) by letting the components be the real and imaginary parts, respectively. Vectors in mathematics provide another example; one-dimensional arrays in programming languages are yet another.

Another important idea based upon the idea of n-tuples is the notion of cartesian products of sets.

Definition For any two sets S and T, the *cartesian product* of S and T, written $S \times T$, is the set of all ordered pairs of the form $\langle s, t \rangle$ where $s \in S$ and $t \in T$. In predicate notation, $S \times T = \{\langle s, t \rangle | s \in S \land t \in T\}$.

Again using the complex numbers for an example, we can say that $C = R \times R$, where the two components are the real and imaginary parts, respectively, of complex numbers.

To summarize the main points of this section, we recall that sets are collections in which repetition and ordering are unimportant. Sets can be constructed from other sets by using operations such as unordered pairing, union, intersection, set difference, set complement, and forming the power set. We also can define sets through use of predicates that are true for elements in the set and are false otherwise. Collections in which repetition is important are multisets, and collections in which order is important are n-tuples.

The reader interested in further pursuit of the study of sets will find interesting reading in the books by Suppes (1957) and Halmos (1960). The axioms used by most set theoreticians today are called the Zermelo-Fraenkel axioms after two philosophers who developed the system in the early part of this century. Halmos (1960) gives an excellent discussion of axiomatic set theory based on the Zermelo-Fraenkel axioms.

Exercises

1.2.1 Let X, Y, and Z be arbitrary sets. Show that the following identities are valid:
  (a) $\varnothing \cup X = X$;
  (b) $\varnothing \cap X = \varnothing$;
  (c) $X \cap (Y \cup Z) = (X \cap Y) \cup (X \cap Z)$;
  (d) $X \cup (Y \cap Z) = (X \cup Y) \cap (X \cup Z)$.
  Construct Venn diagrams for (c) and (d).

1.2.2 Describe each of the following sets with an English phrase:
  (a) $\{x | x \in Z \land \exists y \in Z \ (x = y^2)\}$;
  (b) $\{x | x \in \mathscr{P}(S) \land x \neq S\}$;
  (c) $\{x | x \in R \land x^2 = 1\}$;
  (d) $\{x | x \in C \land \exists n \in P \ (x^n = 1)\}$.

1.2.3 Use Venn diagrams to show that
  (a) $\overline{S \cap T} = \overline{S} \cup \overline{T}$;
  (b) $\overline{S \cup T} = \overline{S} \cap \overline{T}$;
  (c) $(S - T) \cup (T - S) = (S \cup T) - (S \cap T)$.
  For relative complements, assume $S, T \subseteq R$ and take complements in R.

1.2.4 Construct predicates that define each of the following sets:
  (a) the prime numbers;
  (b) the real roots of the polynomial $f(x) = f_0 + f_1 x + \cdots + f_n x^n$;
  (c) the positive factors of 42;
  (d) the positive multiples of 42;
  (e) the perfect squares less than 112.

1.2.5 A survey of a group of 43 students revealed the following statistics:
  (a) 18 of the males in the group are coffee drinkers;
  (b) 16 students are 21 or over;
  (c) 18 students are males under 21;
  (d) three males 21 or older are not coffee drinkers;
  (e) four of the 17 female students are not coffee drinkers, and all four are under 21.
  Determine how many students are in each of the subsets formed from the eight combinations of the dichotomies male/female, 21 or over/under 21, and coffee drinker/coffee nondrinker.

1.2.6 Let R, S, and T be arbitrary sets. Show that we can construct the set $\{R, S, T\}$ by using the set constructions given in this section. HINT: $\{R, S, T\} = \{R\} \cup \{S, T\}$.

1.2.7 Use mathematical induction to prove that we can construct the set $\{S_1, S_2, \ldots, S_n\}$ from the n sets $S_1, S_2, \ldots, S_n$.

1.2.8 Prove that we can define $\langle x, y \rangle$ to be the set $\{\{x\}, \{x, y\}\}$ and obtain the correct characteristics for ordered pairs.

1.2.9 Prove that we cannot define the ordered triple $\langle x, y, z \rangle$ to be the set $\{\{x\}, \{x, y\}, \{x, y, z\}\}$ because this definition does not satisfy the properties of n-tuples.

1.2.10 Prove that the appropriate definition of $\langle x, y, z \rangle$ in Ex. 1.2.9 is

$$\langle x, \langle y, z \rangle \rangle = \{\{x\}, \{\{x\}, \{\{y\}, \{y, z\}\}\}\}.$$

Prove by induction that n-tuples can be defined entirely in terms of sets.

1.2.11 Show by example that the predicate $\neg \, (\forall \, x \in X \, p(x))$ is identical to the predicate $\exists \, x \in X \, (\neg p(x))$. Thus a negation can be moved to the right of $\forall$ by changing $\forall$ to $\exists$. Similarly, show by example that the predicate $\neg \, (\exists \, x \in X \, p(x))$ is identical to the predicate $\forall \, x \in X \, (\neg p(x))$.

1.3 Graphs In the following sections we make frequent use of abstract notions as we develop the tools of discrete mathematics for our computer applications. To make these notions more concrete, we use graphs to provide informative pictorial representations of the mathematical structures. In this section, we give a brief review of the bare essentials of graphical notation.

We define a *graph* in set theoretical terms as an ordered pair, $\langle V, E \rangle$. V is a set of *vertices* (or nodes) and E is a set of *edges* (or arcs). Each edge is an ordered pair of vertices $\langle x, y \rangle$ (where $x, y \in V$), indicating the presence of a connection directed from vertex x to vertex y. The vertex x is called the *initial vertex* and y the *terminal vertex* of the edge. E is thus a subset of $V \times V$.

We can represent a graph by a drawing with dots denoting the vertices and arrows denoting the edges; each arrow points from the initial vertex toward the terminal vertex of the edge. We also can represent a graph by a binary *interconnection matrix* M (sometimes called an *adjacency matrix*) of dimension $|V| \times |V|$. M contains a one in the ith row and jth column if $\langle i, j \rangle \in E$, and contains a zero otherwise. Fig. 1.3.1 shows both the drawing and the matrix that represent the graph

$$G = \langle \{a, b, c, d\}, \{\langle a, c \rangle, \langle a, d \rangle, \langle b, b \rangle, \langle c, b \rangle, \langle c, d \rangle, \langle d, c \rangle \} \rangle.$$

Definition   A *complete graph* has every possible edge; that is, $E = V \times V$.

Definition   A *path of length n* in G is an ordered set, P, of n edges in E, such that each edge in P, except for the first, has as its initial vertex the terminal vertex of the preceding edge. The *initial vertex of the path* is the initial vertex of the first edge in the path; similarly, the *terminal vertex of the path* is the terminal vertex of the last edge.

In graph G (Fig. 1.3.1), a path from a to c is $P = \langle \langle a, c \rangle, \langle c, d \rangle, \langle d, c \rangle \rangle$, where P has initial vertex a and terminal vertex c. Note that we could define a path in an equivalent way as an ordered set of vertices, such that each successive pair of vertices is an edge in E. With this definition, path P just described would be denoted as $\langle a, c, d, c \rangle$.

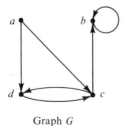

	a	b	c	d
a	0	0	1	1
b	0	1	0	0
c	0	1	0	1
d	0	0	1	0

Graph G Interconnection Matrix for G

FIG. 1.3.1   A Graph and Its Interconnection Matrix

Definition   A *cycle* is a path whose initial vertex is the same as its terminal vertex. A cycle of length 1 is called a *self-loop*.

In graph G, node b has a self-loop, and there is another cycle containing the nodes c and d.

These rather elementary definitions suffice to describe the discrete mathematical structures of interest to us. We use graphs principally to illuminate mathematical concepts, but graphs also are useful to describe computer-related topics that have little in common with discrete mathematics. For example, Knuth (1968) describes at some length the use of various data structures in computer programs; the subject of graphs is quite central to his discussion, both for illustrative purposes and as a tool for analysis. The class of graphs known as *trees* is particularly relevant to the study of data structures, for such graphs have application to compilers, search algorithms, and a wide variety of other programs. However, our applications are far different from this, and a discussion of trees falls outside the scope of this text.

Graph theory in the abstract is a fascinating and challenging study in and of itself, quite apart from any applications that we can make of the theory. The interested reader may pursue this topic in depth in books such as those by Berge (1962) and Harary (1969).

Exercises

1.3.1   An *acyclic graph* is a graph without cycles. Prove that no acyclic graph with n vertices can have more than $n(n - 1)/2$ edges. Prove that there exists a graph that attains this bound for any n.   HINT: Prove that the interconnection matrix of a graph with cycles must contain a one on the main diagonal or a pair of ones above and below the main diagonal. Also, for some ordering of the vertices of an acyclic graph, all of the ones of its interconnection matrix lie on one side of the main diagonal.

1.3.2   An *undirected graph* is a graph in which there exists an edge $\langle y, x \rangle$ for every edge $\langle x, y \rangle$. A *cycle in an undirected graph* is a self-loop or a path visiting vertices $x_1, x_2, \ldots, x_n, x_1$, where $n \geq 3$, and vertex x_i is distinct from vertex x_j for $i \neq j$. Prove that an acyclic undirected graph with n vertices can have no more than $n - 1$ edges.

1.3.3   Consider an arbitrary graph G with n vertices. Prove that if there exists a path between vertices x and y, then there exists a path between x and y of length no greater than $n - 1$.

1.3.4   The familiar family tree is a directed graph in which an edge is directed from each parent to each of his offspring. Prove that a family tree must be acyclic.

1.3.5 Fig. 1.3.2 shows a graph G, its interconnection matrix M, and M^2 (the square of its interconnection matrix under ordinary matrix multiplication). Prove that, for every graph, the entry in the ith row and jth column of M^2 is the number of distinct paths of length 2 from vertex i to vertex j. Similarly, prove that M^k gives the paths of length k.

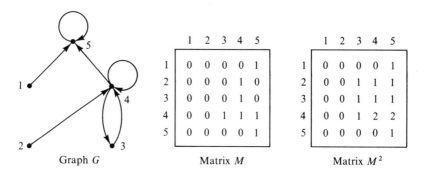

FIG. 1.3.2   Graph G with Interconnection Matrix M (Ex. 1.3.5)

	1	2	3	4	5
1	0	0	0	0	1
2	0	0	1	1	1
3	0	0	1	1	1
4	0	0	1	1	1
5	0	0	0	0	1

FIG. 1.3.3   Matrix $M^{(2)}$ for the Graph of Fig. 1.3.2

1.3.6 For a graph G with interconnection matrix M, we define $M^{(2)}$ to be the matrix obtained from M by *logical matrix* multiplication. More precisely, the (i,j)th element of $M^{(2)}$ (denoted $[M^{(2)}]_{ij}$) is

$$[M^{(2)}]_{ij} = \bigvee_r ([M]_{ir} \wedge [M]_{rj}),$$

where $\vee$ and $\wedge$ are logical connectives defined on $\{0,1\}$ analogously to their definitions on $\{false,\ true\}$. Here 0 is analogous to false and 1 to true, so that

$$0 = 0 \vee 0 = 0 \wedge 0 = 0 \wedge 1 = 1 \wedge 0,$$
$$1 = 1 \wedge 1 = 1 \vee 1 = 1 \vee 0 = 0 \vee 1.$$

In general, $M^{(2)}$ has zeros where M^2 has zeros, and $M^{(2)}$ has ones everywhere else. Fig. 1.3.3 shows $M^{(2)}$ for the graph in Fig. 1.3.2. Prove that the (i,j)th entry of $M^{(2)}$ is one if and only if there is a path of length 2 from vertex i to vertex j. Let $M \vee I$ be the matrix derived from

M by setting its main diagonal to ones. Prove that $(M \vee I)^{(2)}$ has a one in the (i,j)th position if there is a path of length 2 or less from vertex i to vertex j.

1.3.7 We extend the definition of logical matrix multiplication to higher powers by defining $M^{(r)}$ for $r \geq 2$ to be

$$\left[M^{(r)}\right]_{ij} = \bigvee_k \left([M]_{ik} \wedge \left[M^{(r-1)}\right]_{kj}\right).$$

Prove that the (i,j)th entry of $M^{(r)}$ is a one if there exists a path of length r from vertex i to vertex j. Prove that $(M \vee I)^{(r)}$ has a one in its (i,j)th entry if there is a path of length r or less from vertex i to vertex j.

1.3.8 Prove that $(M \vee I)^{(r)} = (M \vee I)^{(n-1)}$ for all $r \geq n$, if M is the interconnection matrix of a graph with n vertices.

1.4 Relations Perhaps one of the most powerful attributes of a digital computer is its ability to perform comparisons and to take different actions depending on the outcome of the comparisons. Programmers make use of this ability throughout most programs; in addition, the computer hardware may make several comparisons in executing each program instruction. In this section, we formalize the notion of comparison through the mathematical structure known as *relation*. Relations commonly found in computers are magnitude comparisons such as $x < y$, $x = y$, and $x > y$. However, within the framework that we develop for relations, we find many other forms of relations that are widely applicable to computer science. Among these are *equivalence relations*, which tell us when one item is indistinguishable from another. In a computer application, for example, these items may be computer components or program subroutines.

We develop our study of relations from set-theoretical grounds beginning with the next definition.

Definition A subset of $S \times T$ is said to be a *binary relation* on $S \times T$. A subset of $S \times S$ is said to be a binary relation on S. We usually mean binary relation when we use the term *relation* alone.

Most readers are familiar with such examples of relations as $<$, $\leq$, and $=$ on $\mathbf{N}$ and on $\mathbf{Z}$. If R is a relation on S, and $\langle s, t \rangle$ is an ordered pair in the set R, then we say s R t. Because $\langle 3, 5 \rangle$ is in the relation $<$, we say $3 < 5$, which is a familiar notation.

Graphs are useful ways to visualize relations. If R is a relation on S, then the graph G determined by R has a vertex for each element of S, and has an edge from x to y in G if and only if x R y. For example, the relation $\leq$ on $S = \{1, 2, 3, 4\}$ has the graph

$$G = \langle\{1, 2, 3, 4\}, \{\langle 1, 1 \rangle, \langle 2, 2 \rangle, \langle 3, 3 \rangle, \langle 4, 4 \rangle,$$
$$\langle 1, 2 \rangle, \langle 1, 3 \rangle, \langle 1, 4 \rangle, \langle 2, 3 \rangle, \langle 2, 4 \rangle, \langle 3, 4 \rangle\}\rangle,$$

as shown in Fig. 1.4.1.

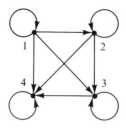

FIG. 1.4.1 Graph of the Relation $\leq$ on $S = \{1, 2, 3, 4\}$

If R is a relation on $S \times T$, we can graph the relation as exemplified in Fig. 1.4.2 for

$$R = \{\langle s_1, t_1 \rangle, \langle s_1, t_2 \rangle, \langle s_2, t_2 \rangle, \langle s_3, t_2 \rangle, \langle s_1, t_3 \rangle\}.$$

Because relations are sets of ordered pairs, the set operations of union and intersection of relations are well defined, provided that the relations are defined on the same sets. For example, let G denote the relation "greater than" on $\mathbf{Z}$ and E denote the relation "equals" on $\mathbf{Z}$. Then the union of the two relations is the set

$$G \cup E = \{\langle x, y \rangle | (\langle x, y \rangle \in G) \vee (\langle x, y \rangle \in E)\}$$
$$= \{\langle x, y \rangle | (x, y \in \mathbf{Z}) \wedge (x \geq y)\}.$$

This set is the relation $\geq$ defined on $\mathbf{Z}$. Similarly, $\geq$ intersected with $>$ is $>$, and $>$ intersected with $=$ is the null, or empty, relation. The operations on relations may be carried over to their graphs. Here we abuse notation slightly and define the union of two graphs $G_1 = \langle S, R_1 \rangle$ and $G_2 = \langle S, R_2 \rangle$ to be $G_1 \cup G_2 = \langle S, R_1 \cup R_2 \rangle$. The intersection is $G_1 \cap G_2 = \langle S, R_1 \cap R_2 \rangle$. G_1 is contained in (is a subgraph of) G_2 if and only if $R_1 \subseteq R_2$.

Some relations have properties of special significance. Let R be a relation on set X. Then R is said to be

(1) *reflexive* if x R x for every x in X;
(2) *symmetric* if x R $y \Leftrightarrow y$ R x for every x, y in X;
(3) *antisymmetric* if $(x$ R $y \wedge y$ R $x) \Rightarrow x = y$ for every x, y in X;
(4) *transitive* if $(x$ R $y \wedge y$ R $z) \Rightarrow x$ R z for every x, y, z in X.

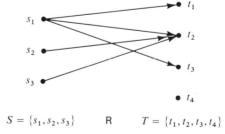

$S = \{s_1, s_2, s_3\}$ R $T = \{t_1, t_2, t_3, t_4\}$

FIG. 1.4.2 Graph of a Relation R on $S \times T$

Consider the relation $<$ on $\mathbf{Z}$. There is no integer x that satisfies $x < x$, so it is not reflexive. Nor is it symmetric, because if $x < y$, then $y \not< x$. However, $<$ is antisymmetric, because we cannot have both $x < y$ and $y < x$ for $x \neq y$. Clearly, $<$ is transitive, for if $x < y$ and $y < z$, then $x < z$.

Note that $\leq$ on $\mathbf{Z}$ is reflexive, antisymmetric, and transitive, but is not symmetric. The relation $=$ on $\mathbf{Z}$ is both antisymmetric and symmetric, as well as transitive and reflexive.

We can deduce many other properties of a relation by determining which of these four properties it has. For example, consider the relation $\leq$ on $\mathbf{Z}$. Because it is reflexive, antisymmetric, and transitive, we can show that there do not exist three distinct integers x, y, and z such that $x \leq y \leq z \leq x$. The relation $=$ on $\mathbf{Z}$ is also reflexive, antisymmetric, and transitive, and it follows that there do not exist three distinct integers such that $x = y = z = x$. Any other relation that also is reflexive, antisymmetric, and transitive will have an analogous property. The remainder of this section we devote to an investigation of some of the characteristics of relations with particular combinations of the four properties just defined.

Definition A relation is said to be a *partial ordering* if it is reflexive, antisymmetric, and transitive.

The relations $=$ and $\leq$ are partial orderings on $\mathbf{Z}$. For any set S, the set inclusion relation $\subseteq$ is a partial ordering on the power set of S, $\mathscr{P}(S)$.

In drawing graphs of partial orderings, we follow the convention of eliminating the self-loops (which are implied by reflexivity) and those edges whose presence is implied by transitivity. Thus if $\langle x, y \rangle$, $\langle y, z \rangle$, and $\langle x, z \rangle$ are pairs in a relation, we do not draw the edge $\langle x, z \rangle$ in the graph of the relation if we have already stated it to be a partial ordering. For example, the relation $\subseteq$ on the subsets of $S = \{0, 1\}$ is drawn as shown in Fig. 1.4.3, rather than including all the implied edges as shown in Fig. 1.4.4.

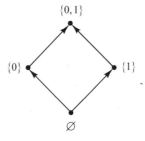

FIG. 1.4.3
Graph of a Partial Ordering
with Implied Edges Eliminated

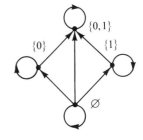

FIG. 1.4.4
Graph of a Partial Ordering
with Implied Edges Included

We now are ready to state in a more general form the important property of partial orderings that we have already mentioned.

Lemma   If R is a partial ordering on X, then for $n \geq 2$, there cannot be $x_1, x_2, \ldots, x_n$ (distinct elements of X) such that $x_1 \, R \, x_2 \, R \ldots R \, x_n \, R \, x_1$.

Proof: Assume that there are distinct $x_1, x_2, \ldots, x_n$ such that $x_1 \, R \, x_2 \, R \ldots R \, x_n \, R \, x_1$. Because a partial ordering is transitive, $x_1 \, R \, x_n$. Then we have both $x_1 \, R \, x_n$ and $x_n \, R \, x_1$, so the antisymmetric property gives $x_1 = x_n$. However, this means that the n elements are not distinct, which is a contradiction of the initial assumption.   □

We can state the lemma in an equivalent way as follows: the graph of a partial ordering contains no cycles (other than the self-loops that we eliminate from the drawing in our convention). Fig. 1.4.5 shows graphs of some partial orderings.

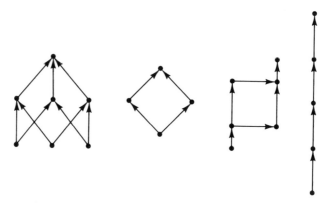

FIG. 1.4.5   Graphs of Some Typical Partial Orderings

Definition   A partial ordering R on a set S is said to be a *total ordering* if either $s \, R \, t$ or $t \, R \, s$ for every s and t in S.

Note that the set inclusion relation on the subsets of $S = \{0, 1\}$ is not a total ordering because $\{0\} \nsubseteq \{1\}$ and $\{1\} \nsubseteq \{0\}$. The relation $\leq$ on $\mathbf{Z}$ is a total ordering. From the definition of total ordering, it follows immediately that the graph of a finite total ordering must be a chain, as illustrated in Fig. 1.4.6. The definition of total ordering guarantees that $s \, R \, t$ or $t \, R \, s$ for all s and t but, because a total ordering is a type of partial ordering, both $s \, R \, t$ and $t \, R \, s$ only if $s = t$.

Partial and total orderings are not the only types of relations that we encounter frequently.

FIG. 1.4.6   The Graph of a Total Ordering

Definition   A relation R on a set S is called an *equivalence relation* if it is symmetric, reflexive, and transitive.

Of the various relations on $\mathbf{Z}$ mentioned thus far, only the equality relation, $=$, is an equivalence relation. An equivalence relation breaks up the set on which it is defined into subsets of equivalent elements. Another definition is useful in showing this.

Definition $S_1, S_2, \ldots, S_n$, a collection of subsets of S, is called a *partition* of S if

    (i)   the subsets are disjoint, so that $(S_i \cap S_j \neq \varnothing) \Rightarrow (i = j)$, and

    (ii) every element of S is in some S_i, so that $\bigcup_i S_i = S$.

In this definition, the notation $\bigcup_i S_i$ denotes $S_1 \cup S_2 \cup \ldots \cup S_n$. The convention we use here is that the i under the $\bigcup$ indicates that the union is to be taken over i, where the set of values for i is implied by the context.

Definition   If R is an equivalence relation on S, then for each s in S, $[s] = \{t \mid s \, \mathsf{R} \, t\}$ is the *set of elements equivalent to s*.

Lemma   If R is an equivalence relation on a set S, then the collection of subsets $\{[s] \mid s \in S\}$ is a partition of S.

Proof: **(i)** *The subsets are disjoint.* Suppose that there exist s and t in S such that $[s] \cap [t] \neq \varnothing$. We prove that $[s] = [t]$. If $[s] \neq [t]$, there is some element in one subset that is not in the other. Let this element be called x, and without loss of generality assume $x \in [t]$. Let $y \in ([s] \cap [t])$. Then $s \, \mathsf{R} \, y$ and $t \, \mathsf{R} \, y$, and by symmetry, $y \, \mathsf{R} \, t$. Now we have $s \, \mathsf{R} \, y \, \mathsf{R} \, t$, so by transitivity, $s \, \mathsf{R} \, t$. Because $x \in [t]$, we have $t \, \mathsf{R} \, x$. Then $s \, \mathsf{R} \, t \, \mathsf{R} \, x$, and transitivity yields $s \, \mathsf{R} \, x$, so $x \in [s]$. This is a contradiction and establishes that $[s] = [t]$.

(ii) *The union of the subsets is S.* To prove this, we must show that every $s \in S$ is in at least one subset in the collection. But for every s, $[s]$ is the subset in the collection containing s, because $s \, \mathsf{R} \, s$ by the reflexive property. This proves $S \subseteq \bigcup_s [s]$. But $[s] \subseteq S$ for all s, so $\bigcup_s [s] \subseteq S$.

Therefore, $S \subseteq \bigcup_s [s] \subseteq S$, and $S = \bigcup_s [s]$. $\square$

In the proof of the lemma, note that we use the properties of symmetry and transitivity to prove the disjointedness of the subsets, and the reflexitivity property to prove that the subsets cover S. If a relation does not satisfy all three properties, then the sets $[s]$ do not partition S.

As we mentioned earlier, the relation $=$ on $\mathbf{Z}$ is an equivalence relation. In the partition induced by $=$, each integer of $\mathbf{Z}$ is an equivalence class by itself. In general, for every finite set, there is an equality equivalence relation in which every element is equivalent only to itself.

The universal relation—which, by definition, relates every element to every other element—is an equivalence relation for every finite set.

The partition induced by this equivalence relation contains only one subset, which is the entire set on which the relation is defined.

A nontrivial example of an equivalence relation is the relation of congruence modulo n.

Definition An integer x is said to be *congruent to an integer y modulo n* if $x - y$ is an integral multiple of n.

For example, the even integers are congruent to 0 modulo 2, and the odd integers are congruent to 1 modulo 2. We use the notation $x \equiv y \bmod n$ to denote that x is congruent to y modulo n.

For every $n \in \mathbf{P}$, congruence modulo n is an equivalence relation on $\mathbf{Z}$ (see Ex. 1.4.4). Consider the relation of congruence modulo 5, for example. All of the multiples of 5 are congruent to one another, so they form one set of the equivalence partition. If we add one to each element of this set, we obtain the set of integers congruent to 1 mod 5. This forms another set of the partition. The five sets in the equivalence partition for congruence modulo 5 are

$$\{0, 5, -5, 10, -10, 15, -15, \ldots\},$$
$$\{1, 6, -4, 11, -9, 16, -14, \ldots\},$$
$$\{2, 7, -3, 12, -8, 17, -13, \ldots\},$$
$$\{3, 8, -2, 13, -7, 18, -12, \ldots\}, \text{ and}$$
$$\{4, 9, -1, 14, -6, 19, -11, \ldots\}.$$

Applications of congruence modulo n arise frequently in our discussions of computer arithmetic.

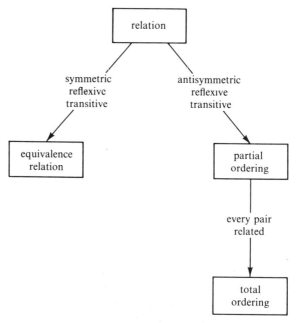

FIG. 1.4.7 Properties Satisfied by Special Types of Relations

Fig. 1.4.7 summarizes the various types of relations discussed in this section. Before closing the section, we mention briefly that it is convenient on occasion to generalize the definition of relation to *n*-tuples (a generalization from our definition based on ordered pairs).

Definition An *n-ary relation* defined on the cartesian product set

$$S_1 \times S_2 \times \cdots \times S_n$$

of the sets $S_1, S_2, \ldots, S_n$ is a subset of $S_1 \times S_2 \times \cdots \times S_n$.

Intuitively speaking, we can say that an *n*-ary relation is analogous to a binary relation, but the *n*-ary relation involves *n* items whereas the binary relation involves two items. For example, a 3-ary (or ternary) relation on $\mathbf{Z} \times \mathbf{Z} \times \mathbf{Z}$ is the relation that contains the ordered triple $\langle x, y, z \rangle$ if and only if $x + y = z$. The properties of symmetry, transitivity, and reflexivity for binary relations seldom are generalized to *n*-ary relations for $n > 2$, and in this text we restrict the usage of these properties to binary relations.

Exercises

1.4.1 The *transitive closure* of a relation R on a finite set S is defined to be the smallest transitive relation on S that contains R. Show that, for every relation R on S, there is some transitive relation R′ such that $R \subseteq R'$. Show that the transitive closure is uniquely defined. That is, if R_1 and R_2 are transitive relations such that $R_1, R_2 \supseteq R$ and $|R_1| = |R_2|$, and if there are no smaller transitive relations that contain R, then $R_1 = R_2$. NOTE: We use "smaller" in the sense that R_1 is smaller than R_2 if $|R_1| < |R_2|$.

1.4.2 Let G be a graph with interconnection matrix M. Show that M^2 gives the number of paths of length 2 between any two vertices of G. How is M^3 related to the paths in G? If G is the graph of a relation R on S, then how do you construct the graph of R′, the transitive closure of R? How is the interconnection matrix of this graph related to M?

1.4.3 Prove the assertion that the graph of a partial ordering contains no cycles other than self-loops.

1.4.4 Prove that congruence modulo *n* is an equivalence relation.

1.4.5 Consider the relation $|$ on $\mathbf{Z}$, where $x|y$ (read as "*x* divides *y*") by definition if *x* is a factor of *y*. Which of the four properties of relations does this relation have?

1.4.6 Two integers *x* and *y* are said to be relatively prime if their greatest common divisor is 1. What are the properties of the relation "*x* is relatively prime to *y*"?

1.4.7 Let R be a partial ordering on a set S. Given any subset T of S, we say that $x \in S$ is a *lower bound* of T if $x \, R \, t$ for every $t \in T$. If x is a lower bound of T, and if $y \, R \, x$ for every lower bound y of T, we call x a *greatest lower bound* of T. Show that every subset of S has at most one greatest lower bound. For the relations $\leq$ and $|$ on $\mathbf{Z}$, what are the greatest lower bounds of two integers?

1.4.8   A relation R on S is said to be a *well ordering* if it is a total ordering, and if every nonempty subset of S has a smallest element (where the smallest element of a subset T is an element x such that $x \, R \, t$ for every $t \in T$). Show that, if R is a total ordering on a finite set S, then it is a well ordering. Show that the nonnegative real numbers are not well ordered by $\leq$. Is $\leq$ on $\mathbf{N}$ a well ordering? Is $\leq$ on the nonnegative elements of $\mathbf{Q}$ a well ordering?

1.4.9   The following algorithm (Warshall 1962) computes the transitive closure of a relation R on n objects by computing $(R \vee I)^{(n-1)}$. (See Exs. 1.3.6 and 1.3.7.)

Step 1:   Set the matrix M equal to $R \vee I$.
Step 2:   Repeat step 3 for $i = 1, 2, \ldots, n$; then terminate.
Step 3:   Repeat step 4 for $j = 1, 2, \ldots, n$.
Step 4:   If $[M]_{ij} = 1$, then repeat step 5 for $k = 1, 2, \ldots, n$.
Step 5:   Set $[M]_{jk}$ to $[M]_{jk} \vee [M]_{ik}$.

Prove that the algorithm is correct.

1.4.10   Modify the algorithm of Ex. 1.4.9, replacing step 1 by the following:

Step 1:   Set the matrix M equal to R.

Show that the modified algorithm also computes the transitive closure of R.

1.5 Functions

Discrete functions essentially are transformations that carry one set into another set. Because the input and output signals of a typical computer are all discrete (in some sense), the computer is an instrument for transforming a discrete set of inputs into a discrete set of outputs. The transformation is the computation performed by the computer. Computer programs also fit this model. For example, a compiler is a computer program that transforms programs written in a high-level language into programs that are executable by the computer; thus a compiler fits our informal definition of a function. In this section we discuss the set-theoretical definition of function and examine several important classes of functions.

Definition   A relation f on $X \times Y$ is said to be a *function* from X to Y if **(i)** for every $x \in X$ there is a $y \in Y$ such that $\langle x, y \rangle \in f$, and **(ii)** for every $x \in X$ and $y_1, y_2 \in Y$,

$$((\langle x, y_1 \rangle \in f) \wedge (\langle x, y_2 \rangle \in f)) \Rightarrow (y_1 = y_2).$$

We call X the *domain* of the function and Y the *range* (or codomain) of the function.

Informally, we may summarize the definition by saying that the function f associates some particular element of Y with each element of X. The first part of the definition guarantees that every element of the function domain is associated with some element in the range; the second part of the definition guarantees that the function is single-valued. If $\langle x, y \rangle \in f$, we say that f *maps x into y*, and we denote this by $f(x) = y$. We often use the term *mapping* as a synonym for function. We commonly use the notation $f: X \to Y$ to state explicitly the range and domain of a function.

The following are examples of functions:

$f: \mathbf{Z} \to \mathbf{Z}, f(x) = 2x;$
$f: \mathbf{Z} \to \{0\}, f(x) = 0;$
$f: \mathbf{Z} \to \mathbf{Z}, f(x) = x + 1;$ and
$f: \mathbf{R} \to \mathbf{C}, f(x) = \sqrt{x}$ (the positive square root of x).

In the graph of a function from X to Y, the definition of function requires that exactly one edge be directed from each vertex in X to some vertex in Y. The graphs in Fig. 1.5.1 are all graphs of functions.

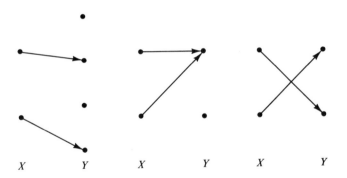

| X | Y | X | Y | X | Y |

FIG. 1.5.1   Graphs of Functions from X to Y

When we define functions on finite sets, we usually specify a value of the function for each of its domain values. Such a specification is simply a tabulation similar to those used in our definition of the logical connectives.

Definition   The *inverse* of a function $f: X \to Y$ is the set of ordered pairs

$$\{\langle y, x \rangle \mid \langle x, y \rangle \in f\}.$$

We denote the inverse of f by the notation f^{-1}.

As the definition suggests, we can graph the inverse of a function f by reversing the edges in the graph of f. The inverse of a function is always a relation on $Y \times X$, but it is not necessarily a function (see Ex. 1.5.1).

We classify functions according to certain properties they may have. The next definition describes some of the more important of these properties.

Definition The function $f: X \to Y$ is said to be

    (i)   an *injection*, or *one-to-one mapping*, if $f(x_1) = f(x_2)$ implies $x_1 = x_2$ for every x_1 and x_2 in X;

    (ii) a *surjection*, or *onto mapping*, if, for every $y \in Y$, there is an $x \in X$ such that $\langle x, y \rangle \in f$; or

    (iii) a *bijection*, or *one-to-one onto mapping*, if f is both one-to-one and onto.

If $f: X \to Y$ is a surjection, then we say that f maps X *onto* Y; otherwise we say that f maps X *into* Y. The graphs of Fig. 1.5.2 illustrate examples of the types of functions.

Note that, if $f: X \to Y$ is a bijection, then $f^{-1}: Y \to X$ is a function and, in fact, is also a bijection.

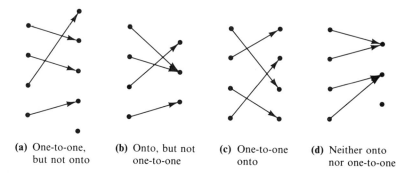

(a) One-to-one, but not onto   **(b)** Onto, but not one-to-one   **(c)** One-to-one onto   **(d)** Neither onto nor one-to-one

FIG. 1.5.2 Examples of Various Types of Functions

Definition For a function $f: X \to Y$, the *image of X under f*, written $f(X)$, is the set $\{y \mid \exists\, x \in X\, (f(x) = y)\}$.

From this definition, we note that $f(X)$ is the subset of elements of Y onto which some element of X is mapped by f. If f is onto, $f(X) = Y$. If f is one-to-one, f^{-1} is a function from $f(X)$ to X.

The following theorem, sometimes known as the *pigeonhole principle*, demonstrates a very important use of properties of functions.

Theorem Let X and Y be finite sets, and let $|X| = |Y|$. Then $f: X \to Y$ is an injection if and only if it is a surjection.

    Proof: **(i)** If f is an injection, then $|X| = |f(X)| = |Y|$. From the definition of f, we have $f(X) \subseteq Y$. But $|f(X)| = |Y|$, and, because $|Y|$ is finite, $f(X) = Y$. Hence, "f is an injection" implies "f is a surjection." **(ii)** If f is a surjection, $f(X) = Y$ by definition of surjection, so that

$|X| = |Y| = |f(X)|$. Because $|X| = |f(X)|$ and $|X|$ is finite, f is an injection. Hence, "f is a surjection" implies "f is an injection." □

Thus, if $|X| = |Y|$ and X and Y are finite sets, then any function from X to Y that is either an injection or a surjection is automatically also a bijection. It is tempting to assume that this theorem also is true for infinite sets, but unfortunately it is not. For example, the function $f: \mathbf{Z} \to \mathbf{Z}$, where $f(x) = 2x$, maps the integers onto the even integers (clearly a proper subset of the integers). In this case, f is one-to-one, yet it is not onto. The proof of the pigeonhole principle breaks down in this case because $f(X) \subset Y$, yet $|f(X)| = |Y| = \infty$.

Definition The *set of all functions from X to Y* is denoted Y^X. The notation is suggested by the fact that the number of functions in this set is $|Y|^{|X|}$ when both Y and X are finite sets.

We mention several discrete functions in preceding sections, without identifying them as such. The logical connectives $\wedge$, $\vee$, $\Rightarrow$, and $\Leftrightarrow$ as defined in Section 1.1 are functions on $\{false, true\} \times \{false, true\}$ into $\{false, true\}$. During syntax checking, a FORTRAN compiler computes a function from the set of FORTRANlike programs into the set $\{incorrect, correct\}$ as it computes whether or not the syntax is correct. Even more central to computer applications is the use of functions to transform the changing physical states of devices within the computer into information that people can comprehend. A typical example is a printing device, which maps electrical signals into mechanical responses.

The reader who is familiar with some elementary aspects of computer operation will know about the various encodings of integers and instructions into groups of two-valued signals. These encodings usually are bijections, because a unique binary representation is associated with each integer or instruction, and a unique integer or instruction is associated with each binary representation.

Exercises

1.5.1 Of the functions graphed in Fig. 1.5.1, which have inverses that are functions?

1.5.2 For finite sets S and T, show that
   (a) a one-to-one map from S to T exists if and only if $|T| \geq |S|$;
   (b) a map from S onto T that is not one-to-one exists if and only if $|T| < |S|$.

1.5.3 Determine whether each of the following functions is an injection, a surjection, a bijection, or none of these:
  (a) $f: \mathbf{Z} \to \mathbf{Z}, f(x) = -x$;
  (b) $f: \mathbf{R} \to \mathbf{C}, f(x) = \sqrt{x}$;
  (c) $f: \mathbf{C} \to \mathbf{C}, f(x) = \sqrt{x}$;
  (d) $f: \mathbf{Z} \to \mathbf{Z}, f(x) = \begin{cases} x/2 & \text{if } x \text{ is even,} \\ (x-1)/2 & \text{if } x \text{ is odd;} \end{cases}$
  (e) $f: \mathbf{Z} \to \mathbf{Z}, f(x) = x + 6$.

1.5.4 Let X and Y be finite, nonempty sets. Prove that Y^X has $|Y|^{|X|}$ elements. How many functions are there in Y^X if either Y or X is empty?

1.5.5 If it is a bijection, a function $p: X \to X$ is said to be a *permutation* of X. Let $p^2: X \to X$ be the function $p^2(x) = p(p(x))$, where p is a function from X to X. Show that p^2 is a permutation if and only if p is a permutation.

1.5.6 Consider the permutation p of the set X graphed in Fig. 1.5.3. Note that p^6 is the identity mapping on X, which carries each element of X into itself. Prove that, for every finite set X and every permutation p on X, there is some integer n such that p^n is the identity mapping on X.

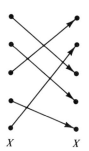

X X

FIG. 1.5.3   A Permutation of the Set X

1.5.7 Consider the integers modulo p, that is, the set

$$\mathbf{Z}_p = \{0, 1, \dots, (p-1)\}.$$

For each integer r, where $1 \le r < p$, the map $f_r: \mathbf{Z}_p \to \mathbf{Z}_p$ is the map such that $f_r(x) = (r \cdot x) \bmod p$. Show that, for each p, every f_r is a bijection if and only if p is prime.   HINT: Examine cases for which $(r \cdot x) \bmod p = (r \cdot 0) \bmod p = 0$, and then use the pigeonhole principle.

1.5.8 Show that there exist two people in the world with identical numbers of hairs on their heads.   NOTE: Very crude estimates are sufficient to establish the validity of this statement.

1.5.9 One astrologer claims to be able to give accurate personalized horoscopes by making use of both the subject's birthday and the subject's mother's birthday. For example, he claims that those born on June 5 whose mothers were born on March 20 have keen, logical minds. Noting that there are well over a million births each year in the United States, show that at least two babies are born each year with the same horoscope. If this astrologer were to include the subject's father's birthday in his determination of the horoscope, prove that at least two people now alive in the United States would have the same horoscope.

1.6 Composition of Functions

Consider the pair of functions $f: X \to Y$ and $g: Y \to Z$. Because the range of f is the domain of g, every element of X can be mapped first by f into an element of Y, and then by g into an element of Z. In other words, by applying the functions f and then g to $x \in X$, we can obtain $g(f(x)) \in Z$. In doing so, we have constructed a new function h, where $h: X \to Z$ and $h(x) = g(f(x))$.

Definition   The *left composition* of a function $g: Y \to Z$ with a function $f: X \to Y$ is the function $g \circ f: X \to Z$, which maps x into $g(f(x))$.

Definition   A function $g: W \to Z$ is said to be *composable on the left* of a function $f: X \to Y$ if $f(X)$, the image of X under f, is a subset of W, the domain of g.

Fig. 1.6.1 graphs the left composition of a particular pair of functions. For our present discussion, the left composition of two functions is the natural way to denote the composable function, because it follows the usual convention of functional notation. The functional notation $g(f(x))$ denotes that we apply f to x and then g to $f(x)$, thus evaluating functions from right to left. Thus the left composition of g with f, $g \circ f$, is evaluated from right to left as well. In some other chapters of this text, we find it convenient to deal with the *right composition* of functions, which is merely a different notational convention in which functions are evaluated from left to right. Thus for right composition, the notation $g \circ f$ means to apply g first and then to apply f. Analogous

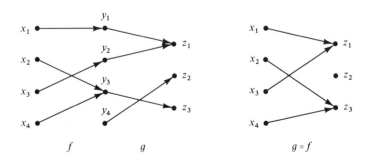

FIG. 1.6.1   Left Composition of a Function g with a Function f

to right composition is the functional notation in which the application of g to x, followed by the application of f to the result, is denoted as $((x)g)f$. In this text, we follow the more common convention of functional notation, and we use the notation $g \circ f$ to denote $g(f(x))$ except in those cases where we explicitly state otherwise.

Lemma The composition of two functions is a function.

    Proof: Let $g: W \to Z$ be left composable with $f: X \to Y$ so that $f(X) \subseteq W$. To prove that $g \circ f$ is a function, we must show **(i)** that $g \circ f$ maps every x in X into some z in Z, and **(ii)** that if z_1 and z_2 are images of x under $g \circ f$, then $z_1 = z_2$.

    To prove **(i)**, note that for every x in X there is a unique ordered pair $\langle x, y \rangle$ in f and a unique ordered pair $\langle y, z \rangle$ in g such that $y = f(x)$ and $z = g(y)$. By definition, $g \circ f(x)$ is the ordered pair $\langle x, z \rangle$. Hence, for each x, we can find a z such that $z = g(f(x))$.

    To prove **(ii)**, suppose that $g \circ f$ contains the ordered pairs $\langle x, z_1 \rangle$ and $\langle x, z_2 \rangle$. This can occur only if there exist y_1 and y_2 in Y such that $\langle x, y_1 \rangle$ and $\langle x, y_2 \rangle$ are in f, and $\langle y_1, z_1 \rangle$ and $\langle y_2, z_2 \rangle$ are in g. Because f is a function, $y_1 = y_2$, and we must have both $\langle y_1, z_1 \rangle$ and $\langle y_1, z_2 \rangle$ in g. But g is a function and hence $z_1 = z_2$.   □

    Assume $g \circ f$ is a function. The inverse of $g \circ f$, written $(g \circ f)^{-1}$, is a well-defined relation, although it may not be a function. Because $g \circ f$ is completely defined in terms of g and f, it is not surprising to find that $(g \circ f)^{-1}$ also is completely defined in terms of g and f.

Lemma Let $f: X \to Y$ and $g: f(X) \to Z$. If $(g \circ f)^{-1}$ is a function from $g \circ f(X)$ onto X, then **(i)** f and g are injections, and **(ii)** $(g \circ f)^{-1} = f^{-1} \circ g^{-1}$.

    Proof: See Ex. 1.6.3.   □

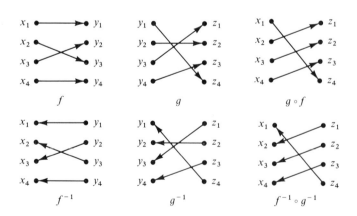

FIG. 1.6.2 Left Composition of Composable Functions f and g and Their Inverses

A graphic example of this lemma appears in Fig. 1.6.2, which shows that the graph of $(g \circ f)^{-1}$ is obtained from the graph of $g \circ f$ by reversing the arrows. Hence, if there is a path in $g \circ f$ from x to y to z—where $y = f(x)$ and $z = g(y)$—then the path is reversed in $(g \circ f)^{-1}$. This reversed path is directed from z to $y = g^{-1}(z)$ to $x = f^{-1}(y)$. In less abstract terms, the process of putting on a shirt and then putting on a tie is reversed by removing the tie and then removing the shirt.

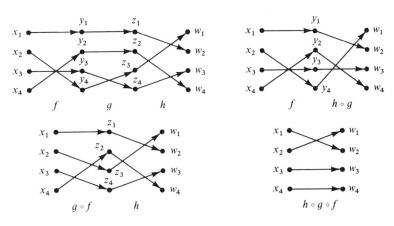

FIG. 1.6.3   Composite Functions of h, g, and f

Consider three functions f, g, and h, where g is composable on the left with f, and h is composable on the left with g. We can construct the composite functions $h \circ (g \circ f)$ and $(h \circ g) \circ f$. Because these two composites always are the same function, we call the resulting function $h \circ g \circ f$, and we can safely ignore the parentheses.

Definition   The composition of functions f, g, and h is said to be *associative* if $h \circ (g \circ f) = (h \circ g) \circ f$. That is, if X is the domain of f, then

$$h(g \circ f(x)) = h \circ g(f(x))$$

for every $x \in X$.

Lemma   The composition of functions is associative.
   Proof: See Ex. 1.6.6.   □

For an informal proof of the lemma, we use the graphs in Fig. 1.6.3. If there is a path $\langle x, y, z, w \rangle$ in the graph of $h \circ g \circ f$, then clearly there is a path $\langle x, z, w \rangle$ in $h \circ (g \circ f)$ and a path $\langle x, y, w \rangle$ in $(h \circ g) \circ f$. Pairwise composition of functions in $h \circ g \circ f$ does not eliminate any of the paths from X to W; it merely shortens each path by eliminating intermediate nodes. Hence, each path that exists from X to W in $h \circ g \circ f$ also exists in $(h \circ g) \circ f$ and in $h \circ (g \circ f)$. Thus we conclude that, for all $x \in X$,

$$h(g(f(x))) = h(g \circ f(x)) = h \circ g(f(x)) = h \circ g \circ f(x).$$

Fig. 1.6.4 graphically illustrates the property of associativity. Each edge of the graph corresponds to the application of a function; a path from the initial node to the final node corresponds to the composition of a sequence of functions. The property of associativity guarantees that every path from the initial node to the final node represents the same function.

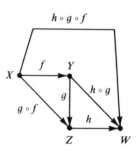

FIG. 1.6.4   Graph Depicting All Possible Ways of Composing Three Functions

Exercises

1.6.1 Let $f_i: X_i \to X_{i+1}$, $i = 1, 2, \ldots, n$, be a family of functions for $n > 3$. Show that the composite function

$$f_n \circ f_{n-1} \circ \cdots \circ f_1 = f_n \circ \left(f_{n-1} \circ \left(f_{n-2} \circ (\ldots \circ f_1) \ldots \right)\right)$$

is equal to

$$\left(\ldots \left((f_n \circ f_{n-1}) \circ f_{n-2}\right) \circ \ldots \right) \circ f_1.$$

1.6.2 List the necessary and sufficient conditions on f and g so that
  (a) $f \circ g$ is an injection;
  (b) $f \circ g$ is a surjection;
  (c) $f \circ g$ is a bijection.

1.6.3 Show that, if $(g \circ f)^{-1}$ is a function, then
  (a) f and g are injections; and
  (b) $(g \circ f)^{-1} = f^{-1} \circ g^{-1}$.

1.6.4 A function $g: S \to T$ is said to be a *left inverse* of a function $f: T \to S$, if $g(f(t)) = t$ for every $t \in T$. If g is a left inverse of f, then f is said to be a *right inverse* of g. Prove that
  (a) $f: T \to S$ has a left inverse if and only if it is an injection;
  (b) $f: T \to S$ has a right inverse if and only if it is a surjection;
  (c) if $g: S \to T$ is both a left and a right inverse of $f: T \to S$, then f is a bijection and $g = f^{-1}$.

1.6.5 Let X, Y, and Z be finite, nonempty sets, and let $f: X \to Y$ and $g: Y \to Z$ be functions on these sets. How many composite functions $g \circ f$ are there? NOTE: Not all of these composite functions are distinct.

1.6.6 Prove that the composition of any three composable functions is associative.

1.6.7 Let f, g, and h be the functions $f(x) = x^2$, $g(x) = e^x$, and
$$h(x) = (1 + x)/(1 - x).$$
What are the sets $f(\mathbf{R})$ and $g(\mathbf{R})$? Why cannot $\mathbf{R}$ be the domain of h? What is the largest subset of $\mathbf{R}$ on which h is defined? What are the composite functions $f \circ g$, $g \circ f$, $f \circ h$, $h \circ f$, $g \circ h$, and $h \circ g$? Which of the composite functions are defined on $\mathbf{R}$?

1.6.8 For the functions f, g, and h of Ex. 1.6.7, what are f^{-1}, g^{-1}, and h^{-1}? Which of these are functions on $\mathbf{R}$? What are $(f \circ g)^{-1}$, $(f \circ h)^{-1}$, $(g \circ f)^{-1}$, and $(g \circ h)^{-1}$?

1.7 Binary Operations

In Section 1.6 we investigate the construction of new functions from old ones through function composition. In this section we treat an important class of functions that lend themselves naturally to a construction similar to function composition.

Definition A function f is said to be a *binary operation* (or *composition*) on a set X if it is a function from $X \times X$ into X.

The usual arithmetic operations $+$, $-$, and $\cdot$ are binary operations on $\mathbf{R}$ because $x + y$, $x - y$, and $x \cdot y$ are real numbers for every pair of real numbers x and y. Note that $x/0$ is usually undefined, so $/$ is not a binary operation on $\mathbf{R}$. Because we can define $x/0$ to be real, but not finite, we shall hereafter assume that $/$ is a binary operation on $\mathbf{R}$. We generally use *infix* notation for binary operations. In this notation the binary operation $\circ$ on X carries the pair $\langle x_1, x_2 \rangle$ into the element denoted $x_1 \circ x_2$.

Computer arithmetic offers many examples of binary operations on finite sets. Every computer has a repertoire of integer arithmetic operations, normally including addition, subtraction, multiplication, and division. Because of the inherent finiteness of the computer, however, only a finite subset of the integers can be manipulated. Hence the arithmetic operations usually are operations on $\mathbf{Z}_n$ for some n, and consequently the arithmetic is done modulo n.

Many of the devices within computers perform binary operations other than the arithmetic operations we have mentioned. Because signal values lie in the set $\mathbf{Z}_2 = \{0, 1\}$, any device with two inputs and one output is a binary operation on $\mathbf{Z}_2$. Binary operations on the set $\{ false, true \}$, including the operations $\wedge$, $\vee$, $\Rightarrow$, and $\Leftrightarrow$, are discussed in Section 1.1. As we have mentioned, computer engineers sometimes use $\{ false, true \}$ rather than $\{0, 1\}$ as the names for the signal values; consequently, the devices that operate on these signals are called *logic devices*.

We specify a binary operation by tabulating its value for each ordered pair in the domain. Such a table often is called a *composition table*; most readers are familiar with examples such as addition and multiplication tables. Fig. 1.7.1 shows the table for the binary operation $*$ on $\{1, 2, 3\}$. To determine the element $x_1 * x_2$, find the row labeled x_1 and the column labeled x_2. The value of $x_1 * x_2$ is found at the intersection of this row and column. For example, $1 * 2 = 2$, and $2 * 1 = 3$.

$*$	1	2	3
1	3	2	1
2	3	3	2
3	3	3	3

FIG. 1.7.1   Composition Table for $*$ on $\{1, 2, 3\}$

Although binary operations are not composable in the same sense that is discussed in the preceding section, there is a natural way to combine them. If both $\circ$ and $*$ are binary operations on X, then for any $x_1, x_2, x_3 \in X$, two possible ways of combining the operations are $(x_1 \circ x_2) * x_3$ and $x_1 \circ (x_2 * x_3)$. Both of these ways are well defined because $x_1 \circ x_2$ and $x_2 * x_3$ are elements of X. A binary operation always can be combined with itself, as exemplified by $x_1 + (x_2 + (x_3 + x_4))$ and $x_1 \cdot (x_2 \cdot (x_3 \cdot x_4))$. Fig. 1.7.2 illustrates the composition of binary operations in comparison with the composition of functions. In the illustrated examples, note that the range of f must be a subset of the domain of g, but that the range of $+$ is not a subset of the domain of $*$ because X is not a subset of $X \times X$.

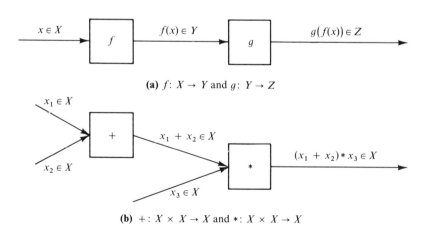

(a) $f: X \to Y$ and $g: Y \to Z$

(b) $+: X \times X \to X$ and $*: X \times X \to X$

FIG. 1.7.2   Composition of Functions and of Binary Operations

These operations are called "binary" to indicate that an ordered pair of elements of X is mapped into a single element of X by the operation. When a binary operation is combined with any other binary operation on the same set or with itself, as in the function $x_1 \circ (x_2 \circ x_3)$, the resulting function maps $X \times X \times X = X^3$ into X. Such a function maps ordered triples of elements of X into single elements of X, and therefore is called a *ternary operation*. More generally, when we combine $n - 1$ binary operations on X, we create a function that maps n-tuples from X into elements of X. In other words, the function is a map from X^n into X. Functions from X^n into X are called *n-ary operations*. As these remarks suggest, some (but possibly not all) n-ary operations can be constructed from binary operations. By simple combinatorial arguments, we can in fact demonstrate the existence of n-ary functions that cannot be constructed from binary functions (Ex. 1.7.9). Because we deal so frequently with binary operations, we refer to them by their more compact name *compositions* whenever there is no possibility for confusion with the composition of functions.

Most computer components have three or four inputs and a single output, so such devices are ternary or quaternary operations on the set $\{0, 1\}$. Although the components have more than two inputs, very few of the functional operations of a computer have more than two operands. For example, the arithmetic operations are binary operations; it is extremely unusual to build a computer with a ternary arithmetic operation such as $f(x, y, z) = x + y + z$.

We can extend to binary operations the definition of the associative property of composable mappings. However, unlike composable mappings, binary operations are not necessarily associative. Moreover, binary operations may have special properties other than associativity.

Definition   Let $\circ$ and $*$ be compositions on a set X. Then,

(i) $\circ$ is said to be *associative* if, for all $x, y, z \in X$,

$$(x \circ y) \circ z = x \circ (y \circ z);$$

(ii) $\circ$ is said to be *commutative* if, for all $x, y \in X$,

$$x \circ y = y \circ x;$$

(iii) $\circ$ is said to be *distributive* over $*$ if, for all $x, y, z \in X$,

$$x \circ (y * z) = (x \circ y) * (x \circ z) \quad \text{and} \quad (y * z) \circ x = (y \circ x) * (z \circ x).$$

The compositions $+$ and $\cdot$ on $\mathbf{R}$ are associative and commutative, but $-$ and $/$ have neither property. Note that $\cdot$ is distributive over $+$ on $\mathbf{R}$, but $+$ is not distributive over $\cdot$.

Elements of sets on which compositions are defined also may have special properties worthy of discussion. For the following definitions, let $\circ$ be a composition on X.

Definition   An element e of X is said to be an *identity* with respect to $\circ$ if, for all $x \in X$,

$$e \circ x = x \circ e = x.$$

Definition   An element z of X is said to be a *zero* with respect to $\circ$ if, for every $x \in X$,

$$z \circ x = x \circ z = z.$$

Definition   An element i of X is said to be an *idempotent* with respect to $\circ$ if $i \circ i = i$.

Consider the compositions $+$ and $\cdot$ on $\mathbf{R}$. The identity for $+$ is 0, and the identity for $\cdot$ is 1. There is no zero for $+$, but 0 is a zero for $\cdot$. From the definitions it is clear that an identity is automatically an idempotent, but an idempotent is not necessarily an identity. The operation $+$ on $\mathbf{R}$ has no idempotent other than its identity, but both 0 and 1 are idempotents for $\cdot$ on $\mathbf{R}$.

Lemma   If $\circ$ is a composition on X, then there is at most one identity in X with respect to $\circ$.

Proof: If x and y are both identities with respect to $\circ$, then $x = x \circ y$ because y is an identity, and $y = x \circ y$ because x is an identity, so $x = x \circ y = y$.   ☐

The notion of inverse extends to compositions with an identity element.

Definition   Let $\circ$ be a composition on X with identity e. Then y is said to be the *inverse* of x with respect to $\circ$ if

$$x \circ y = y \circ x = e.$$

We usually denote the inverse of x as x^{-1}.

For $+$ on $\mathbf{R}$, the inverse of a number x is the number $-x$. For $\cdot$ on $\mathbf{R}$, the inverse of every nonzero number x is the number $1/x$.

Lemma   If $\circ$ is an associative composition on X with an identity e, then every $x \in X$ has no more than one inverse.

Proof: Let $x \in X$ have the inverses y_1 and y_2. We show that $y_1 = y_2$:

$$y_1 = y_1 \circ e = y_1 \circ (x \circ y_2) = (y_1 \circ x) \circ y_2 = e \circ y_2 = y_2.$$ ☐

The compositions $+$ and $\cdot$ on $\mathbf{R}$ satisfy the hypothesis of the lemma, so it is no surprise that every real number has a unique additive inverse, and every nonzero real number has a unique multiplicative inverse.

A property similar to invertibility, but less restrictive, is a property known as cancellability.

Definition   An element $x \in X$ is said to be *cancellable* with respect to a composition $\circ$ on X if, for every $y_1, y_2 \in X$,

$$((x \circ y_1 = x \circ y_2) \vee (y_1 \circ x = y_2 \circ x)) \Rightarrow (y_1 = y_2).$$

The composition $\cdot$ on $\mathbf{Z}$ exemplifies the reason that the property of cancellability is of interest. Because $\mathbf{Z}$ does not contain the fractions $\{1/n\}$, it does not contain the multiplicative inverses of the integers. Hence, no integer other than 1 is invertible with respect to $\cdot$ on $\mathbf{Z}$.

However, every nonzero integer is cancellable with respect to $\cdot$ on $\mathbf{Z}$. For example, if $3 \cdot x = 3 \cdot y$, we can cancel the 3 and conclude that $x = y$.

We can detect cancellable elements through use of a table such as that given in Fig. 1.7.3., for the composition $\circ$ on $X = \{1, 2, 3, 4\}$. In fact, if a row contains any two entries that are equal, the corresponding element is not cancellable. For example, 1 is not cancellable because $1 \circ 2 = 1 \circ 4$. Hence the only candidate for a cancellable element is 4, because only this element heads a row in which each entry is distinct. Cancellability also requires that the column entries be distinct; for example, $1 \circ 1 = 4 \circ 1$ implies that 1 is not cancellable. Note that both row 4 and column 4 have distinct entries. Thus 4 is cancellable with respect to $\circ$ and is the only such element.

$\circ$	1	2	3	4
1	4	2	3	2
2	1	1	3	4
3	4	2	2	1
4	4	1	2	3

FIG. 1.7.3   Composition Table for $\circ$ on $X = \{1, 2, 3, 4\}$

Note that each row is a mapping from X into X. In particular, row i is the map that carries j into $i \circ j$. Similarly, the columns are maps from X into X such that column i carries an element j into $j \circ i$. An element is cancellable if and only if its row and column maps both are one-to-one maps. For finite sets, the pigeonhole principle immediately tells us that the maps also will be onto. However, as we have seen for multiplication on $\mathbf{Z}$, the maps need not be onto for infinite sets.

Exercises

1.7.1 Consider the compositions $\cap$ and $\cup$ on the power set of X, where X is an arbitrary set. Determine if these compositions have any of the following properties: **(a)** associativity; **(b)** commutativity; **(c)** one composition distributes over the other; **(d)** idempotents; **(e)** identities; or **(f)** invertible elements.

1.7.2 How many compositions are there on a set X? How many are commutative?

1.7.3 How many compositions on X have an identity element? Of these, how many are commutative?

1.7.4 For each of the compositions on $X = \{0, 1\}$, determine if the composition is associative or commutative, and if it has an identity element. If it has an identity, determine inverses of the elements that are invertible.

1.7.5   What are the necessary and sufficient conditions for a composition on $X = \{0, 1\}$ to be associative?

1.7.6   In this section we prove that, if an element x has an inverse with respect to an associative composition, then the inverse is unique. If the composition is not associative, does this uniqueness property still hold?

1.7.7   For every $n \geq 2$, *addition modulo n* (denoted $+_n$) is defined to be the composition on $\mathbf{Z}_n$ that maps $\langle x_1, x_2 \rangle$ into x_3, where $x_3 \equiv (x_1 + x_2)$ mod n. The table for $+_3$ is given in Fig. 1.7.4. Show that $+_n$ is associative and commutative.

$+_3$	0	1	2
0	0	1	2
1	1	2	0
2	2	0	1

FIG. 1.7.4   Composition Table for $+_3$

1.7.8   Multiplication modulo n, denoted $\cdot_n$, is the composition on $\mathbf{Z}_n$ that carries $\langle x_1, x_2 \rangle$ into $x_3 = (x_1 \cdot x_2)$ mod n. The table for $\cdot_3$ is given in Fig. 1.7.5. Show that $\cdot_n$ is associative and commutative.

$\cdot_3$	0	1	2
0	0	0	0
1	0	1	2
2	0	2	1

FIG. 1.7.5   Composition Table for $\cdot_3$

1.7.9   How many ternary operations are there on a set X? Prove that sets exist for which there are some ternary operations that cannot be composed from two binary operations.   HINT: Let $f(x_1, x_2, x_3)$ be a ternary operation on X. If f is composable from two binary operations, then

$$f(x_1, x_2, x_3) = x_{i_1} + (x_{i_2} * x_{i_3}),$$

where $+$ and $*$ are some binary operations on X, and where x_{i_1}, x_{i_2}, and x_{i_3} are the elements x_1, x_2, and x_3, possibly in a different order. If B is the number of binary operations on X, then there can be no more than $6 \cdot B^2$ different ternary functions created by composing binary operations.

1.7.10   Extend the argument in Ex. 1.7.9 to show that, for all $n > 3$, there exist sets for which some n-ary operations are not composed from binary operations.

1.8 Algebraic Structures and Structure-Preserving Maps

Everyone is familiar with the systematic rules and properties that characterize arithmetic on the reals. High-school algebra is, in fact, the study of these properties. Arithmetic is just one example of what we call an algebraic structure, and modern algebra is the study of many such structures. In this section we develop the notion of an algebraic structure; in other chapters we study particular structures of practical importance.

From this study of algebraic structures emerges the notion of homomorphism, one of the central notions of this text. We discover that two algebraic structures may exhibit similar characteristics and, in fact, one structure may be used to simulate the other. For example, as we have mentioned several times, computer arithmetic is arithmetic on a finite set of numbers, whereas conventional arithmetic is arithmetic on an infinite set. The two arithmetic systems are inherently different, yet computer arithmetic is a rather acceptable imitation of arithmetic on **R** for many purposes. The imitation is good because it satisfies certain criteria, guaranteeing that many of the crucial properties of arithmetic on **R** are preserved in computer arithmetic, although it is impossible to preserve all of arithmetic on **R**. The study of homomorphisms reveals the conditions under which one algebraic structure can simulate another.

We begin this study by making precise the notion of algebraic structure.

Definition   An *algebraic structure*, written

$$\langle X, Y, \ldots, \circ, *, \ldots, R_1, R_2, \ldots, x, y, \ldots \rangle,$$

is an *n*-tuple that has elements drawn from among the following: **(i)** sets such as X and Y; **(ii)** functions and *n*-ary operations such as $\circ$ and $*$, whose domains and ranges are cartesian products of the sets; **(iii)** relations such as R_1 and R_2 on the sets; and **(iv)** distinguished elements of the sets such as x and y.

Among the algebraic structures already studied in this chapter are the following:

$\langle \mathbf{Z}, \leq \rangle$, a total ordering;

$\langle \mathbf{Z}, = \rangle$, an equivalence relation;

$\langle \mathscr{P}(X), \cup, \varnothing \rangle$, a set with an associative composition $\cup$ that has an identity element $\varnothing$;

$\langle \mathscr{P}(X), \cap, X \rangle$, a set with an associative composition $\cap$ that has an identity element X;

$\langle \mathbf{R}, +, \cdot, 0, 1 \rangle$, a set with two associative compositions $+$ and $\cdot$, and with elements 0 and 1 that are identities with respect to $+$ and $\cdot$, respectively.

In many cases, two different algebraic structures share similar characteristics. When one structure is defined on a set X and a similar structure is defined on a set Y, we can exhibit the nature of the similarity by finding a mapping from X into Y that preserves some characteristics of the structure.

Definition   Let $\langle X, \circ \rangle$ and $\langle Y, * \rangle$ be algebraic structures with compositions $\circ$ and $*$, respectively. Then a function $f: X \to Y$ is said to be a *homomorphism* from $\langle X, \circ \rangle$ to $\langle Y, * \rangle$ if, for every $x_1, x_2 \in X$,

$$f(x_1) * f(x_2) = f(x_1 \circ x_2).$$

A homomorphism preserves some, but possibly not all, of the characteristics of the domain structure. The definition says that composing two elements of the domain and then applying the homomorphism mapping gives you the same result as that obtained by mapping each of the elements first and then composing them in the range. Fig. 1.8.1 illustrates this definition graphically.

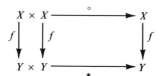

FIG. 1.8.1   Diagram Illustrating a Homomorphism

As an example of a homomorphism, consider the tables for $+_4$ and $+_2$ (Fig. 1.8.2). If we replace 0 and 2 in $+_4$ by E (for even) and replace 1 and 3 by O (for odd), we obtain the composition $+_4$ on $\{E, O\}$ (the evens and odds). This composition is identical to $+_2$ on $\{0, 1\}$. In other words, there exists a homomorphism from $\langle \mathbf{Z}_4, +_4 \rangle$ to $\langle \mathbf{Z}_2, +_2 \rangle$. This homomorphism is the function f such that $f(0) = f(2) = 0$, and $f(1) = f(3) = 1$.

Because homomorphisms are functions, they may also be injections, surjections, or bijections. We give special names to homomorphisms with any of these properties.

$+_4$	0	1	2	3
0	0	1	2	3
1	1	2	3	0
2	2	3	0	1
3	3	0	1	2

$+_4$	E	O
E	E	O
O	O	E

$+_2$	0	1
0	0	1
1	1	0

FIG. 1.8.2   Composition Tables for $+_4$ on $\mathbf{Z}_4$, $+_4$ on $\{E, O\}$, and $+_2$ on $\mathbf{Z}_2$

Definition Let f be a homomorphism from $\langle X, \circ \rangle$ to $\langle Y, * \rangle$. Then f is said to be **(i)** an *epimorphism* if it is onto Y; **(ii)** a *monomorphism* if it is one-to-one into Y; **(iii)** an *isomorphism* if it is one-to-one onto Y; **(iv)** an *endomorphism* if it is into itself (that is, if $Y \subseteq X$); or **(v)** an *automorphism* if it is one-to-one onto itself (that is, if $Y = f(X) = X$).

Fig. 1.8.3 graphically illustrates the relationships among the various definitions.

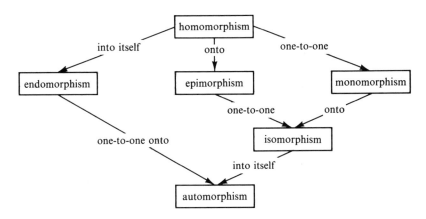

FIG. 1.8.3 Various Types of Homomorphisms and Their Properties

If there exists a homomorphism $f\colon X \to Y$ between two algebraic structures $\langle X, \circ \rangle$ and $\langle Y, * \rangle$, we say that the structure $\langle Y, * \rangle$ is a *homomorphic image* of $\langle X, \circ \rangle$. When the homomorphism is an isomorphism, the two structures are identical except for the labels that we give to the elements of X and Y and the names of the compositions $\circ$ and $*$. For example, consider the compositions multiplication modulo 3 on $\mathbf{Z}_3^+ = \mathbf{Z}_3 - \{0\}$ and addition modulo 2 on $\mathbf{Z}_2$. The tables for these compositions (Fig. 1.8.4) show that the function $f\colon \mathbf{Z}_3^+ \to \mathbf{Z}_2$ for which $f(2) = 1$ and $f(1) = 0$ maps $\langle \mathbf{Z}_3^+, \cdot_3 \rangle$ isomorphically into $\langle \mathbf{Z}_2, +_2 \rangle$. We say that $\langle \mathbf{Z}_3^+, \cdot_3 \rangle$ and $\langle \mathbf{Z}_2, +_2 \rangle$ are *identical to within isomorphism*.

Now suppose that we have to build a box to perform addition modulo 2, and that all we have available are black boxes that perform multiplication modulo 3 and black boxes that rename their inputs by applying f or f^{-1}. Fig. 1.8.4 shows how to construct a box for addition modulo 2 from the black boxes on hand. Here we make use of the fact that, if

$$f(x +_2 y) = f(x) \cdot_3 f(y),$$

then

$$x +_2 y = f^{-1}\big(f(x +_2 y)\big) = f^{-1}\big(f(x) \cdot_3 f(y)\big).$$

Because f is a bijection, f^{-1} is a function (in fact, a bijection) and the

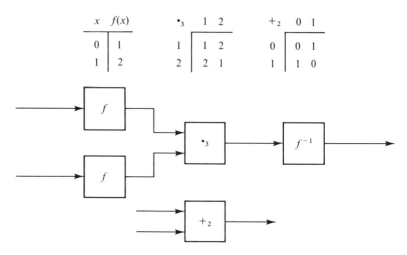

FIG. 1.8.4 A Modulo-2 Adder Constructed from a Modulo-3 Multiplier

right-hand side of the preceding equation is well defined. The use of f and f^{-1} in Fig. 1.8.4 is such that the modulo 3 multiplier sees its inputs labeled properly for its domain, while the inputs and output of the entire network are labeled correctly for the domain of addition modulo 2.

The notion of homomorphism extends to all algebraic structures. For example, consider the algebraic structure $\langle X, +, \cdot, x \rangle$, where $+$ and $\cdot$ are compositions on X, and x is a distinguished element of X. If the structure has a homomorphic image $\langle Y, \oplus, \odot, y \rangle$, then $\oplus$ and $\odot$ are compositions on Y with the same properties as $+$ and $\cdot$, respectively, and y is a distinguished element of Y, the image of x, and with the same properties as x. The homomorphic image of an algebraic structure has sets, n-ary operations, relations, and distinguished elements that correspond one-to-one to each of the elements of the algebraic structure. Moreover, all of the special properties of the original structure hold in the image structure. Hence, if $+$ is associative and commutative in $\langle X, +, \cdot, x \rangle$, then so is $\oplus$ in $\langle Y, \oplus, \odot, y \rangle$. Similarly, if x is an identity with respect to $+$, then so is y with respect to $\oplus$. Thus the map between X and Y preserves some of the structure of $\langle X, +, \cdot, x \rangle$. We discuss examples of homomorphisms of rather complex algebraic structures in other chapters.

The mappings we have been discussing thus far are structure-preserving maps. Analogously, there exist equivalence partitions that preserve the structure of a composition on a set. The following definition makes this notion more precise.

Definition Let R be an equivalence relation on an algebraic structure $\langle X, \circ \rangle$. R is said to be a *congruence* if, for every $x_1, x_2, y_1, y_2 \in X$,

$$((x_1 \text{ R } x_2) \wedge (y_1 \text{ R } y_2)) \Rightarrow (x_1 \circ y_1) \text{ R } (x_2 \circ y_2).$$

Congruence modulo n is a congruence relation on $\langle \mathbf{Z}, \cdot \rangle$ and $\langle \mathbf{Z}, + \rangle$. Because a congruence relation is an equivalence relation, it induces a partition of a set into equivalence classes. The definition of congruence suggests that we can replace any $x \in X$ with any other x', provided that x' is in $[x]$, the set of elements equivalent to x. More precisely, a congruence relation R on the structure $\langle X, \circ \rangle$ induces a *quotient structure* $\langle X/\text{R}, * \rangle$ in which X/R (read as X modulo R) is the set of equivalence classes induced by R, and the composition $*$ is defined by $[x] * [y] = [x \circ y]$ for all $x, y \in X$.

Is the meaning of $[x] * [y]$ unambiguous under this definition? For example, if x_1 and x_2 are in $[x]$, and y_1 and y_2 are in $[y]$, then the definition says $[x] * [y] = [x_1 \circ y_1] = [x_2 \circ y_2]$. But are we guaranteed that $x_1 \circ y_1$ is equivalent to $x_2 \circ y_2$, no matter which x_1, x_2, y_1, and y_2 we select? The definition of congruence does indeed guarantee that $x_1 \circ y_1$ is equivalent to $x_2 \circ y_2$, because x_1 R x_2 and y_1 R y_2, so $(x_1 \circ y_1)$ R $(x_2 \circ y_2)$. Hence there is no ambiguity in the definition of $[x] * [y]$. We say that the operation $*$ on the quotient structure X/R is *well defined*.

Congruence modulo n on $\langle \mathbf{Z}, +, \cdot \rangle$ induces the quotient structure $\langle \mathbf{Z}_n, +_n, \cdot_n \rangle$. In the quotient structure, the elements $0, 1, 2, \ldots, (n - 1)$ of $\mathbf{Z}_n$ actually represents the sets

$$[0] = \{0, \pm n, \pm 2n, \ldots\}, \quad [1] = \{1, 1 \pm n, 1 \pm 2n, \ldots\},$$

through

$$[n - 1] = \{n-1, n-1 \pm n, n-1 \pm 2n, \ldots\}.$$

The tables for $\circ$ and $*$ in Fig. 1.8.5 show compositions on

$$X = \{a, b, c, d, e\} \quad \text{and} \quad X/\text{R} = \{[a, b, c], [d, e]\},$$

where R is the equivalence relation that partitions X into $[a, b, c]$ and $[d, e]$. Inspection of the tables reveals that R is a congruence. The reader can verify, for example, that $[a] * [b] = [a \circ b] = [b]$.

Consider the equivalence relation on $\{a, b, c, d\}$ that induces the classes $[a, b]$ and $[c, d]$. This relation is not a congruence with respect

$\circ$	a	b	c	d	e
a	a	b	c	d	e
b	b	c	c	d	e
c	c	b	a	d	d
d	d	e	e	b	b
e	e	d	e	b	a

$*$	$[a, b, c]$	$[d, e]$
$[a, b, c]$	$[a, b, c]$	$[d, e]$
$[d, e]$	$[d, e]$	$[a, b, c]$

FIG. 1.8.5 **A Composition $\circ$ on $X = \{a, b, c, d, e\}$, and a Composition $*$ on the Quotient Structure $X/\text{R} = \{[a, b, c], [d, e]\}$**

to the composition $\oplus$ on $\{a, b, c, d\}$ given in Fig. 1.8.6. For example, $[a \oplus c] \neq [a \oplus d]$. Note that operations on equivalence classes in the quotient structure are not well-defined if we try to use the definition $[x] * [y] = [x \oplus y]$. By this definition, for example, $[a] * [c]$ could be either $[a \oplus c]$ or $[a \oplus d]$, but we have just seen that these are different classes.

$\oplus$	a	b	c	d
a	b	a	a	c
b	d	d	c	a
c	a	b	c	d
d	a	b	d	c

FIG. 1.8.6   A Composition for Which the Equivalence Relation $\{[a, b], [c, d]\}$ Is Not a Congruence

We return for a moment to Fig. 1.8.5 to make an interesting observation. The algebraic structure $\langle X/\mathsf{R}, * \rangle$ is a homomorphic image of the structure $\langle X, \circ \rangle$. The homomorphism here is the mapping that carries each x in X into $[x]$. Note that, for every congruence relation R on a structure $\langle X, \circ \rangle$, the map $f : X \to X/\mathsf{R}$ that sends $f(x)$ into $[x]$ is a homomorphism. Conversely, if f is a homomorphism from a structure $\langle X, \circ \rangle$ to a structure $\langle Y, * \rangle$ we can find a congruence relation on $\langle X, \circ \rangle$ that in some sense preserves the properties of f. This congruence is the relation R such that $x_1 \mathrel{\mathsf{R}} x_2$ if and only if $f(x_1) = f(x_2)$. An equivalence class for this relation is just the subset of X that is mapped on some y in Y. Hence, if $f(x_1) * f(x_2) = f(x_1 \circ x_2)$ for x_1 and x_2 in X, then $[x_1] * [x_2] = [x_1 \circ x_2]$, so that the relation is indeed a congruence.

The notion of homomorphism is central to computers for several reasons. Perhaps the most important reason is that digital computation is done through simulation. When a computer adds the numbers 1183 and 3675, the computation is done only by changing voltages, currents, directions of magnetization, or other physical variables, and not by manipulating the numbers 1183 and 3675 in a mathematical sense. It so happens that a homomorphism exists from the physical variables to the integers, so that the voltage changes can correctly simulate the addition operation. In other chapters we examine several specific examples of homomorphisms in computers.

Exercises

1.8.1 Let $\langle X, \circ \rangle$ and $\langle Y, * \rangle$ be algebraic structures, and let $f: X \to Y$ be an epimorphism. Show that **(a)** if $\circ$ is commutative on X, then $*$ is commutative on Y; **(b)** if $\circ$ is associative on X, then $*$ is associative on Y; **(c)** if $\langle X, \circ \rangle$ has an identity e, then the image of e is an identity on $\langle Y, * \rangle$; **(d)** if x and x^{-1} are a pair of elements inverse with respect to each other in X, then their images are inverses with respect to each other in Y.

1.8.2 Find all isomorphisms (if any exist) of each of the following pairs of structures:
  (a) $\langle \mathbf{Z}_4 - \{0\}, \cdot_4 \rangle$ and $\langle \mathbf{Z}_3, +_3 \rangle$;
  (b) $\langle \mathbf{Z}_5 - \{0\}, \cdot_5 \rangle$ and $\langle \mathbf{Z}_4, +_4 \rangle$;
  (c) $\langle \mathbf{Z}_6 - \{0\}, \cdot_6 \rangle$ and $\langle \mathbf{Z}_5, +_5 \rangle$.

1.8.3 Consider the algebraic structures of Ex. 1.8.1, and let $f: X \to Y$ be a homomorphism, but not an epimorphism.
  (a) Find an example for which $\circ$ is commutative on X, but $*$ is not commutative on Y.
  (b) Find an example for which $\circ$ is associative on X, but $*$ is not associative on Y.
  (c) Find an example for which $\langle X, \circ \rangle$ has an identity e, and $f(e)$ is not an identity on $\langle Y, * \rangle$.
  (d) Find an example for which x and x^{-1} in $\langle X, \circ \rangle$ are mapped into a pair of elements y_1 and y_2 that are not inverses of each other in $\langle Y, * \rangle$.

1.8.4 Show that $\langle \mathscr{P}(X), \cup \rangle$ is isomorphic to $\langle \mathscr{P}(X), \cap \rangle$ for any set X.

1.8.5 Find a nontrivial congruence with respect to the composition table given in Fig. 1.8.7.

$\bullet$	a	b	c	d
a	b	d	a	c
b	c	b	b	c
c	b	c	b	b
d	c	a	a	c

FIG. 1.8.7   Composition Table for Ex. 1.8.5

1.8.6 Find all of the endomorphisms of $\langle \mathbf{Z}_6, +_6 \rangle$.

1.8.7 Consider the composition $*$ defined on $\mathbf{Z}_3 \times \mathbf{Z}_2$ such that
$$\langle x_1, y_1 \rangle * \langle x_2, y_2 \rangle = \langle x_1 +_3 x_2, y_1 +_2 y_2 \rangle.$$
Show that $\langle \mathbf{Z}_3 \times \mathbf{Z}_2, * \rangle$ is isomorphic to $\langle \mathbf{Z}_6, +_6 \rangle$. Show how to

construct a modulo 6 adder from a modulo 3 adder and a modulo 2 adder.

1.8.8   Let the composition $*$ be defined on $\mathbf{Z}_m \times \mathbf{Z}_n$ as in Ex. 1.8.7. Show that $\langle \mathbf{Z}_m \times \mathbf{Z}_n, * \rangle$ is isomorphic to $\langle \mathbf{Z}_{m \cdot n}, +_{m \cdot n} \rangle$ if and only if m is relatively prime to n.

1.8.9   An nth root of unity is one of the n distinct complex roots of the equation $x^n - 1 = 0$. For each n, we call the set of the nth roots of unity E_n. Show that $\langle E_n, \cdot \rangle$ is isomorphic to $\langle \mathbf{Z}_n, +_n \rangle$ where $\cdot$ is multiplication of complex numbers.

1.8.10   Let $\langle X, + \rangle$, $\langle Y, \circ \rangle$, and $\langle Z, * \rangle$ be algebraic structures, and let $f: X \to Y$ and $g: Y \to Z$ be homomorphisms. Show that $g \circ f$ is a homomorphism of $\langle X, + \rangle$ into $\langle Z, * \rangle$.

TWO | Groups

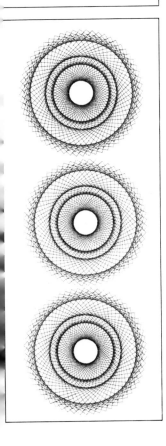

TWO | Groups

In this chapter we take up the study of groups. The results we derive here underlie many facets of computer science; one direct application of group theory is to computer addition. In other chapters we investigate the use of group theory in the design of fast adders and error-correcting adders. Although these applications in themselves would be sufficient to motivate our interest, there are several more subtle applications of equal importance.

The design of data paths for intermodule communication poses important problems in computer design. The objective is to find interconnections that provide a rich set of possible permutations, yet require the least cost. Because every group is isomorphic to a set of permutations, group theory aids us in the design of permutation networks. In Chapter Four we describe a computer memory system with interesting characteristics that result from permutation networks built into the memory.

Group theory also is applied in the analysis of symmetries, where the theory enables us to construct equivalence relations for various classes of objects. For example, when these objects are computer devices, the equivalence classes are classes of interchangeable devices. Hence the use of group theory permits us to reduce the size of a device catalog, because we need only catalog one device in each equivalence class. In the following chapter, we describe the extensive practical results that have been obtained in this area.

The study of groups originally was motivated by the search for methods of solving polynomial equations. Algorithms eventually were found for solving equations of degree less than five, where the algorithms involve only the elementary arithmetic operations and root extraction. However, no such algorithms were found for equations of fifth degree or higher. For several decades, mathematicians unsuccessfully sought algorithms for solving quintic polynomial equations, until N. H. Abel (1829) and Évariste Galois (1832) proved that the search was futile; such algorithms do not exist. Although Abel and Galois were not the first to do research in groups, relatively little had been done before their great advances. The work of Galois in particular has had a great impact in the years since his death. His research in the theory of the roots of algebraic equations led later to solutions to mathematical problems that had remained unsolved by classical methods. For example, Galois theory provides a particularly simple proof that angles cannot be trisected by ruler-and-compass construction; such a proof had evaded perplexed mathematicians since Euclid. In recent years, Galois theory has been applied to computer design and to the design of data-communication systems.

In this chapter, we discuss the mathematical aspects of group theory and prepare for the applications discussed in following chapters. The initial section states the group axioms and derives some of the elementary properties of groups. The following few sections deal with group graphs and permutations, placing the study on a more concrete foundation. The section on subgroups, cosets, and group homomorphisms discusses the central mathematical properties of groups that we use in this text. Following this section, we discuss several important examples of groups, and we summarize briefly some of the classical applications of group theory to problems that are not computer-oriented.

2.1 Group Axioms

We frequently encounter mathematical problems of the type "Given a and b, and the equation $a \circ x = b$, what is x?" A precise statement of such a problem must specify the sets to which a, b, and x belong and also must specify the composition table for the binary operation $\circ$. We can easily concoct example problems of this type, and for these problems there may exist no solutions, many solutions, or a unique solution. The class of problems of immediate interest is that class for which there is a unique solution x for every a and b. The study of groups essentially is a study of this class of problems.

Examining groups from this viewpoint, we can begin to understand why groups have had an impact on computers. For example, we remark in Chapter One that computer arithmetic cannot be identical to arithmetic on **R**, yet we wish if possible to preserve the unique solvability of a computer computation if the corresponding computation on **R** has a unique solution. The property of unique solvability does not quite capture all of the useful aspects of groups; some of our applications require stronger properties. Therefore, we find that the group axioms are slightly stronger than would be necessary only to guarantee unique solvability. The present axiom system has withstood the test of time, and it leads to a rich and widely applicable theory.

We begin our study by stating the group axioms.

Definition

A *group* $\langle G, \cdot \rangle$ is an algebraic structure, where G is a set and $\cdot$ is a composition on that set, such that the following axioms hold:

(i) (*closure*) $g \cdot h$ is in G for all $g, h \in G$;

(ii) (*associativity*) $g \cdot (h \cdot k) = (g \cdot h) \cdot k$ for all g, h, k in G;

(iii) (*identity*) there exists an e in G such that $e \cdot g = g \cdot e = g$ for all $g \in G$;

(iv) (*inverses*) for every $g \in G$ there exists $g^{-1} \in G$ such that
$$g \cdot g^{-1} = g^{-1} \cdot g = e.$$

Note that we have used the multiplication symbol · to denote the group composition. This usage is deliberate in order to emphasize the analogy of group composition with ordinary multiplication. Indeed, $\langle \mathbf{R} - \{0\}, \cdot \rangle$ is a group. We carry the analogy farther by denoting the inverse of g in the group as g^{-1}, which is the usual notation for multiplicative inverses. In keeping with this analogy, we occasionally refer to expressions of the form $g_1 \cdot g_2 \cdot \ldots \cdot g_n$ as *products*. The context always makes clear whether · represents multiplication on $\mathbf{R}$ or another group composition.

In speaking of the structure $\langle G, \cdot \rangle$, technically we should use a name such as "the group $\mathscr{G}$" that distinguishes the structure from the set G. However, here we follow the convention of naming the structure by the set. Thus we refer to the structure $\langle G, \cdot \rangle$ as "the group G." Following this convention, we refer to elements of the set G as "elements of the group G." Careful attention to the context should resolve any confusion that may arise from this minor abuse of notation.

Definition

An *abelian group*, or *commutative group*, is a group for which the commutative axiom holds. That is, $g \cdot h = h \cdot g$ for every g, h in G.

Definition

The *order*, or *cardinality*, of a group G, denoted $|G|$, is the number of elements in the set G.

Several of the structures discussed in Chapter One are groups, including the following:

$\langle \mathbf{Q} - \{0\}, \cdot \rangle$ (multiplication on the nonzero rationals);
$\langle \mathbf{Z}, + \rangle$ (addition of integers);
$\langle \mathbf{R} - \{0\}, \cdot \rangle$ (multiplication on the nonzero reals);
$\langle \mathbf{R}, + \rangle$ (addition on the reals); and
$\langle \mathbf{Z}_n, +_n \rangle$ (addition of integers modulo n).

To check that each of these structures is a group, note that the closure axiom is satisfied, that the compositions are associative, that there is an identity element (1 when the composition is multiplication or 0 when the composition is addition), and that each of the structures contains inverses for all elements.

Not all familiar algebraic structures are groups. For example, $\langle \mathbf{Z}, \cdot \rangle$ and $\langle \mathbf{Z}_n, \cdot_n \rangle$ are not groups because they fail to satisfy the axiom of inverses. Note that $0 \cdot x = x \cdot 0 = 0$ for all x, so we cannot hope to solve the equation $0 \cdot x = 0$ in any multiplicative structure with 0. We can make $\langle \mathbf{Z}_n, \cdot_n \rangle$ into a group for some n by deleting the element 0 from the set; for example, $\langle \mathbf{Z}_n - \{0\}, \cdot_n \rangle$ has an identity for all n. However, it is not always a group because the closure axiom is not satisfied when there exist nonzero x and y such that $x \cdot_n y = 0$.

Next we examine some of the properties that follow directly from the group axioms.

Lemma The identity of a group is unique.

  Proof: We show in Chapter One that, if a composition on a set has an identity, then the identity is unique. ☐

Lemma Every element of a group has a unique inverse.

  Proof: We show in Chapter One that, if an element has an inverse with respect to an associative composition, then the inverse is unique. ☐

Lemma For all a and b in a group $\langle G, \cdot \rangle$, we have $(a \cdot b)^{-1} = b^{-1} \cdot a^{-1}$.

  Proof: Note that $(b^{-1} \cdot a^{-1}) \cdot (a \cdot b) = b^{-1} \cdot (a^{-1} \cdot a) \cdot b = b^{-1} \cdot b = e$. Similarly, $(a \cdot b) \cdot (b^{-1} \cdot a^{-1}) = e$. Hence $b^{-1} \cdot a^{-1}$ is an inverse of $a \cdot b$, and therefore is the unique inverse of $a \cdot b$. ☐

Lemma For any $a, b \in G$, the equation $a \cdot x = b$ has a unique solution x in G.

  Proof: Multiplying both sides of the equation by a^{-1} and then using the group axioms, we obtain

$$a^{-1} \cdot b = a^{-1} \cdot (a \cdot x) = (a^{-1} \cdot a) \cdot x = e \cdot x = x,$$

showing that the equation has a solution x for all a and b. If

$$a \cdot x_1 = a \cdot x_2 = b, \quad \text{then} \quad x_1 = a^{-1} \cdot b = x_2,$$

so the solution is unique. ☐

  The preceding lemma shows that the group axioms guarantee the unique solvability of equations, and its proof demonstrates the existence of a simple calculation by which we can solve equations. If $a \cdot x = b$, then the unique solution is $x = a^{-1} \cdot b$. In particular, observe the use of the axioms of closure, associativity, identity, and inverses in our proof of the lemma.

  The next lemma yields an important result concerning the form of the composition table of a group operation. For this lemma we need an additional definition.

Definition A *permutation* of a set X is a bijection of X onto X.

Lemma Every element of a group is cancellable with respect to the group operation. Consequently, every row and column of a group composition table is a permutation of the elements of the group.

  Proof: By definition, an element g of a group G is cancellable if, for all h_1 and h_2 in G,

$$((g \cdot h_1 = g \cdot h_2) \vee (h_1 \cdot g = h_2 \cdot g)) \Rightarrow (h_1 = h_2).$$

But in a group, if $g \cdot h_1 = g \cdot h_2$, then

$$h_1 = g^{-1} \cdot g \cdot h_1 = g^{-1} \cdot g \cdot h_2 = h_2.$$

Similarly, if $h_1 \cdot g = h_2 \cdot g$, then $h_1 = h_2$. Hence g is cancellable.

  Now consider the row labeled g in the group composition table. To show this row to be a permutation of G, we show that it is one-to-one and onto. If the row is not a one-to-one map from G to G, then there is some element of g (say k) that appears at least twice in the row. That is, there must be h_1 and h_2 such that $g \cdot h_1 = g \cdot h_2 = k$. But we have just shown that this is impossible unless $h_1 = h_2$. Therefore, each row is an injection of G. To show that each row is onto G, we show that each element of G appears at least once in each row. This follows readily, because $g \cdot (g^{-1} \cdot h) = h$ for every h in G, so that h appears in this row in the column labeled $g^{-1} \cdot h$. Hence each row is a permutation of G. Similarly, the column labeled g is a permutation of G, and because g is arbitrary, the proof is complete. $\qquad\square$

Now that we have stated the group axioms and derived some of the basic properties of groups, we turn to the problem of finding all groups. As we gain some experience with groups, we shall see that this problem cannot be solved in practice because there are too many different groups. However, we can make some progress on the related problem of finding all groups with a specified number of elements.

For example, we can easily show that every group with a single element is isomorphic to every other such group. The composition table is shown in Fig. 2.1.1. Any group with a single element must have this table, except that the element e may have a different name.

$$
\begin{array}{c|c}
\cdot & e \\
\hline
e & e
\end{array}
$$

FIG. 2.1.1   Composition Table of a One-Element Group

A group with two elements must have a table of the form shown in Fig. 2.1.2, because e is an identity. The table must be completed in a way that satisfies the axiom of inverses. If the unspecified element in the table is g, then g has no inverse because there is no h in $\langle G, \cdot \rangle$ such that $h \cdot g = g \cdot h = e$. Hence the only possible table for a group with two elements is that shown in Fig. 2.1.3. By construction, the axioms of closure, identity, and inverses hold for the table. You can readily verify that the axiom of associativity holds as well. Therefore $\langle G, \cdot \rangle$ is a group.

$$
\begin{array}{c|cc}
\cdot & e & g \\
\hline
e & e & g \\
g & g & ?
\end{array}
\qquad\qquad
\begin{array}{c|cc}
\cdot & e & g \\
\hline
e & e & g \\
g & g & e
\end{array}
$$

FIG. 2.1.2   Partial Composition Table of a Two-Element Group

FIG. 2.1.3   Composition Table of a Two-Element Group

We have mentioned that $\langle \mathbf{Z}_2, +_2 \rangle$ is a group; its table is shown in Fig. 2.1.4. This group has two elements, but at first glance its table may appear different from that in Fig. 2.1.3. However, the two tables differ only in the way that we have labeled the elements. We can map one group onto the other with a map from $G = \{e, g\}$ onto $\mathbf{Z}_2$ that carries e into 0 and g into 1. This map preserves the structure of the composition table; hence the two groups are isomorphic to each other. Similarly, all groups with two elements are isomorphic. We can say the same thing in another form: to within isomorphism, there is only one group with two elements.

$+_2$	0	1
0	0	1
1	1	0

FIG. 2.1.4   Composition Table for $\langle \mathbf{Z}_2, +_2 \rangle$

Now let us look at groups with three elements. The composition table for such a group must have the form shown in Fig. 2.1.5. We know that rows and columns labeled k and h must be permutations, so it is easy to verify that the table entries can be filled in only one way to create the required permutations. The table must have the form shown in Fig. 2.1.6. You can easily verify that the table satisfies the axioms of closure, identity, and inverses. A tedious check of the table verifies that it is associative, so this is the composition table of a three-element group. Moreover, to within isomorphism, this is the only three-element group.

At this point, we might leap to the hasty conclusion that there is only one group of each order. This conjecture is false; there are two nonisomorphic groups of order 4 (Ex. 2.1.1).

Note that the groups of order 1, 2, and 3 we have derived are all abelian groups. All groups of order 4 and 5 also are abelian. Again, however, we would be wrong to generalize this property to all groups. There is a nonabelian group of order 6 (Ex. 2.1.2); in fact, nonabelian groups of orders larger than 6 are quite plentiful.

$\cdot$	e	k	h
e	e	k	h
k	k	?	?
h	h	?	?

FIG. 2.1.5   Partial Composition Table for a Three-Element Group

$\cdot$	e	k	h
e	e	k	h
k	k	h	e
h	h	e	k

FIG. 2.1.6   Composition Table for a Three-Element Group

Thus far in our study we have successfully discovered all groups (to within isomorphism) of order 3 or less. Unfortunately, the techniques we have used in this search are not sufficiently powerful to be of much help when dealing with reasonably large orders. Can we construct large groups, making use of our knowledge of small groups, but avoiding the tedious steps used thus far? Fortunately, we can do so, and our knowledge of ways to construct large sets from small ones proves quite useful here.

Definition Let G and H be groups with compositions $\circ$ and $*$, respectively. The *direct product* of G and H, written $G \times H$, is the product structure $\langle G \times H, \cdot \rangle$, where $\cdot$ is a composition on $G \times H$ defined by

$$\langle g_1, h_1 \rangle \cdot \langle g_2, h_2 \rangle = \langle (g_1 \circ g_2), (h_1 * h_2) \rangle$$

for all $g_1, g_2 \in G$ and $h_1, h_2 \in H$.

Again we are abusing notation by using $\times$ to denote both cartesian products of sets and direct products of groups. It might be more appropriate to use a different symbol (such as $\otimes$) for direct product, but our use of $\times$ is consistent with our convention of naming the group by the name of its set.

Theorem The direct product of groups G and H is a group.

Proof: We show that the direct product satisfies the axioms of **(i)** closure, **(ii)** associativity, **(iii)** identity, and **(iv)** inverses. **(i)** Because G is closed under $\circ$, and H is closed under $*$, it follows from the definition of direct product that $G \times H$ is closed under $\cdot$. **(ii)** The associativity of groups G and H gives us

$$\langle g_1, h_1 \rangle \cdot (\langle g_2, h_2 \rangle \cdot \langle g_3, h_3 \rangle) = \langle (g_1 \circ g_2 \circ g_3), (h_1 * h_2 * h_3) \rangle$$
$$= (\langle g_1, h_1 \rangle \cdot \langle g_2, h_2 \rangle) \cdot \langle g_3, h_3 \rangle$$

for all $g_1, g_2, g_3 \in G$ and $h_1, h_2, h_3 \in H$. **(iii)** Let e_G and e_H be the identities of G and H. Then $\langle e_G, e_H \rangle$ is the identity of $G \times H$ because

$$\langle g, h \rangle \cdot \langle e_G, e_H \rangle = \langle g, h \rangle = \langle e_G, e_H \rangle \cdot \langle g, h \rangle$$

for all $\langle g, h \rangle \in (G \times H)$. **(iv)** It follows immediately that the inverse of $\langle g, h \rangle$ is $\langle g^{-1}, h^{-1} \rangle$. $\square$

We now can construct the composition table for a group of order 6 as the direct product of groups of order 2 and 3. In Fig. 2.1.7, each element in this group is shown as a pair a, b, where a and b are drawn from the groups of order 2 and 3, respectively.

The direct product is only one way of constructing large groups from smaller ones; there are several other methods, some of which are not discussed in this text. All of these constructions share the property that the structure of the composite group in some way retains the structure of its component groups. The direct-product group $G \times H$ just defined, for example, has two subsets of elements that are groups isomorphic to G and H, respectively. To help us investigate the structure of large

$\cdot$	e_2, e_3	e_2, k	e_2, h	g, e_3	g, k	g, h
e_2, e_3	e_2, e_3	e_2, k	e_2, h	g, e_3	g, k	g, h
e_2, k	e_2, k	e_2, h	e_2, e_3	g, k	g, h	g, e_3
e_2, h	e_2, h	e_2, e_3	e_2, k	g, h	g, e_3	g, k
g, e_3	g, e_3	g, k	g, h	e_2, e_3	e_2, k	e_2, h
g, k	g, k	g, h	g, e_3	e_2, k	e_2, h	e_2, e_3
g, h	g, h	g, e_3	g, k	e_2, h	e_2, e_3	e_2, k

FIG. 2.1.7   Composition Table for One Possible Group of Order 6

groups, we try to find all subsets that are groups. This approach suggests the next definition.

Definition A *subgroup* of a group G is a subset of elements of the set G that forms a group under the composition of the group G.

In the example of a direct-product group of order 6, the sets

$$\{\langle e_2, e_3 \rangle, \langle e_2, k \rangle, \langle e_2, h \rangle\} \quad \text{and} \quad \{\langle e_2, e_3 \rangle, \langle g, e_3 \rangle\}$$

form subgroups of order 3 and 2, respectively.

Theorem Let H be a subgroup of G. Then the identity of H is the same as the identity of G. Moreover, the inverses of the elements of H coincide in G and H.

Proof: Let e_G and e_H be the identities in G and its subgroup H, respectively. We show that $e_G = e_H$. Because e_H is an identity, $e_H \cdot h = h \cdot e_H = h$ for all $h \in H$. Because e_G is an identity on G, we have $e_G \cdot h = h \cdot e_G = h$. Combining these expressions, we have $h \cdot e_H = h \cdot e_G = h$ in G. Then

$$e_H = e_G \cdot e_H = h^{-1} \cdot h \cdot e_H = h^{-1} \cdot h \cdot e_G = e_G \cdot e_G = e_G,$$

so the identity of a subgroup must be the same as the identity of the group. Consequently, for every h in H, its inverse in H is h^{-1} (its inverse in G) because $e_H = h \cdot h^{-1} = h^{-1} \cdot h = e_G$. $\square$

We frequently encounter the problem of showing that particular subsets of groups form subgroups. In general, to show that a set forms a group, we must show that the four group axioms hold, but this procedure can be simplified somewhat for subgroups. If the elements form a subset of a group, the associativity axiom obviously is satisfied. The identity axiom also is trivially satisfied if the subset includes the group identity. Therefore, the nontrivial part of the problem is to show that the axioms of closure and inverses are satisfied. The next theorem simplifies this process.

Theorem Let H be a subset of G. Then H forms a subgroup of the group G if and only if $(h_1 \cdot h_2^{-1}) \in H$ for every h_1, h_2 in H.

Proof: If H is a subgroup, then it follows immediately that $h_1 \cdot h_2^{-1}$ is in H. To prove the theorem, it is sufficient to show that, if $h_1 \cdot h_2^{-1}$ is in H for every h_1, h_2 in H, then H is a group. By picking $h_1 = h_2$, we find that $h_1 \cdot h_1^{-1} = e_G$ is in H, thereby satisfying the identity axiom. For any h, we know that h^{-1} is in H because $e_G \cdot h^{-1} = h^{-1}$ is in H. Hence the axiom of inverses is satisfied. Closure is satisfied because $h_1 \cdot h_2 = h_1 \cdot (h_2^{-1})^{-1}$ is in H. □

If we know that the subset of a group has a finite number of elements, the process of proving that it forms a subgroup becomes even simpler.

Theorem Let H be a finite subset of a group G such that H is closed under the group composition. Then H forms a subgroup of G.

Proof: By the previous theorem, H is a group if $h_1 \cdot h_2^{-1}$ is in H for all $h_1, h_2 \in H$. Because H is closed with respect to $\cdot$, it is sufficient to prove that h^{-1} is in H for all h in H, because then h_2^{-1} is in H if h_2 is, and $h_1 \cdot h_2^{-1}$ is in H because of closure, so H is a group.

Now examine the sequence h, $h^2 = h \cdot h$, $h^3 = h \cdot h \cdot h$, . . . , up to h^{n+1}, where H is assumed to have n elements. Because closure is satisfied, the $n + 1$ elements in the sequence all lie in H; therefore they cannot all be distinct. Hence $h^i = h^j$ for some i and j, where $i < j$. Then $h^{j-i} = e$. From this we conclude that $h^{j-i-1} = h^{-1}$, because $h \cdot h^{j-i-1} = h^{j-i-1} \cdot h = h^{j-i} = e$. Then h^{-1} is in H if h is. □

Summary The axioms of associativity, closure, identity, and inverses are valid for groups. To determine if an algebraic structure is a group, it is sufficient to check if these axioms hold for the structure. All elements of a group are invertible, so that we can solve equations of the form $a \cdot x = b$ in a group.

A group composition table has the interesting property that every row and column is a permutation of the elements of the group. We used this property here to show that the groups of orders 1, 2, and 3 are unique to within isomorphism; in following sections we make further use of this property.

The direct product of two groups G and H is a group derived from the cartesian product of G and H. This group essentially has two dimensions, one for each of G and H. The group structures of both G and H are captured in the direct-product group $G \times H$, in that both G and H are isomorphic to subgroups of $G \times H$.

A subgroup H of a group G is both a subset of G and a group for which the group composition is the same as that of G. To prove that a subset H of a group G forms a subgroup of G, we can dispense with some of the machinery that we normally use to prove that an algebraic structure is a group. For example, in this situation we need not prove that H is closed. Here it is sufficient to show only that, for every pair h_1 and h_2 in H, the element $h_1 \cdot h_2^{-1}$ is also in H.

Exercises

2.1.1 Construct the composition tables for the two nonisomorphic groups of order 4. Show that there can be no other groups of order 4.

2.1.2 Construct the composition table for a nonabelian group of order 6. As subgroups it has the group of order 2 and the group of order 3.

2.1.3 Consider the set of rotations of the unit square (Fig. 2.1.8) into itself. Let I be the rotation of $0°$, and let R, D, and L be rotations of $90°$, $180°$, and $270°$, respectively. Show that this set of rotations forms a group, with the group operation being composition of rotations.

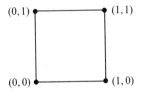

FIG. 2.1.8 The Unit Square

2.1.4 Let X be the set $\{1, 2, 3\}$. Let P be the set of permutations of X.
   (a) Show that the set of permutations, P, forms a group with respect to function composition.
   (b) Show that every set of permutations of a finite set X that is closed with respect to function composition forms a group.

2.1.5 Let X be the set $\{1, 2, 3, 4, 5\}$. Let f_1 and f_2 be bijections from X onto X, as shown in Table 2.1.1. What is the smallest set of functions that we must include with f_1 and f_2 to form a group under function composition?

TABLE 2.1.1 Bijections from X onto X (Ex. 2.1.5)

x	$f_1(x)$	$f_2(x)$
1	2	1
2	3	5
3	4	4
4	5	3
5	1	2

2.1.6 For $n \geq 2$, show that $\langle \mathbf{Z}_n - \{0\}, \cdot_n \rangle$ is a group if and only if n is a prime number.

2.1.7 Prove that, if G is an abelian group, then $(g \cdot h)^n = g^n \cdot h^n$ for all g and h in G and for any integer $n \geq 1$.

2.1.8 Prove that the intersection of two subgroups of a group G is also a group.

2.1.9 Discover whether the following statement is true or false: The union of two subgroups of a group G is also a group.

2.1.10 Consider the subset of $\mathbf{Z}_n$ that consists of the integers in $\mathbf{Z}_n$ that are relatively prime to n. (Two integers are said to be *relatively prime* if their greatest common divisor is 1.) Show that this set forms a group under multiplication modulo n.

2.1.11 Show that, if every g in a group G satisfies the relation $g^2 = e$, then the group is abelian. HINT: Consider $g^2 = e$ and replace g by $h \cdot k$.

2.1.12 Show that, if every g and h of a group G have the property $(g \cdot h)^2 = g^2 \cdot h^2$, then G must be abelian.

2.1.13 An element g of a group G is said to be *self-inverse* if $g = g^{-1}$; that is, if $g^2 = e$. Show that every group of even order has at least one self-inverse element other than the identity.

2.1.14 Show that, if G is a finite set with a closed associative composition $\cdot$ and if every element of G is cancellable with respect to $\cdot$, then G is a group.

2.1.15 Let G be a group with subgroups H_1 and H_2. Define $H_1 \cdot H_2$ to be $\{h_1 \cdot h_2 | (h_1 \in H_1) \wedge (h_2 \in H_2)\}$, where $\cdot$ is the group operation of G. Is $H_1 \cdot H_2$ necessarily a group?

2.2 Generators and Group Graphs

Thus far, we have been able to specify the composition of a group only by specifying each of the entries in the composition table. This method becomes rather cumbersome for larger groups, because the number of entries in the composition table grows as the square of the order of the group. A much more compact specification is possible with relations known as generator relations, which we describe in this section. Occasionally we encounter applications of group theory that arise directly from generator relationships, rather than from any of the several other notions associated with groups. One application of this type is the analysis of data-switching networks that can permute input data by any one of several possible permutations. We perform such an analysis in Chapter Four.

Generator relations lend themselves naturally to graphical descriptions. Consequently, we devote a large portion of this section to graphs of groups. The graph of a group provides a great deal of insight about the structure of the group, and it does so in a concrete and concise way.

In fact, a group graph describes a group completely because it determines uniquely the group composition table.

We begin this section with a discussion of generators of a group, then describe the use of a set of relations on the generators to define a group uniquely. From the set of defining relations, we construct group graphs using a method described in the final part of the section.

Speaking intuitively, we can say that a set of generators for a group G is a subset of group elements from which the entire group can be constructed. By specifying the interactions of the generators under the group composition, we can infer how arbitrary group elements interact under group composition. First, we make the notion of a generator more precise.

Definition   A subset of elements of a group G is said to be a set of *generators* of G if every element of G can be expressed as a product of the subset elements and their inverses.

In this definition, a product of group elements is simply the composition of those elements under the group operation. Thus, if g and h are generators of a group $\langle G, \cdot \rangle$, then $g \cdot h \cdot g^{-1}$ and $g \cdot h \cdot g \cdot h$ are both products in the sense used in the definition.

The group of order 2 has two elements, e and g. A generator set for this group is $\{g\}$ because $g \cdot g = g^2 = e$. The group of order 3 has a generator set $\{h\}$ because the group is $\{h, h^2 = k, h^3 = e\}$. Note that $\{k\}$ also is a generator set. The group of order 6 that is the direct product of groups of order 2 and 3 has a generator set $\{g, h\}$. The elements of this group are $\{e, g, h, g \cdot h, h^2 = h^{-1}, g \cdot h^2 = g \cdot h^{-1}\}$.

Next we consider construction of the group graph from a set of generators for the group. To construct the graph we must first determine the number of nodes in the graph and give each node a distinct label. Then we determine which nodes are connected by edges.

1. Place one node in the graph for each group element. Label each node with the name of a group element in a one-to-one fashion.
2. Let x be a generator of the group. Then for each group element y in G, we draw an edge from the node labeled y to the node labeled z, where $z = y \cdot x$.

Fig. 2.2.1 shows group graphs for groups of order 2, 3, and 6. The labels of the nodes correspond to the names for these elements that we used in Section 2.1. Each edge of a group graph is associated with a particular generator of the group, and in Fig. 2.2.1 we have used solid or dashed lines to distinguish the edges of the graph and make this association clear; for each group graph we indicate the group generators represented by each type of edge. Thus the group graph for the group of order 2, which has the generator set $\{g\}$, has two

Group Graphs   Types of Generators

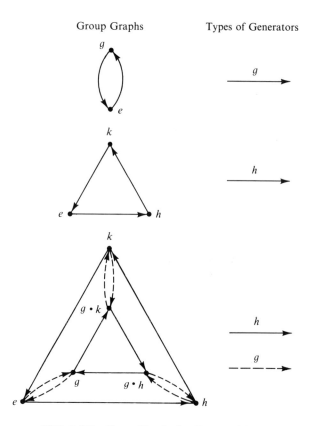

FIG. 2.2.1   Group Graphs for Groups of Order 2, 3, and 6

nodes—labeled e and g, respectively—and two edges, each of type g. The group graph of order 6 has both g and h edges, with the g edges drawn dashed and the h edges solid. By a tedious calculation, we can verify that this group graph correctly represents the composition table in Fig. 2.2.2.

The examples make evident the following property of group graphs: if g is a generator of a finite group, then every node of a group graph has a generator edge of type g entering it and another leaving it. From the construction, we know that an edge of type g leaves each node. No node has two edges of type g entering it for, if there were two such edges—one from node x to node z and one from node y to node z—then $z = x \cdot g = y \cdot g$. But every element of a group is invertible, so this expression gives us $z \cdot g^{-1} = x = y$, so that x and y are not distinct nodes, and the two edges are not distinct edges. Thus, every node in the graph has exactly one edge of each type entering it.

The construction of the group composition table from the group graphs is relatively straightforward. For example, the first graph tells us that $g^2 = e$. To see this, start at node e and traverse two successive

$\cdot$	e	g	h	$g \cdot h$	k	$g \cdot k$
e	e	g	h	$g \cdot h$	k	$g \cdot k$
g	g	e	$g \cdot h$	h	$g \cdot k$	k
h	h	$g \cdot h$	k	$g \cdot k$	e	g
$g \cdot h$	$g \cdot h$	h	$g \cdot k$	k	g	e
k	k	$g \cdot k$	e	g	h	$g \cdot h$
$g \cdot k$	$g \cdot k$	k	g	e	$g \cdot h$	h

FIG. 2.2.2   Composition Table for Group of Order 6 in Fig. 2.2.1

g edges. By construction, this takes us to the node $e \cdot g \cdot g$, which is the node $g \cdot g = g^2$ because e is the group identity. But traversing two successive g edges starting at e takes us back to e. Hence $e = g \cdot g = g^2$. Similarly, we note that $e \cdot g = g$, because the g edge leaving node e enters node g. How can we determine $g \cdot e$ from the graph? We do this by starting at node g and traversing a path that corresponds to e. Because $g^2 = e$, we see that such a path consists of two g edges. Starting at node g and traversing two successive g edges leaves us at node g, so $g \cdot e = g^3 = g$. We have now determined three of the four entries of the group composition table; to complete the table we must find the element equal to $e \cdot e$. Obviously, this is the element e, and we can deduce this fact from the group graph. Thus the composition for the group of order 2 is that shown in Fig. 2.2.3.

$\cdot$	e	g
e	e	g
g	g	e

FIG. 2.2.3   Composition Table for Group of Order 2

The table corresponding to the group of order 6 is somewhat more complex to construct from the graph. Because e is an identity, the row and column labeled e are determined. The columns labeled g and h can be read directly from the graph, because the edges show the effect of right multiplication by g and h. For example, the column labeled h is the map that sends each element x into $x \cdot h$. Therefore, the column is the map shown in Fig. 2.2.4.

To determine the column labeled k in the composition table, we must find the map that sends each x into $x \cdot k$. Because $k = h^2$, this is the map that sends x into $x \cdot h \cdot h$. This map can be read from the graph by tracing a path from each node x along two successive h edges

x	$x \cdot h$
e	h
g	$g \cdot h$
h	k
$g \cdot h$	$g \cdot k$
k	e
$g \cdot k$	g

FIG. 2.2.4   Map for Column Labeled h in Group Composition Table of Order 6

to $x \cdot h \cdot h$. Because $h^2 = h^{-1}$, we also can determine the map by traveling backward along one h edge. The column labeled $g \cdot h$ is the map that sends each x into $x \cdot g \cdot h$. We can read this map from the graph by tracing a path from each node along first a g edge and then an h edge. In similar fashion, we can determine each of the columns of the table; from the graph with six nodes (Fig. 2.2.1) we obtain the composition table given in Fig. 2.2.2.

In the remainder of this section, we pursue the question of how much we have to know about a group in order to construct a graph for the group. We shall see that a group graph and group composition table are determined completely by a set of relations on the generators.

Definition   For a group G with generator set $\{g_1, g_2, \ldots, g_r\}$, a *generator relation* is a formula that equates a product of the generators and their inverses to e, the group identity.

For a generator set $\{g, h, k\}$, examples of generator relations are $g^3 = e$, $g \cdot h^{-1} \cdot k \cdot g^{-1} = e$, and $(h^{-1})^5 \cdot g^3 \cdot (k^{-1})^2 \cdot h = e$. Every group has many generator relations, and certain sets of these generator relations determine completely the group graphs. Before discussing methods for constructing group graphs, let us examine the information contained in generator relations.

Although each group has many generator relations associated with it, not all of these relations are independent. For example, if $g^2 = e$ is a generator relation for a group, then $g^4 = g^2 \cdot g^2 = e \cdot e = e$ is another generator relation. In fact, $g^{2n} = e$ is a generator relation for all $n \geq 1$. In this case, we say that the generator relation $g^2 = e$ *implies* the relations $g^4 = g^6 = g^8 = \cdots = e$. In this case, we use the axiom of identity to show that one generator relation implies another, but more generally we can use any combination of the group axioms to show implications among generator relations.

Such implications can cause difficulty in finding the composition table for a group from a set of generator relations. Let G be a group generated by the element g, and let the relation $g^2 = e$ hold for this group. Because G has only one generator, every element of the group is a product of g and g^{-1}. Because $g^2 = e$, by definition $g = g^{-1}$.

Therefore, every element of G is some power of g. Because the relation $g^2 = e$ implies $g^{2n} = e$ for all $n \geq 1$, we have $g^{2n+1} = g^{2n} \cdot g = e \cdot g = g$ for all $n \geq 1$. Thus, there can be at most two elements in this group, namely g and e. However, the relation $g^2 - e$ yields no further information about the group G. We have not yet determined G uniquely, because it could have either one or two elements.

Suppose that G has one element. In this case, the relation $g = e$ holds. Note that $g = e$ implies $g^2 = e$, but $g^2 = e$ does not imply $g = e$. Because it conveys more information, the relation $g = e$ is said to be *stronger* than the relation $g^2 = e$. If the only information given about G is that $g^2 = e$, we should assume that no stronger relation holds. Therefore, we conclude that G is not the group with one element, but must be the group with two elements.

Carrying this argument one step farther, suppose that a group G is generated by g, and that the relation $g^4 = e$ holds. Using the same line of reasoning as before, we can readily show that G contains no more than the four elements g, g^2, g^3, and e. The groups of order 1 and 2 both have generators that satisfy $g^4 = e$, but in these two cases the generators also satisfy one or both of the relations $g = e$ and $g^2 = e$. Because $g^4 = e$ implies neither of the other two relations, the groups of order 1 and 2 satisfy stronger relations than the one given. Hence we say that the relation $g^4 = e$ defines the group $G = \{g, g^2, g^3, e\}$ rather than any smaller group.

The preceding examples suggest a general procedure for constructing a composition table for a group from a set of generator relations. The procedure is to examine all of the products of the generators and their inverses, and to determine from the generator relations how many of these products must be distinct. The weakest assumption we can make is to let each of the distinct products be an element of the group, so that the number of elements equals the number of distinct products. A stronger set of generator relations will equate two or more of these products. For example, the generator relation $g^4 = e$ yields four distinct products: g, g^2, g^3, and e. The generator relation $g^2 = e$ equates g to g^3 and g^2 to e, thus reducing the number of distinct products. Hence $g^2 = e$ is stronger than $g^4 = e$.

The complete construction of a group table or graph from a set of generator relations becomes quite simple after we have associated one group element with each distinct product of the generators. Every product of the generators corresponds to some group element. Therefore, given an element x and a generator g, we can use the generator relations to discover which group element is $x \cdot g$. Hence we can construct both the group graph and the composition table for the group. We turn next to the details of this construction.

Definition   A set of generator relations R is said to *imply* a generator relation r if r can be derived from the relations in R through use of the group axioms. A set of relations R is said to imply another set of relations R'

if each relation in R' is implied by R. Two sets R and R' are said to be *equivalent* if they imply each other. R is said to be *stronger* than R' if R implies R', but R' does not imply R.

You can easily verify that equivalence of generator relations is an equivalence relation as defined in Chapter One (see Ex. 2.2.4).

Definition   Let R be a set of generator relations for a group G. R is said to be a set of *defining relations* for G if, for every set of generator relations R' such that R' is stronger than R, R' fails to hold for G.

The notion of defining relations is brought out in the examples earlier in this section. For example, the relation $g^2 = e$ is a defining relation for a group of order 2; the relation $g = e$ is stronger than $g^2 = e$, but it fails to hold for this group.

Now, how do we construct the composition table for a group from a set of defining relations for the group? The essential idea of the construction is set forth in the preceding discussion; here we work through several examples. First, consider the group defined by the relation $g^2 = e$. We know that $g^{2n} = e$ and $g^{2n+1} = g$ for all $n \geq 1$. Hence this group contains the two elements g and e. The group graph has two vertices and must be the graph shown in Fig. 2.2.5.

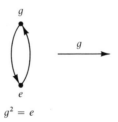

$$g^2 = e$$

FIG. 2.2.5   Group Graph Defined by $g^2 = e$

With similar reasoning, we can show that the defining relation $g^r = e$ generates a group graph with r vertices. Because these vertices are labeled $g, g^2, \ldots, g^r = e$, the edges of the graph form a complete cycle, and the graph must be a polygon with r sides. Fig. 2.2.6 shows group graphs for $r = 3$ and $r = 4$.

Next we turn to a more complex example. Consider the group G with generators $\{g, h\}$ and defining relations $g^2 = e$, $h^3 = e$, and $(g \cdot h)^2 = e$. We can show that this group has six elements: g, h, $h \cdot g$, h^2, $h^2 \cdot g$, and e. Every product of the generators and their inverses can be reduced to one of these six forms through application of the defining relations and the group axioms. The relation $g^2 = e$ implies that all powers of g can be reduced modulo 2; that is, $g^{2n} = e$ and $g^{2n+1} = g$ for all $n \geq 1$. The relation $h^3 = e$ implies that all powers of h can be reduced modulo 3, yielding the elements h, h^2, and e. Next we must consider all products of the generators of the form $g^r \cdot h^s \cdot g^t \cdots$.

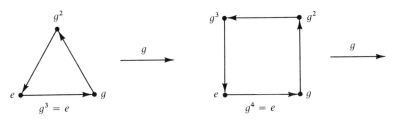

FIG. 2.2.6   Group Graphs Defined by $g^3 = e$ and $g^4 = e$

where the powers of g lie in $\mathbf{Z}_2$ and the powers of h lie in $\mathbf{Z}_3$. In particular, we must show that any such product can be reduced to one of the six elements listed previously. From the defining relations, we know that $g^{-1} = g$, $h^{-1} = h^2$, and $(g \cdot h)^{-1} = g \cdot h$. From these relations, we obtain $g \cdot h = h^{-1} \cdot g^{-1} = h^2 \cdot g$, a formula that we can use to interchange the order of h's and g's in a product. For example, we can reduce $g \cdot h \cdot g \cdot h^2 \cdot g \cdot h^2$ in the following fashion:

$$
\begin{aligned}
g \cdot h \cdot g \cdot h^2 \cdot g \cdot h^2 &= g \cdot h \cdot g \cdot h^2 \cdot (g \cdot h) \cdot h \\
&= g \cdot h \cdot g \cdot h^2 \cdot (h^2 \cdot g) \cdot h \\
&= g \cdot h \cdot g \cdot h^4 \cdot g \cdot h \\
&= g \cdot h \cdot g \cdot h \cdot g \cdot h.
\end{aligned}
$$

Continuing this process, we can move all of the g's to the right and all of the h's to the left, while reducing the powers of the g's modulo 2 and the powers of the h's modulo 3 as we make the substitutions:

$$
\begin{aligned}
g \cdot h \cdot g \cdot h \cdot g \cdot h &= g \cdot h \cdot g \cdot h \cdot (g \cdot h) \\
&= g \cdot h \cdot g \cdot h \cdot (h^2 \cdot g) \\
&= g \cdot h \cdot g \cdot h^3 \cdot g \\
&= g \cdot h \cdot g \cdot g \\
&= g \cdot h \cdot g^2 \\
&= g \cdot h \\
&= h^2 \cdot g.
\end{aligned}
$$

Because the reduction process places the h's to the left of the g's, the only possible results of the reduction process are the six group elements given. We can easily show that these six elements cannot be combined further, so the group graph must have six vertices.

Now the construction of the group graph is quite simple. We know how each of the six elements are formed by products of the generators, and we can determine how the g and h edges connect the vertices in the graph by observing the action of multiplication by g and h on the group elements. We obtain the graph shown in Fig. 2.2.7. As an example, note the h edge directed from $h^2 \cdot g$ to $h \cdot g$. This edge appears because $h^2 \cdot g \cdot h = h^2 \cdot (g \cdot h) = h^2 \cdot (h^2 \cdot g) = h \cdot g$.

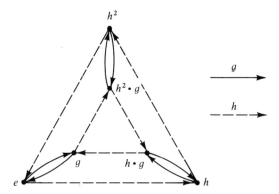

FIG. 2.2.7 Group Graph Defined by $g^2 = h^3 = (g \cdot h)^2 = e$

We can now set forth a general procedure for construction of a group graph from the set of generators and the set of defining relations:

(i)   Find the distinct products of the generators and their inverses, as determined from the defining relations;

(ii) Associate a group element with each of the distinct products;

(iii) Construct the group graph by including one vertex for each group element, and by placing edges between the vertices in positions determined by the defining relations.

We have established that each group graph represents some specific group. We must also consider the important question of the validity of the converse statement: does each group have a unique graph associated with it? Unfortunately, the answer is negative. Several different sets of defining relations may exist for a single group, and the graphs determined from these different sets need not be identical.

For example, the group defined by the generator relation $k^6 = e$ is isomorphic to the group defined by the generator relations $g^2 = e$, $h^3 = e$, and $g \cdot h \cdot g \cdot h^2 = e$. If we substitute k^3 for g and k^2 for h in the set of three relations, we see that the three relations are still satisfied. Hence the relation $k^6 = e$ implies the three relations by implying the existence of the generators k^2 and k^3. Similarly, we can substitute $g \cdot h$ for k in the relation $k^6 = e$ and show that the relation still holds. Thus the two sets of relations imply each other, and therefore they define the same group. Fig. 2.2.8 shows the group graphs corresponding to the two sets of defining relations. Although the two graphs represent the same group, the graphs obviously are different.

Summary The central notion of this section is the following: a group can be specified completely by specifying relations on a set of generators of the group. A set of defining relations usually is much more compact than a composition table, yet it contains the same information; we can construct the entire composition table from the defining relations.

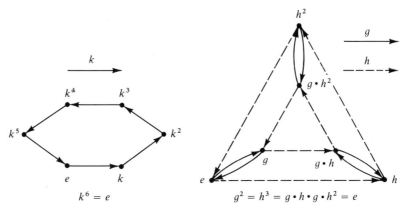

FIG. 2.2.8 Two Different Group Graphs for the Same Group

The group graph is an alternate description of a group; the graph can be determined from the defining relations. The graphical description is valuable because it often makes obvious properties of a group that are not easily observable in the defining relations or in the composition table. The group graph also has the advantage of being more compact than a composition table.

As a matter of historical interest, we mention that the subject of group graphs was developed by Arthur Cayley (1878a, 1878b, 1889). He referred to the graphs as *color groups*, because he used different colors to distinguish the edges associated with each generator (as we have used dashed and solid lines). Group graphs are sometimes called *Cayley diagrams*, to give proper credit to the man who first explored their properties in a systematic fashion. However, some of Cayley's ideas on the subject were anticipated half a century earlier by Augustin-Louis Cauchy (1815) who, for example, represented the relation $g^n = e$ by an n-sided polygon with edges directed around the polygon to show the effect of multiplication by g. Grossman and Magnus (1964) present a thoroughly enjoyable treatment of this subject in a book that most readers will find both easy and highly entertaining. Another graphical representation of groups was developed by Dyck (1882) and is described by Burnside (1911, ch. 18).

Exercises

2.2.1 Construct group graphs from the following sets of defining relations:
(a) $g^2 = e$, $h^4 = e$, $(g \cdot h)^2 = e$;
(b) $g^2 = e$, $h^4 = e$, $g \cdot h \cdot g \cdot h^3 = e$;
(c) $g^2 = e$, $h^2 = e$, $(g \cdot h)^2 = e$.

2.2.2 If G and H are groups, describe the structure of the group graph of $G \times H$ in terms of the group graphs for G and H.

2.2.3 Let H be a subgroup of G, and let R be a set that is a set of defining relations for H and is also a subset of a set of defining relations for G. How is the group graph for H related to the group graph for G, where these graphs are constructed from the given sets of defining relations?

2.2.4 Show that the implication relation for sets of generator relations (one set of relations implying another) is a partial ordering on the collection of sets of generator relations. Show that the equivalence of sets of generator relations is an equivalence relation.

2.2.5 Prove that the group G, defined by the relation $g^4 = e$, has four elements.

2.2.6 Let G be a group defined by the relations $g^2 = h^3 = (g \cdot h)^2 = e$. In this section we show that every product of the generators for this group can be reduced to one of the six forms g, h, h^2, $h \cdot g$, $h^2 \cdot g$, or e. Prove that no further reduction is possible.

2.2.7 Let G be a group defined by the relations $g^2 = h^n = (g \cdot h)^2 = e$ for a given n, where $n \geq 2$.
  (a) Find all of the elements of G in terms of products of the generators g and h.
  (b) Construct the group graph for G.

2.2.8 Let G be a group defined by the relations $g^2 = h^n = g \cdot h \cdot g \cdot h^{n-1} = e$ for a given n, where $n \geq 2$.
  (a) Find all of the elements of G in terms of products of the generators g and h.
  (b) Construct the group graph for G.
  (c) Determine whether this group is isomorphic to the group described in Ex. 2.2.7.

2.3 Permutation Groups

In the latter half of the eighteenth century, Karl Friedrich Gauss and Joseph-Louis Lagrange made numerous observations on the properties of permutations. Several reasons motivated their studies, most notably their interests in solving cubic, quartic, quintic, and higher-degree equations. Because the properties of equations under the actions of permutations of their roots proved essential to their study, they wrote several papers centered on this topic. Eventually, Niels Abel and Évariste Galois solved the problem by using the properties of finite permutation groups.

Just as finite permutations are the central issue in solving polynomial equations, so are they central in the applications of group theory to computers. Permutations arise in such contexts as the state behavior of sequential machines, classes of equivalent devices, data transfers, and computer arithmetic. In this section we concentrate on the basic mathematical tools for manipulating permutations. One of the important results we show is the fact that every group is isomorphic to a permutation group. Hence we lose no generality by limiting our attention to permutation groups.

Our first step is to develop a convenient notation for permutations. We begin by recalling the definition of a permutation.

Definition   A *permutation* of a set X is a bijection from X onto X. The *degree* of a permutation is the cardinality of the set on which it is defined.

As an example, consider the set $\{0, 1\}$. There are two permutations on this set (p_1 and p_2 in Fig. 2.3.1). Both of these permutations are of degree 2.

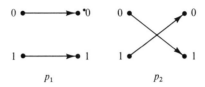

FIG. 2.3.1   The Two Permutations of the Set $\{0, 1\}$

In the early part of the nineteenth century, Cauchy (1815) introduced a shorthand notation for permutations, and his notation has remained useful. Let p be a permutation on the n letters $1, 2, \ldots, n$, and let $p(i)$ be the image of i under the permutation. We can specify p completely by listing the table of pairs $\{i, p(i)\}$ in a form such as Fig. 2.3.2. Cauchy used this tabular notation, but he also improved upon it with a more compact notation. Observe that the permutation p maps 1 into $p(1)$, and it maps $p(1)$ into $p(p(1))$, which we denote as $p^2(1)$. Then it maps $p^2(1)$ into $p(p^2(1)) = p^3(1)$, and $p^3(1)$ into $p^4(1)$, and so forth, until we reach $p^j(1) = 1$ for some $j \leq n$. We represent this action of p by the notation $(1\, p(1)\, p^2(1) \ldots p^{j-1}(1))$, and we call this ordered list a *cycle*. The *length* of a cycle is the number of letters that it contains. In this example, the cycle has length j, and we call it a *j-cycle*. If all n letters appear in the cycle, then we have a complete representation of p. If not, we select any other letter (say i) that does not appear in the cycle, and we construct a new cycle of the form $(i\, p(i)\, p^2(i) \ldots p^{k-1}(i))$. The notation indicates that p maps $p^{k-1}(i)$ onto i, so that $p^k(i) = i$. We continue constructing cycles until every letter appears in some cycle. Thus the identity permutation on the set $\{1, 2, \ldots, 8\}$ has the form $(1)(2) \ldots (8)$, and examples of other permutations on this set are $(1\,2)(3\,4\,5\,6)$, $(1\,3\,5\,7)(2\,8\,6\,4)$, and $(1)(2\,3)(4\,5\,6)(7)(8)$. Notice that in this notation

$$\begin{pmatrix} 1 & 2 & 3 & \ldots\ldots & n \\ p(1) & p(2) & p(3) & \ldots\ldots & p(n) \end{pmatrix}$$

FIG. 2.3.2   Tabular Form for Specifying the Permutation p

each letter appears in one and only one cycle. To simplify the notation, we omit 1-cycles from the listing; thus, if a letter is absent from the notation, it is understood that the permutation leaves that letter fixed.

Any two permutations p_1 and p_2 on the same set can be composed to form a new function $p_1\ p_2$, which also is a permutation on the set. Here we follow the usual convention of left composition; that is, we compose permutations from right to left. Thus, when we write

$$(a\,b)\ (a\,c),$$

we mean to apply $(a\,c)$ first, and then to apply $(a\,b)$. Hence

$$(a\,b)\ (a\,c) = (a\,c\,b), \quad \text{but} \quad (a\,b)\circ(a\,c) \neq (a\,b\,c).$$

Some algebraists prefer to compose permutations from left to right. Obviously, the direction of composition does not alter the underlying concepts, but it is essential that one convention be applied consistently in any given discussion.

Because the composition of two permutations on a set X is itself a permutation of X, the composition of permutations (as functions) is a binary operation on the set of permutations. We have shown that function composition is associative, so $p_1 \circ (p_2 \circ p_3) = (p_1 \circ p_2) \circ p_3$ for all permutations p_1, p_2, and p_3. Thus the binary operation $\circ$ defined on a set of permutations is associative. Because the identity permutation is an identity with respect to the composition of permutations and because every permutation is an invertible function, we can find sets of permutations that are groups with respect to function composition.

Definition   A *permutation group* is a set of permutations that forms a group with respect to function composition.

$\circ$	p_1	p_2
p_1	p_1	p_2
p_2	p_2	p_1

FIG. 2.3.3   Composition Table for Permutation Group of Fig. 2.3.1

The permutations in Fig. 2.3.1 constitute a permutation group with the composition table shown in Fig. 2.3.3. The set of permutations of a three-element set forms a more interesting example of a permutation group (Fig. 2.3.4).

We can prove readily that the set of all possible permutations of a set with n elements forms a permutation group, as exemplified by the six permutations of the three-element set. In general, there are $n!$ distinct permutations of n elements, and the group formed from these permutations is called S_n, the *symmetric group of degree n*. The following are other examples of permutation groups:

$$\{(1)(2)(3), (1\,2\,3), (1\,3\,2)\};$$
$$\{(1)(2)(3)(4), (1\,2\,3\,4), (1\,3)(2\,4), (1\,4\,3\,2)\};$$
$$\{(1)(2)(3)(4), (1\,2)(3\,4), (1\,3)(2\,4), (1\,4)(2\,3)\}.$$

You can verify that each of these sets of permutations is closed under composition, that each contains the identity permutation, and that the inverse of each permutation in the set is also in the set.

The importance of permutation groups stems from the fact that every group is isomorphic to a permutation group. Thus we can focus our study of groups on groups of permutations.

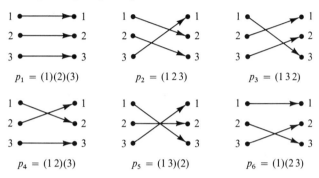

FIG. 2.3.4   The Set of Permutations of a Three-Element Set

Theorem   Every group of order n is isomorphic to a permutation group of degree n.

Proof: Let $\langle G, \cdot \rangle$ be a group of order n with elements $g_1, g_2, \ldots, g_n$. Each row of the composition table of G is a permutation of G, so the composition table provides a natural way to associate each element of G with a permutation of G. We shall construct a permutation group P on the set of n letters $N = \{1, 2\,3, \ldots, n\}$ by associating a permutation p_{g_i} with each group element g_i, where p_{g_i} is essentially just a relabeling of row g_i in the group composition table. Thus, for each $x \in N$, we set $p_{g_i}(x) = y$ if and only if $g_i \cdot g_x = g_y$. Now we show that the map f: $G \to P$, for which $f(g_i) = p_{g_i}$, is an isomorphism of the groups G and P.

Clearly, this map is onto P, and it is one-to-one because no two rows of a group composition table are identical. (If two rows were identical, there would be two distinct solutions to $x \cdot a = b$ for some values of a and b.) To complete the proof, we must show that the map f is a homomorphism. In particular, we must show that, for $g_i, g_j \in G$,

$$p_{g_i \cdot g_j} = f(g_i \cdot g_j) = f(g_i) \circ f(g_j) = p_{g_i} \circ p_{g_j}.$$

Now suppose that $p_{g_i \cdot g_j}(x) = y$, or equivalently that $(g_i \cdot g_j) \cdot g_x = g_y$. By associativity, $g_y = g_i \cdot (g_j \cdot g_x)$, which is true if and only if

$$y = p_{g_i}(p_{g_j}(x)) = p_{g_i} \circ p_{g_j}(x).$$

Thus, for all x,

$$p_{g_i \cdot g_j}(x) = p_{g_i} \circ p_{g_j}(x),$$

and the mapping f is a homomorphism. $\square$

	g_1	g_2	g_3	g_4	g_5	g_6
g_1	g_1	g_2	g_3	g_4	g_5	g_6
g_2	g_2	g_3	g_1	g_5	g_6	g_4
g_3	g_3	g_1	g_2	g_6	g_4	g_5
g_4	g_4	g_6	g_5	g_1	g_3	g_2
g_5	g_5	g_4	g_6	g_2	g_1	g_3
g_6	g_6	g_5	g_4	g_3	g_2	g_1

FIG. 2.3.5   Composition Table for a Six-Element Group G

We can clarify the details of the proof by presenting an example. Consider the six-element group G whose composition table is given in Fig. 2.3.5. To construct a permutation group isomorphic to G, we treat each row of the composition table as a permutation on the set $N = \{1, 2, 3, 4, 5, 6\}$. We obtain the following permutations:

$$p_1 = (1)(2)(3)(4)(5)(6);$$
$$p_2 = (1\,2\,3)(4\,5\,6);$$
$$p_3 = (1\,3\,2)(4\,6\,5);$$
$$p_4 = (1\,4)(2\,6)(3\,5);$$
$$p_5 = (1\,5)(2\,4)(3\,6);$$
$$p_6 = (1\,6)(2\,5)(3\,4).$$

Thus we obtain the permutation group $P = \langle \{p_1, p_2, p_3, p_4, p_5, p_6\}, \circ \rangle$. To verify that P is isomorphic to G, we must show that $g_i \cdot g_j = g_k$ if and only if $p_i \circ p_j = p_k$, or equivalently that their composition tables differ only in the labeling of elements. Although somewhat tedious, this verification is straightforward. For example,

$$p_3 \circ p_4 = (1\,3\,2)(4\,6\,5) \circ (1\,4)(2\,6)(3\,5) = (1\,6)(2\,5)(3\,4) = p_6$$

and $g_3 \cdot g_4 = g_6$.

The permutation group we have just constructed often is called the *regular representation* of a group. Because the columns of a composition table are permutations as well as the rows, we also can construct a permutation representation of a group from the columns. To distinguish the two, the row representation sometimes is called the *left regular representation* and the column representation the *right regular representation*.

Because the composition table can be constructed from the permutations, it is obvious that the regular representation has all of the information contained within the corresponding composition table. The degree of the regular representation always is equal to the order of the group, so that the representation actually is no more compact than the composition table. In both the composition table and the regular representation, we must specify n^2 items for a group of order n, because we must specify n items for each of n group elements. However, we can construct more compact permutation representations for a group, since the degree of a permutation representation may be substantially less than the order of the group. We have mentioned that the set of all permutations of degree n is a group of order $n!$. Thus the regular representation of this permutation group has degree $n!$, which is larger than n for all $n \geq 3$ and is vastly larger than n even for n as small as 10. Hence the regular representation of a group can be too large for practical use, but the same group may have quite useful permutation representations of much smaller degree.

For example, there is a degree-3 permutation representation of the order-6 group whose composition table is given in Fig. 2.3.5. This representation is the following set of permutations: $p_1 = (1)(2)(3)$, $p_2 = (1\,2\,3)$, $p_3 = (1\,3\,2)$, $p_4 = (1\,2)(3)$, $p_5 = (1\,3)(2)$, and $p_6 = (1)(2\,3)$. This is the set of permutations illustrated in Fig. 2.3.4. The groups represented by the sets of permutations of degree 6 and of degree 3 are isomorphic; in fact, their composition tables are identical. For example, observe that $p_2 \quad p_4 = p_5$ in both representations. In the degree-6 representation

$$(1\,2\,3)(4\,5\,6) \circ (1\,4)(2\,6)(3\,5) = (1\,5)(2\,4)(3\,6),$$

and in the degree-3 representation

$$(1\,2\,3) \circ (1\,2)(3) = (1\,3)(2).$$

The permutation representation reveals some interesting properties of a group that are not always evident in a composition table. The following definition mentions one such property.

Definition The *order* of a group element g is the least integer r such that $g^r = e$.

Lemma The order of a permutation in a permutation group is the least common multiple of its cycle lengths.

Proof: Suppose a permutation p includes the k-cycle $(1\,2\ldots k)$. Then p^2 includes a cycle of the form $(1\,3\ldots)$; p^3 has a cycle $(1\,4\ldots)$; p^{k-1} has a cycle $(1\,k\ldots)$; and p^k includes the cycles $(1)(2)\ldots(k)$. More generally, p^i is the identity map on the letters $1, 2, \ldots, k$ if and only if i is a multiple of k. Thus, if p has only one cycle, the theorem is true. If p has two or more cycles, then p^i is an identity map on all of the letters in p's representation if and only if i is a multiple of all the cycle lengths. In particular, if i is the least common multiple of the cycle lengths, p^i is an identity map, and p^i cannot be an identity map for any smaller i. Hence the order of p is the least common multiple of the cycle lengths. $\square$

We can determine quickly from the preceding lemma that the following permutations have order 2: $(1\,2)(3)$; $(1\,2)(3\,4)(5)(6)$; and $(1\,3)(5\,7)(2\,4)(6)$. However, the permutation $(1\,2\,3)(4\,5)$ has order 6.

Lemma Every set of finite-degree permutations that is closed under function composition is a permutation group.

Proof: The axiom of closure holds by hypothesis, and we have shown previously that the axiom of associativity holds for the composition of permutations. The axiom of identity holds because closure implies that p^i is in the set for all $i \geq 1$ if p is a permutation in the set. In particular, if p has order k, then $p^k = e$ is in the set and, for permutations of finite degree, k is finite. The inverse of p is p^{k-1}, because $p^{k-1} \circ p = p \circ p^{k-1} = e$. If p has order k, then p^{k-1} must be in the set, and the axiom of inverses holds. $\square$

As we see from the preceding lemma, the problem of proving that a finite set of permutations forms a group reduces to the problem of proving that the axiom of closure holds. In practice, we often encounter sets of permutations that are not closed under composition, but in such a case the set can be used as a generator set to produce a set of permutations that is closed. Hence, by the preceding lemma, a set of permutations of finite degree is a generator set for a permutation group. Therefore, virtually any application that involves sets of permutations has something to do with permutation groups.

Summary We have developed a notation for permutations in which we describe each permutation as a collection of disjoint cycles. This notation lends itself to manipulation because, if $p_1 = p_2 \circ p_3$, then we can easily compute the cycle notation for p_1 from the notation for p_2 and p_3.

A permutation group is a set of permutations that satisfies the group axioms under the operation of function composition. A major point of this section is the notion that every group is isomorphic to a permutation group. The central idea in the proof of this notion is that the rows and columns of a group composition table are permutations of the elements of the group. Hence we can associate a permutation with each group element by letting each group element correspond to its row (or column) permutation.

The other major point of the section is the notion that a group can have many different representations as a permutation group. In this context, a representation of a group is a set of permutations that have a composition table isomorphic to the group composition table. The degree of a permutation representation is the number of elements in the set on which the permutations act. A group can have permutation representations of many different degrees; the degree of the representation may be less than, equal to, or greater than the order of the group. The composition table of a permutation group depends only on the way in which the permutations interact with each other under function composition; this does not directly depend upon the number of set elements on which the permutations act.

We mention in this section that the cycle notation for permutations was developed by Cauchy (1815). Galois (1832) apparently had a good grasp of the notion of group isomorphism in his work on the solvability of polynomial equations. After a long period of inactivity in group theory, Cauchy reawakened interest in this area with a series of publications that began in 1845. The papers of Galois were republished in 1846 and attracted much attention, although the French Academy of Sciences had rejected them some fifteen years before (an act that played an important part in leading Galois to his death after a duel in 1832 at the age of 20). Cayley (1854) discovered that the structure of a group is determined solely by its composition table, and the central theorem of this section (that every group is isomorphic to a permutation group) is credited to Cayley. For this reason, the regular representation of a group sometimes is called the *Cayley representation.*

By the time Camille Jordan published his excellent monograph on group theory, the idea of the regular representation had been formalized and was well understood (Jordan 1870, p. 60).

Exercises

2.3.1 Find the regular representation of all distinct groups of orders 2, 3, and 4.

2.3.2 Prove that $(1\ 2\ 3 \ldots n) \circ (1\ i)$ is equal to $(1\ i+1\ i+2 \ldots n) \circ (2\ 3 \ldots i-1\ i)$.

2.3.3 Find the permutation $(1\ i) \circ (1\ 2\ 3 \ldots n)$.

2.3.4 Prove that $(1\ 2\ 3 \ldots n) \circ (n\ n+1) = (1\ 2\ 3 \ldots n\ n+1)$.

2.3.5 Prove that $(1\ 3\ 2) \circ (1\ 4\ 7\ 3\ 2\ 8\ 6) \circ (1\ 2\ 3)$ is equal to $(3\ 4\ 7\ 2\ 1\ 8\ 6)$.

2.3.6 Show that, if p is an arbitrary permutation and q is the cycle $(1\ 2 \ldots i)$, then the permutation $q^{-1} \circ p \circ q$ has the same cycle structure as p. In fact, this is the permutation obtained by replacing the letters in the representation of p by their images under the mapping q^{-1}. Is the latter statement still true if q is an arbitrary permutation?

2.3.7 Let $g = (1\ 2\ 3\ 4\ 5\ 6)$. Find the permutations in the group generated by g.

2.3.8 Let $g = (1\ 2\ 3)(4\ 5)$. Find the permutations in the group generated by g. Show that this group is isomorphic to the group g in Ex. 2.3.7.

2.3.9 Find the regular representations of the following groups:
(a) $\langle \mathbf{Z}_4, +_4 \rangle$;
(b) $\langle \mathbf{Z}_5 - \{0\}, \cdot_5 \rangle$;
(c) $\langle \mathbf{Z}_5, +_5 \rangle$;
(d) $\langle \mathbf{Z}_6, +_6 \rangle$;
(e) $\langle \mathbf{Z}_7 - \{0\}, \cdot_7 \rangle$.
Which of these groups are isomorphic to one another?

2.4 Subgroups, Cosets, and Group Homomorphisms

Our applications of group theory require that we solve two types of problems. First, given a particular group, we must find its structure; more precisely, we must determine its subgroups and determine how the group composition table depends on its subgroups. Second, we must be able to determine whether one group can simulate another group, a problem that is equivalent to that of determining if there exists a homomorphism between the two groups.

The first problem is motivated by applications involving the construction of functional computer modules that do group operations. We wish to construct such modules by joining together modules that do operations in subgroups. In Chapter Four, we discover that this construction technique enables us to build computer adders that attain optimal or near-optimal computation speeds. In this section, we pursue the topic through the study of subgroups and cosets. The principal result here is the demonstration that, for each subgroup of a group, there exist two partitions of the group into equal-sized classes called *cosets*. The properties of the cosets enable us to prove that the order of a subgroup must divide the order of the group. In the following three chapters, we make frequent reference to cosets and the results obtained in this section as we examine computer applications.

The second topic of this section concerns group homomorphisms. Here we discover a strong correspondence between special subgroups, called *normal subgroups*, and group homomorphisms. In fact, we show that to each homomorphism of a group G there corresponds a normal subgroup H of G, and to each normal subgroup there corresponds a group homomorphism. For the second part of this statement, we find that the partition of the group into cosets is a homomorphic image of the group.

We begin our study of cosets by introducing a notation that we use frequently in the remainder of this text.

Definition Let K be a subset of elements of a group $\langle G, \cdot \rangle$, and let g be an element of G. Then the set $g \cdot K$ is the set $\{g \cdot k \mid k \in K\}$.

Occasionally we deal with a product such as $g_1 \cdot K \cdot g_2 \cdot K$ in which a set appears two or more times. By convention, we take this to be the set $\{g_1 \cdot k_1 \cdot g_2 \cdot k_2 \mid k_1, k_2 \in K\}$. Each of the two occurrences of K represents all of the elements in K, so the pair of occurrences of K represents all possible pairs of elements in K.

Definition Let H be a subgroup of G. The *left cosets of G relative to H* are defined to be sets of the form $g \cdot H$ where g is in G. These sets are called simply *left cosets* when the groups G and H are known by context. Similarly, *right cosets* are sets of the form $H \cdot g$.

As an example, consider the group S_3, the symmetric group of degree 3. Its degree-3 representation as a permutation group has the following elements: $e, (1\,2), (1\,3), (2\,3), (1\,2\,3), (1\,3\,2)$. Consider the subgroup

$H = \{e, (1\,2)\}$. The left cosets of S_3 relative to H are the following:

$$e \cdot H = \{e, (1\,2)\};$$
$$(1\,2) \cdot H = \{(1\,2), e\};$$
$$(1\,3) \cdot H = \{(1\,3), (1\,2\,3)\};$$
$$(2\,3) \cdot H = \{(2\,3), (1\,3\,2)\};$$
$$(1\,2\,3) \cdot H = \{(1\,2\,3), (1\,3)\};$$
$$(1\,3\,2) \cdot H = \{(1\,3\,2), (2\,3)\}.$$

Observe that there are three distinct cosets, and that the cosets form a partition of the group. This is not a coincidence.

Theorem   The left cosets of a group G relative to a subgroup H form a partition of G. Moreover, all of the left cosets of G relative to H have equal numbers of elements.

   Proof: First we prove that the left cosets form a partition of G. **(i)** *Distinct cosets are disjoint.* If two cosets $g_1 \cdot H$ and $g_2 \cdot H$ are not disjoint, then there is some g such that g is in both $g_1 \cdot H$ and $g_2 \cdot H$. Then there exist h_1 and h_2 in H such that $g = g_1 \cdot h_1 = g_2 \cdot h_2$. Because H is a subgroup, $h \cdot H = H$ for all h in H, from which we find that

$$g_1 \cdot H = g_1 \cdot (h_1 \cdot H) = (g_1 \cdot h_1) \cdot H = g \cdot H$$
$$= (g_2 \cdot h_2) \cdot H = g_2 \cdot (h_2 \cdot H) = g_2 \cdot H.$$

Therefore the left cosets $g_1 \cdot H$ and $g_2 \cdot H$ must be identical.   **(ii)** *Every element of G is in some left coset.* Because H contains e (the identity of G), $g \cdot e = g$ is in the left coset $g \cdot H$ for every g in G. Thus the cosets are a partition of G.

   Now we show that each coset has the same number of elements as any other coset. In fact, each has $|H|$ elements. For any group element g, left multiplication by g is an injection. That is, the map $L_g : G \to G$, such that $L_g(g') = g \cdot g'$ for each g' in G, is one-to-one, because $(g \cdot g' = g \cdot g'') \Rightarrow (g' = g'')$. Hence $g \cdot H$ has precisely the same number of elements as H, for all g in G. $\square$

   Thus we see that the left cosets of G with respect to H always form a partition of G. The right cosets similarly form a partition. We call these two partitions the *left coset partition* and the *right coset partition*, respectively. These partitions need not be identical, although the subgroup H is always an equivalence class in both partitions. For the group S_3, the left and right coset partitions with respect to the subgroup $\{e, (1\,2)\}$ are the following:

left cosets: $\{e, (1\,2)\}$, $\{(1\,3), (1\,2\,3)\}$, $\{(2\,3), (1\,3\,2)\}$;
right cosets: $\{e, (1\,2)\}$, $\{(1\,3), (1\,3\,2)\}$, $\{(2\,3), (1\,2\,3)\}$.

   Because cosets are partitions of a group, it is natural to find equivalence relations on the group that induce these partitions. Obviously, one way to state such an equivalence relation is to say that $g_1 \text{ R } g_2$ if

g_1 is in the left coset $g_2 \cdot H$. It is easy to verify that this relation is symmetric, reflexive, and transitive. The more usual way of defining the equivalence relations is given in the following definition.

Definition Let G be a group with subgroup H. The *left coset relation* on G with respect to H is the relation R with the property that g_1 R g_2 if and only if $g_1^{-1} \cdot g_2$ is in H. Similarly, the *right coset relation* is the relation R such that g_1 R g_2 if and only if $g_1 \cdot g_2^{-1}$ is in H.

Lemma The left coset relation is an equivalence relation on G, and the equivalence classes are the left cosets of G with respect to H.

Proof: To prove that the left coset relation is an equivalence relation, we must show that it is symmetric, reflexive, and transitive. **(i)** *Symmetric.* Suppose that $g_1^{-1} \cdot g_2 = h$, where h is in H. Then, because H is a subgroup, $h^{-1} = g_2^{-1} \cdot g_1$ is in H. This proves that symmetry holds. **(ii)** *Reflexive.* For all g, we have $g^{-1} \cdot g = e$, which is in H. **(iii)** *Transitive.* If $g_1^{-1} \cdot g_2$ and $g_2^{-1} \cdot g_3$ are in H, then the closure of H implies that $g_1^{-1} \cdot g_2 \cdot g_2^{-1} \cdot g_3 = g_1^{-1} \cdot g_3$ is in H, so transitivity holds. Therefore, the relation is an equivalence relation.

The equivalence classes are the left cosets because, if $g_1 \cdot H = g_2 \cdot H$, then

$$H = e \cdot H = (g_1^{-1} \cdot g_1) \cdot H = g_1^{-1} \cdot (g_1 \cdot H)$$

$$= g_1^{-1} \cdot (g_2 \cdot H) = (g_1^{-1} \cdot g_2) \cdot H,$$

so $g_1^{-1} \cdot g_2$ is in H. Conversely, if $g_1^{-1} \cdot g_2$ is in H, then $g_1 \cdot H$ and $g_2 \cdot H$ have the element $g_1 \cdot g_1^{-1} \cdot g_2 = g_2$ in common, and they must be the same coset. □

In Chapter One, we discuss a right coset relation without calling it that in the context of congruence modulo n. Note that the multiples of an integer n form a subgroup of $\langle \mathbf{Z}, + \rangle$. The set of right cosets relative to this subgroup is $\mathbf{Z}_n$, the integers modulo n. We say that two integers x and y are in the same right coset (that is, they are congruent modulo n) if their difference $x - y$ is a multiple of n. But this is just another way of saying that x is equivalent to y if $x + (y^{-1})$ is in the subgroup of the multiples of n. The right coset relationship is quite apparent in the latter formulation. It is not difficult to verify that the right coset partition is the same as the left coset partition for this example.

Now we know that a subgroup of G induces a partition of G into equal-sized subsets. This is an extremely powerful tool for discovering properties relating the sizes of G and H.

Definition The number of left cosets of G relative to H is called the *index of H under G*, and is written $[G:H]$.

Theorem $|G| = |H| \cdot [G:H]$.

Proof: G is the union of the left cosets, and they form a partition of G. Thus $|G|$ is the total number of elements in the left cosets. Each left coset has $|H|$ elements, and there are $[G:H]$ distinct cosets. □

Theorem   (Lagrange 1771)   If H is a subgroup of G, then $|H|$ divides $|G|$.
Proof: Follows trivially from $|H| = |G|/[G:H]$.   □

Lagrange's theorem is a fundamental one, and we use it frequently. Note how it follows directly from the development of the properties of cosets. The theory of cosets also is a useful tool, quite apart from its use in the proof of Lagrange's theorem. We have seen one such application in showing that two numbers x and y are congruent modulo n (for a specified n) if and only if they are in the same coset relative to Z_n. Several applications of the results thus far appear in following chapters.

The second portion of this section is concerned with group homomorphisms. The key result here is the derivation of the correspondence between a normal subgroup and a group homomorphism. We begin this study with the formal definition of a normal subgroup.

Definition   A subgroup H of a group G is said to be a *normal subgroup* if the left coset partition induced by H is identical to the right coset partition induced by H. Equivalently, H is normal if $g \cdot H = H \cdot g$ for all g in G.

If G is an abelian group, then every subgroup of G is a normal subgroup. The commutativity axiom that holds for abelian groups actually is stronger than needed to demonstrate this fact because, if $g \cdot h = h \cdot g$ for all h and g in G, then certainly $g \cdot H = H \cdot g$ for all g in G and for any subgroup H. However, if a group G has a subgroup H for which $g \cdot H = H \cdot g$, we cannot conclude that the group is abelian. The following example shows a normal subgroup of order 3 of the group S_3. Note that $(1\,2) \cdot H = H \cdot (1\,2)$, but the permutation $(1\,2)$ does not commute with every element of H. In particular, you can verify that $(1\,2) \cdot (1\,3\,2) \neq (1\,3\,2) \cdot (1\,2)$.

Left cosets:
$$H = e \cdot H = (1\,2\,3) \cdot H = (1\,3\,2) \cdot H = [e, (1\,2\,3), (1\,3\,2)];$$
$$(1\,2) \cdot H = (1\,3) \cdot H = (2\,3) \cdot H = [(1\,2), (1\,3), (2\,3)].$$

Right cosets:
$$H = H \cdot e = H \cdot (1\,2\,3) = H \cdot (1\,3\,2) = [e, (1\,2\,3), (1\,3\,2)];$$
$$H \cdot (1\,2) = H \cdot (1\,3) = H \cdot (2\,3) = [(1\,2), (1\,3), (2\,3)].$$

Because the left cosets are identical to the right cosets for normal subgroups, we drop the modifier and simply call them *cosets*. We denote the partition of G into cosets (induced by the normal subgroup H) as G/H (read "G modulo H"), and we refer to G/H as a quotient structure. We let $[g]$ denote the coset containing g.

Now we can make an interesting observation about the cosets of a group with respect to a normal subgroup. If G is a group with respect to the binary operation $\cdot$, then the operation $\cdot$ is well defined on the quotient structure G/H. That is, we can define the composition of cosets to be $[g] \circ [h] = [g \cdot h]$, and we discover that, if we pick any g_1 in $[g]$ and h_1 in $[h]$, then the product $g_1 \cdot h_1$ is always in $[g \cdot h]$. The following example illustrates this property; a proof of the property follows the example.

Fig. 2.4.1 shows the composition table for the group S_3 and illustrates that the composition is well defined on the quotient structure with respect to the normal subgroup $\{e, (1\,2\,3), (1\,3\,2)\}$.

$\cdot$	e	$(1\,2\,3)$	$(1\,3\,2)$	$(1\,2)$	$(1\,3)$	$(2\,3)$
e	e	$(1\,2\,3)$	$(1\,3\,2)$	$(1\,2)$	$(1\,3)$	$(2\,3)$
$(1\,2\,3)$	$(1\,2\,3)$	$(1\,3\,2)$	e	$(1\,3)$	$(2\,3)$	$(1\,2)$
$(1\,3\,2)$	$(1\,3\,2)$	e	$(1\,2\,3)$	$(2\,3)$	$(1\,2)$	$(1\,3)$
$(1\,2)$	$(1\,2)$	$(2\,3)$	$(1\,3)$	e	$(1\,3\,2)$	$(1\,2\,3)$
$(1\,3)$	$(1\,3)$	$(1\,2)$	$(2\,3)$	$(1\,2\,3)$	e	$(1\,3\,2)$
$(2\,3)$	$(2\,3)$	$(1\,3)$	$(1\,2)$	$(1\,3\,2)$	$(1\,2\,3)$	e

$\circ$	$[e] = [e, (1\,2\,3), (1\,3\,2)]$	$[(1\,2)] = [(1\,2), (1\,3), (2\,3)]$
$[e]$	$[e]$	$[(1\,2)]$
$[(1\,2)]$	$[(1\,2)]$	$[e]$

FIG. 2.4.1   The Composition Tables for S_3 and for S_3 Modulo a Subgroup of Order 3

The dashed lines in the group composition table partition the table into four areas. The vertical dashed line divides the columns into the two cosets of the group; similarly, the horizontal dashed line divides the rows of the table into the two cosets of the group. In the table for the quotient structure, each of the four entries corresponds to one of the areas of the group table. Note that all of the entries in a single area of the group belong to a single coset. Thus, for example, the coset composition $[e] \circ [(1\,2)]$ is well defined and is equal to $[(1\,2)]$ because all of the elements in the upper-right quadrant of the table belong to the coset $[(1\,2)]$.

Another interesting property of this table is that the quotient structure of cosets of S_3 is a group with respect to coset composition. Later in this section, we show that the quotient structure *always* is a group with respect to the composition of the original group and, in fact, is a homomorphic image of the original group.

Fig. 2.4.2 presents a graphical example to illustrate a well-defined operation on the quotient structure of a group. In this example, the group is a particular group of order 8 that is called the *quaternion group*. This group is generated by the following relations: $g^2 \cdot h^2 = e$; $g^3 \cdot h \cdot g \cdot h = e$; $g^4 = e$; $h^4 = e$. Although every subgroup of this group is normal, the group is not an abelian group. Obviously, every subgroup of an abelian group is a normal subgroup, but it is surprising to find a nonabelian group for which this is true also.

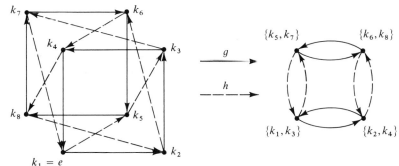

Group Graph of the Quaternion Group

Graph of the Quotient Structure
Relative to the Subgroup $\{e, g^2\}$

**FIG. 2.4.2   Graphs of the Quaternion Group and its Quotient Structure
Relative to a Normal Subgroup**

In Fig. 2.4.2 the nodes of the group graph are labeled $k_1, k_2, \ldots, k_8$ to simplify the construction of the regular representation of the group. The left regular representation is the following (with the symbol $\cong$ denoting "is isomorphic to"):

$$k_1 = \quad e \quad \cong (1)(2)(3)(4)(5)(6)(7)(8);$$
$$k_2 = \quad g \quad \cong (1\,2\,3\,4)(5\,6\,7\,8);$$
$$k_3 = \quad g^2 \quad \cong (1\,3)(2\,4)(5\,7)(6\,8);$$
$$k_4 = \quad g^3 \quad \cong (1\,4\,3\,2)(5\,8\,7\,6);$$
$$k_5 = \quad h \quad \cong (1\,5\,3\,7)(2\,8\,4\,6);$$
$$k_6 = g \cdot h \cong (1\,6\,3\,8)(2\,5\,4\,7);$$
$$k_7 = \quad h^3 \quad \cong (1\,7\,3\,5)(2\,6\,4\,8);$$
$$k_8 = h \cdot g \cong (1\,8\,3\,6)(2\,7\,4\,5).$$

You can easily verify that $\{e, g^2\}$ is a normal subgroup of the quaternion group. The cosets relative to this subgroup are $[e, g^2]$, $[g, g^3]$, $[h, h^3]$, and $[(g \cdot h), (h \cdot g)]$.

The second graph in Fig. 2.4.2 depicts the group composition on the quotient structure. The quotient-structure graph is constructed by joining pairs of nodes in the group graph, where each pair is in the same coset. Hence nodes k_1 and k_3 of the group graph are joined, as are nodes k_2 and k_4, nodes k_5 and k_7, and nodes k_6 and k_8. The graph of the quotient structure retains the edges of the original graph, with duplications eliminated. For example, because there is a g-type edge between nodes k_1 and k_2 and another between nodes k_3 and k_4 in the group graph, a single edge is retained from nodes $\{k_1, k_3\}$ to nodes $\{k_2, k_4\}$ in the quotient graph. Note that there is exactly one generator edge of each type leaving each node in the graph of the quotient structure. Therefore the group composition in the quotient structure carries each coset into a unique coset, which is another way of saying that the group composition is well defined on the quotient structure.

If we attempt to construct a quotient-structure graph for cosets relative to a nonnormal subgroup, we find that the group operation is not well defined on the quotient structure. In the quotient graph, we find that some node has two or more edges of the same generator type leaving it. Consider the cosets of S_3 relative to the subgroup $\{e, (1\,2)\}$. This subgroup is not normal in S_3. For example,

$$(1\,3) \circ [e, (1\,2)] = [(1\,3), (1\,2\,3)] \neq [e, (1\,2)] \quad (1\,3) = [(1\,3), (1\,3\,2)].$$

In Fig. 2.4.3, the composition table for the group has rows and columns partitioned into left cosets by dashed lines. In this case, the left cosets are $[e, (1\,2)]$, $[(1\,3), (1\,2\,3)]$, and $[(2\,3), (1\,3\,2)]$. Notice that we can find several instances in the table of a square region whose entries do not all belong to one left coset. Fig. 2.4.4 shows the group graph for S_3 constructed from the generator relations $g^3 = e$, $h^2 = e$, and $(g \cdot h)^2 = e$,

$\circ$	e	$(1\,2)$	$(1\,3)$	$(1\,2\,3)$	$(2\,3)$	$(1\,3\,2)$
e	e	$(1\,2)$	$(1\,3)$	$(1\,2\,3)$	$(2\,3)$	$(1\,3\,2)$
$(1\,2)$	$(1\,2)$	e	$(1\,3\,2)$	$(2\,3)$	$(1\,2\,3)$	$(1\,3)$
$(1\,3)$	$(1\,3)$	$(1\,2\,3)$	e	$(1\,2)$	$(1\,3\,2)$	$(2\,3)$
$(1\,2\,3)$	$(1\,2\,3)$	$(1\,3)$	$(2\,3)$	$(1\,3\,2)$	$(1\,2)$	e
$(2\,3)$	$(2\,3)$	$(1\,3\,2)$	$(1\,2\,3)$	$(1\,3)$	e	$(1\,2)$
$(1\,3\,2)$	$(1\,3\,2)$	$(2\,3)$	$(1\,2)$	e	$(1\,3)$	$(1\,2\,3)$

FIG. 2.4.3   Composition Table for S_3, Showing Left Cosets Relative to $\{e, (1\,2)\}$

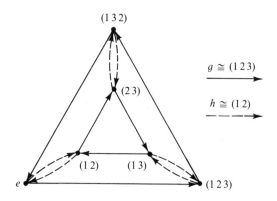

FIG. 2.4.4   Group Graph for S_3

where $g \cong (1\,2\,3)$ and $h \cong (1\,2)$. Fig. 2.4.5 shows the graph of the quotient structure of the left cosets. Note that two g-type edges leave the node corresponding to the coset $[e, (1\,2)]$, showing that, when the members of this coset are composed with $g \cong (1\,2\,3)$, the resulting elements do not all lie in the same left coset. Hence the group composition is not well defined on this quotient structure.

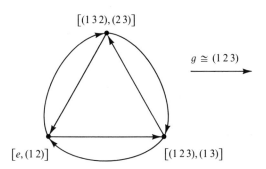

$[(1\,3\,2), (2\,3)]$

$g \cong (1\,2\,3)$

$[e, (1\,2)]$ $[(1\,2\,3), (1\,3)]$

FIG. 2.4.5 **Quotient-Structure Graph for Left Cosets of S_3 Relative to $[e,(1\,2)]$**

Now we must prove that the properties observed in the examples hold in general. First we show that the group composition is well defined on the cosets of a normal subgroup. You will recall from Chapter One that such a demonstration is equivalent to showing that the coset relation is a congruence with respect to the group operation. Then we prove that the converse statement is true—that is, if a group composition is well defined on the cosets of a group relative to a subgroup, then the subgroup is a normal subgroup.

Theorem The coset relation on the elements of a group G with respect to a normal subgroup H is a congruence relation with respect to the group composition.

Proof: We must show that, for any two cosets $[g]$ and $[h]$, if g' is in $[g]$ and h' is in $[h]$, then $g' \cdot h'$ is in $[g \cdot h]$. In set notation, we must show that $(g \cdot H) \cdot (h \cdot H) \subseteq g \cdot h \cdot H$. Recall that, because H is normal, $[h] = h \cdot H = H \cdot h$. Then,

$$(g \cdot H) \cdot (h \cdot H) = g \cdot (H \cdot h) \cdot H = g \cdot (h \cdot H) \cdot H = g \cdot h \cdot (H \cdot H).$$

Because H is a subgroup, $H \cdot H = H$. Then $g \cdot H \cdot h \cdot H = g \cdot h \cdot H$. $\square$

We mention earlier that the converse of this theorem holds as well: if a coset relation with respect to a subgroup H is a congruence relation, then H is a normal subgroup. Thus the following three statements are equivalent: **(i)** H is a normal subgroup of G; **(ii)** the left cosets induced by H are identical to the right cosets induced by H; and **(iii)** the left or right coset relation with respect to H is a congruence relation. We can prove the converse of the preceding theorem, but first we need an easy way of proving that a subgroup H is a normal subgroup. The following lemma provides this tool.

Lemma A subgroup H of a group G is a normal subgroup of G if and only if $g^{-1} \cdot H \cdot g \subseteq H$ for all g in G.

Proof: If H is normal in G, then $g \cdot H = H \cdot g$ for all g. Then $g^{-1} \cdot H \cdot g = g^{-1} \cdot g \cdot H = H$, which proves the implication in one direction. Conversely, if $g^{-1} \cdot H \cdot g \subseteq H$ for all g in G, then $H \cdot g \subseteq g \cdot H$, and replacing g by g^{-1} yields $H \cdot g^{-1} \subseteq g^{-1} \cdot H$. Manipulation of the latter relation yields

$$g \cdot H = g \cdot (H \cdot g^{-1}) \cdot g \subseteq g \cdot (g^{-1} \cdot H) \cdot g = H \cdot g.$$

Thus $g \cdot H \subseteq H \cdot g \subseteq g \cdot H$, so $g \cdot H = H \cdot g$ for all g in G. □

Now we can prove the converse of the preceding theorem.

Theorem Let H be a subgroup of a group G. If the left or right coset relation with respect to H is a congruence relation, then H is a normal subgroup of G.

Proof: We prove here only that, if the left coset relation is a congruence relation, then H is normal in G. The proof for the right coset relation follows similarly.

If the left coset relation is a congruence, then $g' \cdot H \cdot g \cdot H \subseteq g' \cdot g \cdot H$ for all g and g' in G. In particular, for the group identity $g' = e$, we obtain $H \cdot g \cdot H \subseteq g \cdot H$ for all g in G. Then $H \cdot g \cdot h \subseteq g \cdot H$ for all h in H, which implies that

$$(H \cdot g \cdot h) \cdot h^{-1} \subseteq (g \cdot H) \cdot h^{-1} = g \cdot (H \cdot h^{-1}) = g \cdot H.$$

Thus for all g in G, we have $H \cdot g \subseteq g \cdot H$, or $g^{-1} \cdot H \cdot g \subseteq H$, and H is normal in G. □

At this point we have proved that normal subgroups correspond one-to-one to coset quotient structures for which group composition is well defined. It so happens that the quotient structures are groups themselves; in fact, they are homomorphic images of the groups on which they are defined. In Fig. 2.4.1, note that the quotient-structure composition table is a group composition table. In Fig. 2.4.2, note that the quotient-structure graph is a group graph. In each case, the homomorphism is the map that sends a group element g of the original group into the coset containing g in the quotient structure.

One of the important results in the theory of groups is that every group homomorphism is associated with a normal subgroup. Thus to each normal subgroup there corresponds a group homomorphism, and to each group homomorphism there corresponds a normal subgroup. This result is of practical importance because it states the precise conditions under which one group can simulate another. For example, addition in computers must be addition on a finite set, yet that addition must be in some sense imitate addition on an infinite set. To imitate addition on $\langle \mathbf{Z}, + \rangle$, we must find a finite homomorphic image of

$\langle \mathbf{Z}, + \rangle$. The theory of group homomorphisms tells us that every finite homomorphic image of $\langle \mathbf{Z}, + \rangle$ must be a group of the form $\langle \mathbf{Z}_n, +_n \rangle$. Therefore, addition of integers in computers is always isomorphic to addition modulo n for some n, because there are no other finite homomorphic images of addition on $\mathbf{Z}$.

Now we proceed to the main results of this section. First we show that the quotient structure is a group.

Theorem   If H is a normal subgroup of G, then the quotient structure $\langle G/H, \circ \rangle$ is a group, where $\circ$ is composition of cosets defined by $[g] \circ [h] = [g \cdot h]$.

Proof: We show that the four group axioms hold for the quotient structure.   **(i)** *Associativity.* Because composition in G is associative, $([g] \circ [h]) \circ [k] = [g \cdot h \cdot k] = [g] \circ ([h] \circ [k])$ for all g, h, and k in G.   **(ii)** *Closure.* The composition of the cosets $[g]$ and $[h]$ is the coset $[g \cdot h]$. Because G is closed, $g \cdot h$ is in G, and the coset $[g \cdot h]$ is in the quotient structure.   **(iii)** *Identity.* The coset $[e]$ is an identity for all $[g]$, because $[e] \circ [g] = [g] \circ [e] = [g \cdot e] = [e \cdot g] = [g]$.   **(iv)** *Inverses.* For all $[g]$, we have $[g]^{-1} = [g^{-1}]$ because

$$[g] \circ [g^{-1}] = [g \cdot g^{-1}] = [e] = [g^{-1} \cdot g] = [g^{-1}] \circ [g]. \qquad \square$$

We have seen that the normal subgroup of order 3 in S_3 induces a quotient structure with two cosets. Examination of the composition table for the quotient structure (Fig. 2.4.1) shows that it is a group composition table. Note that each of the group axioms holds for this table, as the previous theorem demonstrates. In particular, observe that $[(1\,2)]^{-1} = [(1\,2)^{-1}] = [(1\,2)]$ because $(1\,2)^{-1} = (1\,2)$. Also note that $[(1\,2\,3)]^{-1} = [(1\,2\,3)^{-1}] = [(1\,3\,2)]$, which is consistent because $(1\,2\,3)$ and $(1\,3\,2)$ are both elements of $[e]$.

Definition   The group $\langle G/H, \circ \rangle$ is called the *quotient group* (or *factor group*) of G relative to the normal subgroup H.

We have mentioned previously that the quotient group is a homomorphic image of the group G. The natural map between G and G/H is the map that sends each g in G into its coset in G/H. The following theorem verifies that this mapping is a homomorphism.

Theorem   If H is a normal subgroup of G, then the mapping $f: G \to G/H$, $f(g) = [g]$, is a group epimorphism.

Proof: The mapping is a homomorphism because, for all g and g' in G,

$$f(g) \circ f(g') = [g] \circ [g'] = [g \cdot g'] = f(g \cdot g').$$

The mapping is onto G/H because each g in G is in the coset $[g]$. $\square$

Returning to the examples in Fig. 2.4.1 and Fig. 2.4.2, note that the property of homomorphism is guaranteed because the operations on the quotient structures are well defined. In Fig. 2.4.1, the coset composition captures the structure of the group composition table, because each region of the group table is replaced by a corresponding coset in the quotient-structure table.

We next wish to prove the converse of the preceding theorem: to every group homomorphism there corresponds a normal subgroup. The following definition will be of use.

Definition   The *kernel* of a group homomorphism is the set of domain elements that is mapped onto the identity element in the range.

Returning to the example of Fig. 2.4.1, note that the kernel of the group epimorphism is the set of elements mapped into $[e]$, or the set $\{e, (1\,2\,3), (1\,3\,2)\}$, which is the normal subgroup that determines the group epimorphism. More generally, the epimorphism from G onto G/H maps the subgroup H onto $[e]$, the identity in G/H. Thus, in all of the examples studied in this section, the kernel of a group homomorphism is a normal subgroup. The next theorem shows that the kernel of a group homomorphism is a normal subgroup in every case. This theorem provides the basis for the one-to-one correspondence between group homomorphisms and normal subgroups. It is often called the *fundamental theorem of group homomorphisms*.

Theorem   Let $f: G \to G'$ be any group homomorphism. Then the kernel of the homomorphism is a normal subgroup of G.

Proof: We must show first that the kernel K is a subgroup of G and then that it is a normal subgroup. K is a subgroup if and only if, for all $h_1 \cdot h_2$ in K, $h_1 \cdot h_2^{-1}$ is also in K. That is, if $f(h_1) = f(h_2) = e_{G'}$, then $f(h_1 \cdot h_2^{-1}) = e_{G'}$. Because f is a homomorphism, we have

$$f(h_1 \cdot h_2^{-1}) = f(h_1) \cdot f(h_2^{-1}) = f(h_1) \cdot f(h_2)^{-1} = e_{G'} \cdot e_{G'} = e_{G'}.$$

Hence, $h_1 \cdot h_2^{-1}$ is in the kernel, and the kernel is a subgroup of G.

Now we show that K is normal in G. For all g in G, we have

$$f(g^{-1} \cdot K \cdot g) = f(g^{-1}) \cdot f(K) \cdot f(g) = f(g)^{-1} \cdot e_{G'} \cdot f(g) = e_{G'}.$$

Hence, $g^{-1} \cdot K \cdot g \subseteq K$, and K is normal in G.   □

Theorem   Let $f: G \to G'$ be a group epimorphism, and let H be the normal subgroup that is the kernel of the epimorphism. Then G' is isomorphic to G/H.

Proof: The isomorphism $\theta: G' \to G/H$ maps $f(g)$ into $[g]$. It is a simple matter to verify that every element in G' corresponds to a unique coset in G/H, and that composition in G' is the same as composition in G/H under the isomorphism. The proof is left as an exercise (Ex. 2.4.6).   □

Summary   We have found that every subgroup of a group is associated with two partitions of the group: the left and right coset partitions. Lagrange's theorem (the order of a subgroup of a group G divides the order of G) follows from the fact that all cosets relative to a subgroup are of equal size.

A normal subgroup is a subgroup for which the left and right coset partitions are the same. The group composition is well defined on a coset partition if and only if that partition is relative to a normal subgroup. Moreover, the coset quotient structure is a group with respect to the group composition if and only if the cosets are relative to a normal subgroup. In this case, the quotient-structure group is a homomorphic image of the original group. Conversely, if there exists a homomorphism from a group G to a group G', then the kernel of the homomorphism is a normal subgroup of G.

In some sense, this section traces the first century in the history of group theory. Lagrange's theorem resulted from his investigation of third and fourth order equations, particularly with respect to the effect of permutations of the roots of the equations. K. F. Gauss and Augustin-Louis Cauchy, among others, made various contributions to group theory in the late eighteenth and early nineteenth centuries, but their contributions were of lesser importance, and a true theory of groups did not emerge until later. N. H. Abel (1829) and Évariste Galois (1832) achieved significant advances in their demonstration that equations of fifth degree and higher are unsolvable. Galois, in particular, established many of the ideas from which group theory evolved. Galois viewed groups as algebraic structures and grasped the notion of group isomorphism for different representations of a group. Unfortunately, Galois left only sixty pages of mathematical writings when he was mortally wounded in a duel at age 20. Most of his major contributions stem from a letter hastily written the night before the duel.

Research in group theory lay dormant for a decade after the deaths of Abel and Galois. Then Cauchy (1845) cast the study of groups into a cohesive theory in a series of classic papers. He formalized the notion of left and right cosets and of normal subgroups, and he made many other contributions that awakened interest in the area. Arthur Cayley, J. A. Serret, Camille Jordan, and many others focused their attention on various problems of group theory. The notion of group homomorphism and the fundamental theorem of group homomorphisms are credited largely to Jordan (1870, p. 56), although these ideas were anticipated by earlier mathematicians including Galois.

Herstein (1964) has written a good modern textbook on group theory and other algebraic structures. The book by Van der Waerden (1931) is an unusually thorough text of historic importance. Other books of interest on group theory are those by Burnside (1911), Carmichael (1937), and Hall (1959).

Exercises

2.4.1 Find a subgroup of order 6 in $\langle \mathbf{Z}_{24}, +_{24} \rangle$. Display the cosets of $\mathbf{Z}_{24}$ relative to this subgroup.

2.4.2 Show that the intersection of two normal subgroups is a normal subgroup.

2.4.3 Let H and K be two subgroups of a group G. The *product set* of H and K, written $H \cdot K$, is the set of all products of the form $h \cdot k$, for h in H and k in K. Show that the product set is a normal subgroup of G, if H and K are normal in G.

2.4.4 Show that, if $[G:H] = 2$, then H is a normal subgroup of G.

2.4.5 Let H and K be subgroups of G, and let K be normal. Show that $H \cdot K$ is a subgroup of G.

2.4.6 Let $f: G \to G'$ be a group epimorphism, and let H be the normal subgroup that is the kernel of the epimorphism. Prove that G' is isomorphic to G/H.

2.4.7 For each g in a group G, the set $N_g = \{h \mid h \cdot g \cdot h^{-1} = g\}$ is called the *normalizer of g*. Show that N_g is a subgroup of G for every g.

2.4.8 In a group G, an element h is said to be a *conjugate* of an element g if there is an element x such that $h = x \cdot g \cdot x^{-1}$. Show that the number of conjugates of g is equal to $[G:N_g]$ (see Ex. 2.4.7).

2.4.9 An element g of a group G is said to be a *commutator* if there exist h and k in G such that $g = h^{-1} \cdot k^{-1} \cdot h \cdot k$. Show that every conjugate (see Ex. 2.4.8) of a commutator is a commutator.

2.4.10 If K is a normal subgroup of H, and H is a normal subgroup of G, then K is not necessarily normal in G. Find three groups, H, K, and G that demonstrate the validity of this statement. HINT: Let G be the group with defining relations $g^4 = h^2 = h \cdot g \cdot h \cdot g^{-1} = e$.

2.4.11 Show that, if H and K are subgroups of G and if K is normal in G, then $H \cap K$ is normal in H.

2.4.12 Prove that, if H is a subgroup of a group G, then $g^{-1} \cdot H \cdot g$ is also a subgroup for all g in G.

2.4.13 Let H be the unique subgroup of order $|H|$ in G, and let G be finite. Show that H is a normal subgroup. HINT: Use the preceding exercise.

2.4.14 In a group G, the *center* $C(G)$ of the group is the subset of elements of G that commute with every element of G. That is,

$$C(G) = \{h \mid g \cdot h = h \cdot g \text{ for all } g \text{ in } G\}.$$

Prove that $C(G)$ is a normal subgroup of G.

2.4.15 For each element g of G, the map $F_g : G \rightarrow G$ is the map $F_g(h) = g \cdot h \cdot g^{-1}$.
   (a) Prove that each F_g is an automorphism of G.
   (b) Prove that the set of maps $\{F_g | g \in G\}$ is a group under function composition.
   (c) Prove that the group of maps $\langle \{F_g\}, \rangle$ is isomorphic to $G/C(G)$.

2.4.16 Let H be a normal subgroup in G. Prove that G/H is abelian if and only if $g_1 \cdot g_2 \cdot g_1^{-1} \cdot g_2^{-1}$ is in H for all g_1 and g_2 in G.

2.4.17 Let H and K be normal subgroups of a group G. Prove that H is a normal subgroup of $H \cdot K$.

2.4.18 Let H and K be normal subgroups of a group G. Prove that $H \cap K$ is normal in K.

2.4.19 Prove that the groups $H \cdot K/H$ of Ex. 2.4.17 and $K/H \cap K$ of Ex. 2.4.18 are isomorphic.

2.5 Special Families of Groups

In this section we examine several different families of groups, families that we encounter frequently in applications. The simplest family of all is the class of cyclic groups. We can obtain directly the exact structures of the groups in this family, and we discover that a cyclic group of order n has a cyclic group of order m as a subgroup if and only if m divides n. Moreover, a cyclic group of order n is the direct product of smaller cyclic groups whose orders are the prime divisors of n and their powers.

An abelian group, in general, is the direct product of cyclic groups, but we cannot determine completely the structure of an abelian group from its order as we can the structure of a cyclic group. We do, however, obtain some information about the structure of abelian groups in the second part of this section.

Other families of interest are the dihedral groups, symmetric groups, and alternating groups. For the latter family, we derive the classic result that the alternating group of degree n has no nontrivial normal subgroups for all $n \neq 4$. Galois implicitly used this result to prove that equations of degree 5 and higher cannot be solved algebraically.

2.5.1 Cyclic Groups

The least complex groups are those in the family of cyclic groups. A cyclic group has only one generator. Specifically, a group G is said to be a *cyclic group* generated by the element g if every element in G is a power of g. To within isomorphism, there is only one cyclic group of order n for each $n \geq 1$. This group is called C_n, the *cyclic group of order n*.

Because a cyclic group has only one generator, it is defined completely by a single generator relation that has the form $g^n = e$. Hence the group graph of a cyclic group is a polygon. (For $n \leq 2$, the polygon is degenerate.) Fig. 2.5.1 shows the group graphs for C_4, C_5, and C_6 (the cyclic groups of orders 4, 5, and 6, respectively).

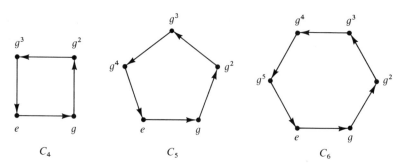

FIG. 2.5.1   Group Graphs for C_4, C_5, and C_6

As the graphs show, powers of g greater than the order of the group can be expressed as powers of g less than the order of the group. For example, in C_4 we see that g^6 is equal to g^2 because composition with g^4 in this group carries a group element around the cycle in the graph and back to itself. We can extend this same result to all groups of finite order, cyclic or not. In the following theorem, recall that the *order* of a group element g is the least integer n such that $g^n = e$.

Theorem   Let g be an element of any finite group G, and let g have order n. Then **(i)** $g^m = e$ if and only if n divides m, and **(ii)** n divides the order of G.

   Proof: For all i, we have $g^i = g^n \cdot g^{i-n} = e \cdot g^{i-n} = g^{i-n}$. Hence we can always replace the exponent of g^i by $i \bmod n$ without changing the group element. But $g^{i \bmod n} = e$ if and only if $i \equiv 0 \bmod n$, because g has order n. Then $g^i = e$ if and only if i is a multiple of n. This proves part **(i)** of the theorem.

   To prove **(ii)**, notice that the powers of any element g of G form a cyclic subgroup of G whose order is equal to the order of g. But the order of any subgroup of G divides the order of G.   □

Using this theorem, we can immediately discover all subgroups of C_n for every n. In fact, these are all cyclic groups of order m, where m divides n. We can informally deduce this fact from the group graph for C_n by looking at every kth node around the polygon for $k = n/m$. These nodes are separated from each other by k generator edges, and hence these nodes are nodes of a subgroup generated by g^k. If m does not divide n evenly, then the number of edges between nodes of the subgraph is not uniform, and we cannot construct a subgroup graph. The following theorem presents a more formal derivation of these results.

Theorem   Every subgroup of a finite cyclic group is a cyclic group whose order divides the order of the group.

Proof: Let G be a cyclic group of order n generated by an element g. Then every subgroup of G contains elements of the form g^i, where $0 \le i < n$. Let H be a subgroup of G, and let m be the smallest nonzero exponent of g in H. Then H contains all powers of g^m including $g^{-m} = (g^m)^{-1}$. Suppose there is some element g^r in H such that r is not a multiple of m. We know that $r = pm + q$ for some $p > 0$ and $0 < q < m$. Then, we find that $(g^{-m})^p \cdot g^r = g^{r-pm} = g^q$ is in H. But this contradicts the hypothesis that m is the smallest exponent of a power of g in H. Hence all other elements in H are of the form g^{im}, and H is a cyclic group generated by g^m. $\qquad\square$

Notice that every cyclic group is abelian. The preceding theorem shows that the order of a cyclic group completely characterizes the structure of the group and all of its subgroups. To be more specific, C_n is isomorphic to the direct product of cyclic subgroups whose orders divide n. The following theorem shows one of many ways of expressing this structure.

Theorem   Let C_n be a cyclic group of order $n = p_1^{x_1} p_2^{x_2} \ldots p_k^{x_k}$, where the p_i are distinct primes. Then

$$C_n \cong C_1 \times C_2 \times \cdots \times C_k,$$

where C_i is a cyclic group of order $p_i^{x_i}$.
Proof: See Ex. 2.5.3. $\qquad\square$

2.5.2 Abelian Groups   Next we explore the structure of abelian groups, because this topic depends heavily on the structure of cyclic groups. In particular, consider an abelian group with generators $g_1, g_2, \ldots, g_n$. Because $g_i \cdot g_j = g_j \cdot g_i$ for all i and j, every element in the group is equal to a product of powers of the generators of the form $g_1^{x_1} \cdot g_2^{x_2} \cdot \ldots \cdot g_n^{x_n}$. Every product of the generators that is not in this form can be placed in this form by using the axiom of commutativity. Because inverses of generators can appear in products, we let negative exponents denote powers of inverses, using the relation $(g^{-1})^r = g^{-r}$. Thus, as $x_1, x_2, \ldots, x_n$ range over $\mathbf{Z}$, the product $g_1^{x_1} \cdot g_2^{x_2} \cdot \ldots \cdot g_n^{x_n}$ ranges over all group elements in the abelian group generated by $g_1, g_2, \ldots, g_n$.

If the order of each of the generators g_i is an integer r_i, then the abelian group is finite. Moreover, it clearly is isomorphic to the direct product of the cyclic groups generated by each of its generators. That is,

$$G \cong C_{r_1} \times C_{r_2} \times \ldots \times C_{r_n}.$$

Thus every finite abelian group is the direct product of cyclic groups. In most cases, the cyclic group structure just indicated for an abelian group is not unique, because we often can find many different generator sets for a group, and such sets can lead to different cyclic groups in the direct product decomposition of the group.

For example, suppose that G is an abelian group that has among its generators the generators g and h of orders 2 and 3, respectively. These two generators account for the groups C_2 and C_3 that appear in the direct-product decomposition of G. Now consider the element $g \cdot h$. Its powers are

$$(g \cdot h)^1 = g \cdot h,$$
$$(g \cdot h)^2 = g^2 \cdot h^2 = h^2,$$
$$(g \cdot h)^3 = g,$$
$$(g \cdot h)^4 = h,$$
$$(g \cdot h)^5 = g \cdot h^2, \text{ and}$$
$$(g \cdot h)^6 = e.$$

Thus $g \cdot h$ generates both g and h, and in fact generates C_6, which is isomorphic to $C_2 \times C_3$. Thus, if a generator set for G contains both g and h, they can be replaced by the single generator $g \cdot h$. This has the effect of replacing $C_2 \times C_3$ by C_6 in the structural decomposition of G. This example illustrates a general principle that is the subject of the following theorem.

Theorem   Let g and h be elements of an abelian group G, and let their orders be relatively prime. Then the order of the element $g \cdot h$ is the product of the orders of g and h.

Proof: Let C_g and C_h be the cyclic subgroups of G generated by g and h, respectively. Let n be the product of the orders of g and h. Then $(g \cdot h)^n = g^n \cdot h^n = e \cdot e = e$. Hence the order of $g \cdot h$ must be a divisor of n. Suppose the order of $g \cdot h$ is m where $m < n$. Then $g^m \cdot h^m = e$, or $g^m = (h^{-1})^m$. But g^m is in C_g and $(h^{-1})^m$ is in C_h, so that the cyclic subgroups generated by g^m in C_g and by $(h^{-1})^m$ in C_h must have the same order. Moreover, their order must be a divisor of both the order of g and the order of h. Because g and h have relatively prime orders, the order of g^m must be 1, so that $g^m = (h^{-1})^m = e$. Also, if $(h^{-1})^m = e$, then $e = h^m \cdot (h^{-1})^m = h^m$. Hence m must be a multiple of both the order of g and the order of h. Because g and h have relatively prime orders, the least common multiple of their orders is n, their product. Therefore if $m < n$, we have a contradiction, and the proof follows.   □

The next definition leads to a powerful result concerning the orders of elements in an abelian group. We show that orders of all of the elements divide the largest order.

Definition   The *exponent* of an abelian group is the maximum of the orders of the elements of the group.

Theorem   The order of every element of a finite abelian group divides the exponent of that group.

Proof: Let g be an element of the abelian group G, and let n be both the order of g and the exponent of G. Suppose there is an element *h* of *G* with order n', where n' is not a divisor of *n*. Then, if *i* is the greatest common divisor of *n* and n', the order of g^i, which is n/i, is relatively prime to n'. Then the order of $h \cdot g^i$ is $n' \cdot n/i$, the product of the orders of *h* and g^i. Because n', by assumption, does not divide *n*, we have $n' > i$, and the order of $h \cdot g^i$ is greater than *n*, which is a contradiction. □

If the order of a finite abelian group *G* is equal to its exponent, the group is cyclic. If not, the group has two or more generators, which we call $g_1, g_2, \ldots, g_n$. Now consider the orders of the various generators g_i and assume that they are pairwise relatively prime. If the orders of the generators g_1 and g_2 are relatively prime, then the order of the element $g_1 \cdot g_2$ is the product of the orders of g_1 and g_2. Similarly, the orders of $(g_1 \cdot g_2)$ and g_3 have no nontrivial divisor in common, so that the order of $(g_1 \cdot g_2) \cdot g_3$ is the product of the orders of g_1, g_2, and g_3. By induction, the order of the element $g_1 \cdot g_2 \cdot \ldots \cdot g_n$ is equal to the product of the orders of all the generators, which is the order of the group. Hence the group is cyclic. Thus a noncyclic abelian group must have two generators whose orders have a nontrivial divisor in common. Hence, the most general structure that can be established for an abelian group is that given by the following theorem.

Theorem Every abelian group *G* has the structure

$$G \cong H_1 \times H_2 \times \cdots \times H_k,$$

where the H_i are cyclic groups whose orders are powers of primes, and the primes are not necessarily distinct. The orders of the H_i uniquely determine *G* to within isomorphism.

Proof: See Ex. 2.5.4. □

2.5.3 Dihedral Groups

We now have shown the explicit structure of cyclic groups in particular and abelian groups in general. It is more difficult to make statements about nonabelian groups. Such groups may have rich and complex structures that are not readily described by sweeping theorems such as those we have encountered for abelian groups. In the remainder of this section, we explore a few important families of nonabelian groups.

The class of *dihedral groups*, of degree *n* and order 2*n*, is just slightly more complex than cyclic groups. A dihedral group is generated from a cyclic group of order *n* by adding a new element of order 2. Specifically, a dihedral group D_n is defined by the relations $g^n = e$, $h^2 = e$, and $h \cdot g \cdot h \cdot g = e$. Fig. 2.5.2 shows the dihedral groups D_2 (of order 4), D_3 (of order 6), and D_4 (of order 8). The dihedral groups have the familiar polygonal shapes, demonstrating that they have cyclic subgroups of order *n*. Note that the third generator relation forces the inner and outer polygons to be reversed in direction in the group graphs.

In the dihedral groups of orders 6 or larger, $g \cdot h = h \cdot g^{n-1} \neq h \cdot g$,

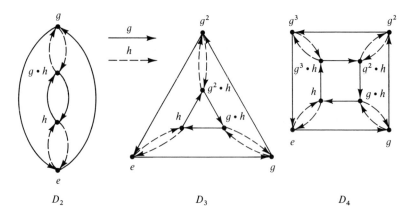

FIG. 2.5.2 Group Graphs for the Dihedral Groups D_2, D_3, and D_4

so all such groups are nonabelian. The dihedral group of order $2n$ arises naturally when we consider all possible rigid transformations of an n-sided regular polygon into itself. For example, consider every possible way to transform the hexagon as a rigid body so that the transformed hexagon can be superimposed on the original hexagon. Two transformations are illustrated in Fig. 2.5.3. Rotation through multiples of 60° clockwise yields five transformations distinct from the original hexagon. Each transformation can be regarded as a permutation of the labels on the nodes, with the original position representing the identity permutation. These six permutations form C_6 under function composition. If we let g represent the operation of clockwise rotation through 60°, this cyclic group is defined by the relation $g^6 = e$. Another six distinct transformations can be produced by reflecting each of the six transformations through a vertical axis. We can represent the operation of reflection by h. Because the reflection operation satisfies $h^2 = e$, the twelve permutations form D_6 under function composition. No other distinct transformations can be produced by any combination of rotations and reflections.

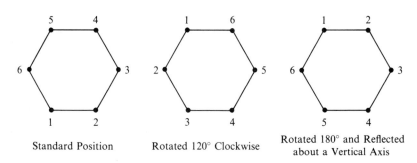

FIG. 2.5.3 Three Transformations of a Hexagon

The transformations can be written directly as permutations on 6 letters. The 120° clockwise rotation is the permutation $(1\,3\,5)(2\,4\,6)$ because node 1 is relabeled as 3, node 3 as 5, and so forth. Following this convention, the permutation corresponding to the 180° rotation with a reflection about a vertical axis is $(1\,5)(2\,4)(3)(6)$.

Because the direct product of two abelian groups must be abelian, $C_2 \times C_n$ cannot be isomorphic to D_n for dihedral groups of orders 6 or larger. Hence we cannot decompose D_n into a direct product of smaller cyclic groups as we did for C_n. However, it is true that C_n is normal in D_n for all n and, because we know the complete structure of C_n, we know a great deal about the structure of D_n.

2.5.4 Symmetric Groups

In this chapter we have settled many issues concerning permutation groups with simple structures. We now turn our attention to permutation groups with complex structures.

We have mentioned the symmetric group S_n on n letters. This group exists for all $n \geq 1$, and it has $n!$ elements. One subset of the elements of S_n is of particular interest to us; this is the subset containing the permutations that interchange two letters. These permutations, known as *transpositions*, take the form $(a\,b)(c)(d)(e)\ldots$, which we usually write simply as $(a\,b)$. Consideration of transpositions leads us to the study of an important subgroup of S_n.

Definition A permutation is called an *odd* permutation if it is a product of an odd number of transpositions; otherwise it is called an *even* permutation.

Obviously, the concept of odd and even permutation is well defined only if no permutation can be written both as a product of an odd number of transpositions and as a product of an even number of transpositions. It is tedious, but not difficult, to prove this. Because $(a\,b) \circ (a\,c) = (a\,c\,b)$, a 3-cycle such as $(a\,c\,b)$ is an even permutation. Similarly, we can show that 4-cycles are odd, 5-cycles are even, and so forth. A permutation that has an odd number of odd cycles and any number of even cycles is an odd permutation; any permutation that cannot be so represented is an even permutation.

Theorem The even permutations of S_n form a subgroup of S_n of order $n!/2$. This group is a normal subgroup of S_n, and it is called A_n, the *alternating group* of degree n.

Proof: To prove that the even permutations form a group, we need only show that the even permutations are closed under the group composition. This demonstration can be made easily by induction, and this part of the proof is left to the reader (Ex. 2.5.6).

To show that A_n is a normal subgroup of S_n, it is sufficient to show that A_n has $n!/2$ elements, or one-half the number of elements in S_n. Then the two left cosets are A_n and $S_n - A_n$, and these are identical to the right cosets. To establish that A_n has $n!/2$ elements, we find a one-to-one map from the even permutations in S_n onto the odd permutations in S_n. If p is an even permutation, then $(1\,2) \circ p$ is odd, and $p_1 \neq p_2$ implies that $(1\,2) \circ p_1 \neq (1\,2) \circ p_2$ for any p_1 and p_2. Thus the odd permutations are at least as numerous as the even ones. Similarly, left composition with $(1\,2)$ maps the odd permutations injectively into the even ones. Hence the even and odd permutations are equally numerous, and A_n has half as many elements as S_n. Then A_n is normal in S_n, and the proof is complete. $\square$

The permutation representation of A_3 is

$$A_3 = \{e, (1\,2\,3), (1\,3\,2)\},$$

and that of the odd permutations of S_3 is

$$(1\,2) \circ A_3 = \{(1\,2), (2\,3), (1\,3)\}.$$

These two sets form the two cosets of S_3 relative to A_3.

Definition   A group is said to be *simple* if it has no proper normal subgroups other than the identity group e.

We now proceed to show that A_n is a simple group for all n except $n = 4$. This is a classic result of group theory; it was perhaps the first practical application of the theory. The simplicity of A_5 guarantees that no algebraic procedure involving root extraction can solve arbitrary fifth-degree equations. The details of the proof are rather interesting because they exercise some of our notions about generators and normal subgroups. Two lemmas will be useful in our proof of the main theorem.

Lemma   The set of permutations $\{(1\,2\,i) \,|\, 3 \leq i \leq n\}$ generates A_n.
    Proof: Because these permutations are even, the group they generate must contain only even permutations, and therefore it cannot generate permutations that are not in A_n. The proof that every permutation in A_n can be written as a product of permutations in this set is left as an exercise (Ex. 2.5.7). $\square$

Lemma   Let H be a normal subgroup of A_n, where $n > 2$. If H contains a cyclic permutation on three letters, then $H = A_n$.
    Proof: Let H contain the 3-cycle $(1\,2\,3)$. Then $(1\,2\,3)^2 = (1\,3\,2)$ is in H, and also $g \circ (1\,3\,2) \circ g^{-1}$ is in H for all g in A_n because H is normal in A_n. Pick g to be $(1\,2) \circ (3\,i)$ where $i > 3$, an even permutation in A_n. It is easy to verify that $g \circ (1\,3\,2) \circ g^{-1} = (1\,2\,i)$. Hence all three-cycles of the form $(1\,2\,i)$ for $i = 3, 4, \ldots, n$ are in H, but these are the permutations that generate the alternating group A_n. $\square$

Theorem   The alternating group A_n is simple for all $n \neq 4$.
    Proof.   By inspection, the theorem holds for A_1, A_2, and A_3. For $n > 4$, it is sufficient to show that every normal subgroup of A_n contains a 3-cycle permutation. If H is a normal subgroup of A_n that does not

contain a 3-cycle, then every permutation in H except the identity cycle must displace at least 4 letters. (Two-cycles are odd permutations and are not in A_n.) Let h be a permutation in H, not equal to the identity, that displaces the least number of letters. Since h is even, it must displace at least three letters. If h is a 3-cycle then, by the lemma above, H is A_n. Consequently, to prove this theorem it is sufficient to show that h does not displace more than three letters.

Suppose, on the contrary, that h displaces four or more letters. Because H is normal in A_n, $g \cdot h \cdot g^{-1} \cdot h^{-1}$ is in H for all g in A_n. We can make use of this fact to find a contradiction and complete the proof. If h consists only of 2-cycles, then $h = (a\,b)(c\,d)(e\,f) \dots$ or $h = (a\,b)(c\,d)$. By letting $g = (c\,d\,e)$, we find $g \cdot h \cdot g^{-1} \cdot h^{-1} = (c\,e)(d\,f) \dots$ and $(c\,e\,d)$ for the respective cases. In both cases we have found another permutation in H that displaces fewer letters than h, which is a contradiction.

We must still examine the case for which h contains at least one cycle other than a 2-cycle. Because h is even, it cannot be of the form $(a\,b\,c\,d)$, which is odd. Thus h must displace at least five letters and, therefore, must be of the form $h = (a\,b\,c \dots)(d\,e \dots) \dots$ or of the form

$$h = (a\,b\,c\,d\,e).$$

In the former case we let $g = (a\,b\,d)$, and in the latter case we let $g = (a\,b\,c)$, and in both cases we find that $g \cdot h \cdot g^{-1} \cdot h^{-1}$ displaces fewer letters than h. Again we have found permutations in H that displace fewer letters than h. This contradiction forces us to conclude that h is a 3-cycle, and proves the theorem. $\qquad\square$

We conclude this section with a historical note. The structure of abelian groups was studied by Karl Friedrich Gauss around 1800 (see Gauss 1876, v. 1, p. 374, and v. 2, p. 266). The statement and proof of the simplicity of the alternating group was given by Camille Jordan (1870, p. 63), but Galois knew much earlier that A_5 is simple. The term "simple" is rather ill-chosen in this context, because A_5—a group of order 60 with no normal subgroups—is hardly a simple structure in the usual sense. In fact, because of the inherent complexity of A_5, the question of the solvability of quintic equations was unanswered for decades. Arbib (1969, pp. 284–286) gives an interesting survey of the history of simple groups.

Exercises

2.5.1 Find a representation of D_4 as a permutation group of degree 4.

2.5.2 Construct a representation of S_4 as a permutation group on four letters. Show the cosets of S_4 relative to A_4. In A_4, form a normal subgroup of order 4 and show the cosets relative to this subgroup. (This subgroup is sometimes called K_4, or Klein's four-group.)

2.5.3 Prove that every cyclic group of order $n = p_1^{x_1} p_2^{x_2} \ldots p_k^{x_k}$ is isomorphic to the direct product of cyclic groups of order $p_i^{x_i}$.

2.5.4 Prove that every abelian group is isomorphic to a direct product of cyclic groups whose orders are powers of primes, where the primes need not be distinct.

2.5.5 Show that S_n is generated by the permutations $(1\ 2)$ and $(1\ 2 \ldots n)$.

2.5.6 Show that the set of even permutations of S_n forms a subgroup of S_n.

2.5.7 Show that A_n is generated by the set of permutations
$$\{(1\ 2\ i) | i = 3, 4, \ldots, n\} \qquad \text{for } 3 \leq i \leq n.$$

	1	2	3
4	5	6	7
8	9	10	11
12	13	14	15

FIG. 2.5.4   The Well-Known "15" Puzzle

2.5.8 Fig. 2.5.4 shows a well-known puzzle. There are fifteen tiles, numbered 1 through 15, arranged in a square array with one cell empty. A move consists of placing into the empty cell one of the tiles that is adjacent to that cell—that is, one of the tiles that is immediately above or below or to the right or left of the empty cell. In effect, the tiles cannot be moved outside the boundary of the array or in the dimension perpendicular to the square array. If any tile could be transposed with any other tile or with the empty cell, 16! states would be possible for the puzzle. Prove that, under the rules of movement given above, exactly $16!/2$ of these possible states are reachable from the initial state.

2.5.9 Prove that a group G is abelian if $(g \cdot h)^3 = g^3 \cdot h^3$, $(g \cdot h)^4 = g^4 \cdot h^4$, and $(g \cdot h)^5 = g^5 \cdot h^5$ for all g and h in G.

2.5.10 Prove that S_n is nonabelian for all $n \geq 3$.

2.5.11 Determine whether A_4 is abelian or nonabelian.

2.5.12 We define $\phi(n)$ (*Euler's totient function*) to be $\phi(1) = 1$ and $\phi(n)$ for $n > 1$ equals the number of positive integers smaller than n that are relatively prime to n. The first few values of $\phi(n)$ are given in Table 2.5.1. Prove that precisely $\phi(n)$ elements of C_n have order n, and that therefore these $\phi(n)$ elements each generate C_n.

TABLE 2.5.1 Values of $\phi(n)$ for n ≤ 10

n	1	2	3	4	5	6	7	8	9	10
$\phi(n)$	1	1	2	2	4	2	6	4	6	4

2.5.13 Let G be an abelian group with elements g and h of orders n and m, respectively. Show that G has a subgroup whose order is the least common multiple of n and m. HINT: Use the cyclic subgroup decomposition structure of G.

2.5.14 Let M_n be the subset of $\mathbf{Z}_n - \{0\}$ that contains all of the nonzero integers in $\mathbf{Z}_n$ that are relatively prime to n. Show that M_n is a cyclic group under multiplication modulo n.

2.6 Symmetry Groups The analysis of a difficult problem often leads to the discovery of symmetries, and these in turn lead to a simplification of the problem. We have made use of such symmetries already in many proofs, when we use the word "similarly" to avoid the repetition of a previous proof with only minor mechanical changes. The simplification of problems through use of symmetries is an extremely important application of group theory. In this section we treat two practical applications of symmetry groups.

The first part of this section gives the background of the use of permutation groups in solving polynomial equations. The discussion is at an elementary level and just summarizes briefly the principal results. The second part of the section concerns symmetry operations on graphs. Again the discussion is elementary, but it serves as an introduction to the results in the next chapter, where we see how symmetry groups have been used extensively to describe equivalent computer devices.

We begin our discussion by considering the algebraic expression

$$a \cdot b + c \cdot d,$$

where the operations are ordinary multiplication and addition. Because multiplication is commutative, we can interchange a and b to form the expression

$$b \cdot a + c \cdot d,$$

which is equivalent to the original expression. Similarly, we can interchange the addends to form the equivalent expression

$$c \cdot d + a \cdot b.$$

In the first case we transformed the expression by the permutation $(a\,b)$; in the second case we used the permutation $(a\,c)(b\,d)$. Because the permutation $(a\,b)$ leaves the expression unchanged in value, we say that a and b are symmetric variables. The permutation $(a\,c)(b\,d)$ as a symmetry indicates that the pair a and b contributes to the value of the expression in the same way as does the pair c and d, so that we can simultaneously exchange a with c and b with d and leave the value of the expression unaltered. Note that we can apply the two symmetry operations sequentially to obtain the expression

$$d \cdot c + a \cdot b,$$

which also has the same value. This expression corresponds to the permutation $(a\,d\,b\,c) = (a\,c)(b\,d) \circ (a\,b)$. It should be no surprise that, if two permutations each leave the expression unchanged in value, then so does their composite permutation.

If we examine the complete set of symmetries for this expression, we find that the corresponding permutations form a group under function composition. This follows because we have seen that the set of symmetries is closed under composition, and every finite set of permutations that is closed under function composition is a group. For the example, the set of permutations in the symmetry group includes the following elements: $e = (a)(b)(c)(d)$, $(a\,b)$, $(c\,d)$, $(a\,b)(c\,d)$, $(a\,c)(b\,d)$, $(a\,d\,b\,c)$, $(a\,c\,b\,d)$, and $(a\,d)(b\,c)$.

To summarize the point illustrated by this example, we note that, if something is invariant under two permutations p_1 and p_2, then it is invariant under the composition $p_1 \circ p_2$. Thus the property of invariance assures us that the set of symmetry operations is closed under function composition. Closure is the crucial property in this context, because we know from a previous lemma that the set of symmetries is a group if it is closed.

Although the expression that we picked as an example is somewhat contrived, algebraic expressions with symmetries are commonplace. The quadratic equation

$$x^2 + bx + c = 0$$

has hidden symmetries. We know that this equation has two roots in **C**, say r_1 and r_2, such that

$$(x - r_1)(x - r_2) = 0.$$

The latter equation is symmetric in r_1 and r_2. After removing parentheses by multiplying the factors together, we obtain

$$x^2 - (r_1 + r_2)x + r_1 r_2 = 0.$$

Matching coefficients with the original equation yields

$$b = -(r_1 + r_2) \quad \text{and} \quad c = r_1 \cdot r_2.$$

It is no coincidence that the coefficients b and c, when expressed as functions of the roots, are invariant under permutations of r_1 and r_2. Because the factored form of the equation is symmetric in r_1 and r_2, so is the unfactored form. Then all coefficients of powers of x must be symmetric in r_1 and r_2.

This statement obviously extends to polynomials of all degrees. The cubic equation

$$x^3 + bx^2 + cx + d = 0$$

leads naturally to the formulas for the coefficients

$$b = -(r_1 + r_2 + r_3),$$
$$c = r_1 r_2 + r_1 r_3 + r_2 r_3, \text{ and}$$
$$d = -r_1 r_2 r_3,$$

all of which are invariant under permutations in the symmetric group S_3 acting on $\{r_1, r_2, r_3\}$. The first equation has three addends, $-r_1$, $-r_2$, and $-r_3$. Similarly, the second equation has three addends, and the third has   addend. Notice that the addends of the first equation are every possible way of taking one root at a time. The second equation shows every possible way of forming the product of two roots, and the third equation lists every way of forming a product of the three roots. This suggests a general formula that we give after the following definition.

Definition   The *elementary symmetric function* of n variables taken r at a time, denoted $S_{n,r}(x_1, x_2, \ldots, x_n)$, is the sum of every possible distinct product of exactly r of the n variables, where $r \leq n$.

The following examples illustrate the definition:

$S_{3,1}(x_1, x_2, x_3) = x_1 + x_2 + x_3;$
$S_{4,1}(x_1, x_2, x_3, x_4) = x_1 + x_2 + x_3 + x_4;$
$S_{n,1}(x_1, x_2, \ldots, x_n) = x_1 + x_2 + \cdots + x_n;$
$S_{3,2}(x_1, x_2, x_3) = x_1 x_2 + x_1 x_3 + x_2 x_3;$
$S_{4,2}(x_1, x_2, x_3, x_4) = x_1 x_2 + x_1 x_3 + x_1 x_4 + x_2 x_3 + x_2 x_4 + x_3 x_4;$
$S_{4,3}(x_1, x_2, x_3, x_4) = x_1 x_2 x_3 + x_1 x_2 x_4 + x_1 x_3 x_4 + x_2 x_3 x_4;$
$S_{3,3}(x_1, x_2, x_3) = x_1 x_2 x_3;$
$S_{4,4}(x_1, x_2, x_3, x_4) = x_1 x_2 x_3 x_4.$

Let $p(x)$ be a polynomial of degree n, with the coefficient of x^n equal to 1. Then, if the n roots of this polynomial are $r_1, r_2, \ldots, r_n$, the polynomial satisfies the equation

$$p(x) = \sum_{i=0}^{n} (-1)^i S_{n,i}(r_1, r_2, \ldots, r_n) \cdot x^i.$$

We see that we can solve polynomial equations if we can deduce the roots from elementary symmetric functions of the roots. Consequently, the properties of the symmetric group play a determining role in the solvability of such equations. A complete discussion of the solvability of polynomial equations is beyond the scope of this text, but it suffices to say that polynomial equations of order n are solvable by formulas involving the elementary arithmetic operations and root extraction if and only if the following properties hold: **(i)** there is a sequence of groups $G_1, G_2, \ldots, G_n$ such that $G_1 = S_n$ and $G_n = \{e\}$, and such that G_{i+1} is a normal subgroup of G_i for all $1 \leq i \leq n - 1$; and **(ii)** each of the quotient groups G_i/G_{i+1} is cyclic. These properties are satisfied for $n = 2, 3$, and 4. In particular, for $n = 4$, the properties are satisfied by the following sequence of groups: S_4, A_4, K_4, S_2, S_1. Here K_4 is an abelian group of order 4, the only such subgroup of A_4. We know that A_n is simple and not cyclic for $n \geq 5$, so that polynomial equations of degree 5 or greater cannot be solved by formulas.

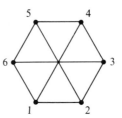

FIG. 2.6.1   A Graph with Several Symmetries

We now turn our attention from algebra to graphs to show that symmetry groups also play an important role in this field. A brief examination of the graph in Fig. 2.6.1 shows that it is invariant under rotations and reflections and therefore is invariant under D_6 acting on its six nodes. (In this graph, each line between two nodes a and b represents two edges, one directed from a to b and the other directed from b to a. The arrowheads and double lines have been omitted to simplify the drawing.)

The graph appears to have perfect symmetry, and we may be tempted to say that it is invariant under not only D_6 but S_6 as well. However, more careful examination shows that this is not true. For example, if we map node 1 onto node 5, we are then not free to map node 2 arbitrarily onto any other node of the graph. Node 2 is connected to node 1 in the original graph; any symmetry operation therefore must map node 2 onto a node connected to the image of node 1. Hence node 2 must be mapped either onto itself or onto node 4 or 6. These three possible mappings for node 2 are shown as partially labeled graphs in Fig. 2.6.2. Looking at the first of these graphs, we see that we can complete it consistently in only two different ways. Nodes 4 and 6 are connected to node 1 on the original graph, and there are only two different ways to place these nodes in the new graph so that they will be connected to the image of node 1. In either case, after nodes 4 and 6 are placed, there is only one consistent mapping for nodes 3 and 5. Thus the first of the graphs shown in Fig. 2.6.2 can be completed only in the two ways shown in Fig. 2.6.3. Similarly, each of the other two graphs in Fig. 2.6.2 can be completed in two different ways, yielding a

total of six distinct graphs that can result from mapping node 1 onto
node 5. In fact, there are six distinct possible mappings for each of the
six nodes onto which we could map node 1, and the total number of
distinct mappings in the symmetry group is 36. This group is much
smaller than S_6, which has order 720.

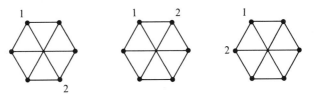

FIG. 2.6.2   Three of the Possible Mappings for Nodes 1 and 2 in Fig. 2.6.1

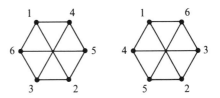

FIG. 2.6.3   Two Ways to Complete the First Graph of Fig. 2.6.2

Now that we know the order of the symmetry group, we can make
at least one nontrivial observation about this graph. Consider the
problem of finding the number of *different* ways to label this graph,
where we say that two labelings are different if there is no symmetry
operation that will carry one into the other. Under this definition, the
36 mappings discussed thus far are the same labeling. However, the
labeling shown in Fig. 2.6.4 is different. For example, note that nodes 1
and 2 are not connected in this graph as they are in the others. Although
there are 6! = 720 ways of placing numbers on the nodes, many of
these ways are not different from each other. It is a simple exercise to
show that there is one labeling for each coset of S_6 relative to the symmetry group of the graph. That is, all of the labelings within a single

FIG. 2.6.4   A Labeling Essentially Different from That of Fig. 2.6.1

coset are equivalent, but they are different from the labelings in the other cosets. Hence there are $720/36 = 20$ different labelings. (We describe this enumeration technique more rigorously in Chapter Three where we determine equivalence of devices for computers.)

Of course, we could solve the labeling enumeration problem without use of symmetries by listing all 720 labelings and identifying those that are truly different. However, this technique obviously is quite tedious and yields very little insight about the graph. By first finding the symmetry group, we are able to solve the enumeration problem almost trivially, and we can use the symmetry group to help us gain an understanding of other properties of the graph.

·	e	h	k
e	e	h	k
h	h	k	e
k	k	e	h

·	e	k	h
e	e	k	h
k	k	h	e
h	h	e	k

FIG. 2.6.5   Composition Tables for C_3

The use of symmetries extends beyond the few examples that we have listed here. For example, it extends to symmetries of elements of groups as exhibited by the composition tables for C_3 in Fig. 2.6.5. The two tables are the same, differing only in the order in which we have labeled the rows and columns. Note that the permutation $(e)(h\,k)$ changes one table into the other. Hence this permutation is an automorphism of C_3, and it shows that the elements h and k are symmetric with respect to each other. Because C_3 is isomorphic to $\langle Z_3, +_3 \rangle$, this means that we may interchange 1 and 2 in Z_3 without altering the operation of addition modulo 3.

The principal idea of this section is the following: two items that can be interchanged by a symmetry operation are indistinguishable. In the case of polynomial equations, the roots are indistinguishable in the sense that a polynomial is invariant under arbitrary permutations of its roots. In a group, the identity is distinguishable from all other elements under every symmetry operation because it is the unique identity in the group. However, in some groups, two or more elements may be indistinguishable in the sense described here.

For further information on the solvability of equations, see the books by Herstein (1964) or Van der Waerden (1931). This topic requires some knowledge of the Galois theory of fields that is beyond the scope of this text.

Exercises

2.6.1 Find the order of symmetry groups of the graphs shown in Fig. 2.6.6.

FIG. 2.6.6   Graphs with Symmetries (see Ex. 2.6.1)

2.6.2 Show that any sum of the elementary symmetric functions of n variables is symmetric in the n variables.

2.6.3 Show that each coset of S_6 relative to the symmetry group of order 36 corresponds to a distinct labeling of the graph shown in Fig. 2.6.1.

2.6.4 Let g be a generator of C_n. Show that g^i is a generator of C_n if and only if i is relatively prime to n.   HINT: Show that the mapping $f(g^j) = g^{i \cdot j}$ is a group automorphism.

2.6.5 A challenging board game is tic-tac-toe played on a three-dimensional board with cells arranged in a $4 \times 4 \times 4$ array. A win occurs when either player succeeds in placing four counters on a straight line in any dimension. How many different lines pass through the board, where two lines are different if there is no symmetry operation that maps one onto the other? What is the order of the symmetry group that you found?

2.6.6 In the game described in Ex. 2.6.5, the board position known as a *double force* (shown in Fig. 2.6.7) is a winning combination for the player who succeeds in reaching it. In the positions shown in the figure, a single move by white cannot stop black from completing one line on his next move. How many different double forces are there on a two-dimensional 4×4 board? How many are there on a three-dimensional $4 \times 4 \times 4$ board?

2.6.7 The *eight queens problem* is a celebrated parlor puzzle in which a player is asked to place eight queens on a chessboard so that no queen is attacking any other queen. This problem in its generalized form is called the *n queens problem*, and it requires that n queens be placed on an $n \times n$ chessboard so that no queen is attacking any other queen.
(a)   Find all of the solutions to the n queens problem for $n = 4$. Use the symmetry of the problem to help reduce the search for solutions.
(b)   Taking into account the natural symmetries of the chessboard, how many different solutions are there to the four queens problem?

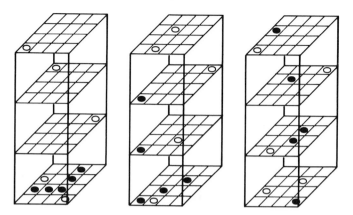

FIG. 2.6.7 Some Double Forces for Black in Three-Dimensional Tic-Tac-Toe (see Ex. 2.6.6)

(c) Prove that there are no solutions to the two and three queens problems.

(d) Write a computer program to solve the seven and eight queens problems. Determine how many different solutions exist for each of these problems, after taking into account the natural symmetries.

2.6.8 The *n rooks problem* is a variant of the *n* queens problem. It requires that *n* rooks be placed on an $n \times n$ chessboard so that no rook is attacking any other rook.

(a) Find all different solutions to the two and three rooks problems, after taking into account the natural symmetries of the problem.

(b) Use a computer program to find all solutions to the six rooks problem. How many of these are truly different after taking into account the symmetries?

THREE | The Pólya Theory of Enumeration

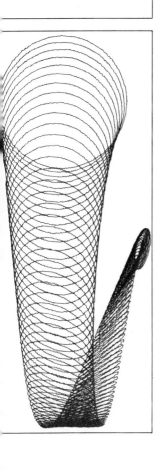

THREE | The Pólya Theory
of Enumeration

In this chapter we discuss a very practical application of the theory of groups, namely the Pólya method for enumerating equivalence classes. This application pertains to situations in which a set has symmetries that form a group under composition. We shall see that the symmetries induce a partition of the set into equivalence classes, where elements that are similar belong to the same class. The Pólya theory greatly simplifies the problem of determining the number of equivalence classes and that of finding a representative of each class.

As an example of the practicality of the results, consider the problem of the detailed design of digital computers. The design problem consists of finding good circuit implementations of specified functional behavior. To be more precise, let $X = \{0, 1\}$, and let a function from X^n into X for each positive n be called a *switching function*. Each computer device implements a switching function of two to four variables, and only about ten or fewer different types of devices typically are used in the construction of a computer. A computer engineer faces the problem of finding a network of devices that implements a given function of n variables, where n may be as small as one or as large as twenty. The design should be a good one in some sense—in its economy, in its speed, or in some other function of the network topology.

Simple combinatorial analysis shows that there are 2^{2^n} switching functions of n variables, a number that grows astronomically as n increases. In principle, it would be possible to catalog the best way to construct a network realization for each switching function. Then a computer designer needing to use a particular function in a new computer could go to the catalog and look up the corresponding network. However, this approach is out of the question for $n \geq 4$ because the catalog would be far too large. For example, there are 65,536 individual 4-variable functions. Fortunately, we can take advantage of symmetries that do not change the cost or the structure of the network implementation, and we can thereby reduce the size of the catalog.

While the first computers were being built, a team of researchers at Harvard carried out the laborious task of enumerating all of the possible 4-variable functions and sorting out those that are equivalent. They eventually constructed a table showing the best ways to design each of 222 nonequivalent 4-variable functions (Harvard 1951). Years later, it was recognized that the Pólya theory is applicable to this problem.

Making use of this theory, it is relatively easy to show that the catalog of 4-variable functions reduces to exactly 222 equivalence classes and to find a representative of each class.

3.1 Counting Equivalence Classes à la Burnside

We begin this section by stating in abstract terms the problem with which we will be dealing. Let S be a set, and let G be a group of permutations of S. We say that two elements s and t of S are equivalent if there is some permutation g in G such that $g(s) = t$. This definition of equivalence defines an equivalence relation, and therefore the action of G on S partitions S into equivalence classes. Our problem is to determine the number of equivalence classes induced by G and to find representatives of each class. Speaking less formally, we can say that G is a group of symmetries that tells whether two or more elements of S are alike. The problem is to find all the elements that are different, after taking into account the symmetries.

To put the central problem in less abstract terms, we look next at a few very specific examples. Each of the following questions fits the abstract problem just outlined.

How many different ways are there to color the faces of a cube with two colors? We will say that two cubes are colored the same if one cube can be oriented in space so that the colors of its faces are oriented exactly as are the colors of the faces of the second cube. The set in this example is the set of faces of the cube; the permutation group is the group of rigid motions of the cube. This group contains all of the rotations of the cube about each of its axes of symmetry.

In how many different ways can we use the integers 0, 1, and 2 to label the vertices of an n-sided regular polygon? Two labelings are considered the same if we can rotate and/or reflect one labeled polygon in such a way that the labelings of its vertices are oriented identically to those of the other polygon. The set in this example is the set of vertices of an n-sided regular polygon; the group is the group of symmetries of the polygon under rotation and reflection (the dihedral group of order $2n$).

How many essentially different kinds of electronic black boxes (each carrying out one switching function) do we need in order to implement all of the n-variable switching functions? Here we are dealing with the set of switching functions from $\{0,1\}^n$ into $\{0,1\}$. Let $f_1(x_1, x_2, \ldots, x_n)$ be a typical function. We will say that the function $f_2(x_1, x_2, \ldots, x_n)$ is equivalent to f_1 if there is a permutation π of the n letters $1, 2, \ldots, n$ such that

$$f_1(x_1, x_2, \ldots, x_n) = f_2(x_{\pi(1)}, x_{\pi(2)}, \ldots, x_{\pi(n)}).$$

The group in question here is S_n. This definition of equivalence is appropriate to the original question. For example, suppose that an electronic black box performs the function f_2, as shown in Fig. 3.1.1.

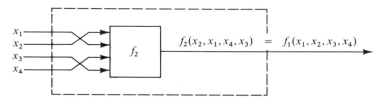

FIG. 3.1.1 A Network for f_1 That Is Constructed from the Network for the Equivalent Function f_2

Then the same black box also can perform the function f_1, provided that we permute the input wires according to the permutation π. In Fig. 3.1.1 we show the network for f_1 within dashed lines; it consists of the network for f_2 preceded by a permutation of the inputs. It costs us nothing extra to permute the input wires when we hook up the black box. Thus one black box may be used to implement a number of different functions. The number of different black boxes needed is the number of equivalence classes induced by S_n acting on the set of n-variable functions.

For each of the three questions just given, the most direct way to find an answer is to examine each element of the set, compare it with the other elements, and thus sort the elements into equivalence classes. Of course, before the end of the chapter we will find that the Pólya theory provides an elegant shortcut to avoid this tedious enumeration and sorting process.

In this section we describe a powerful technique for counting the number of equivalence classes—a technique based heavily on our notions of groups, subgroups, and cosets. This counting technique generally is credited to Burnside (1911, sect. 145), but it appears implicitly in an earlier paper by Frobenius (1896). The essential idea of this technique is to count the number of equivalence classes by performing an enumeration over the symmetry group rather than enumerating the elements of the set on which the group acts. The Burnside method is particularly efficient when the order of the group is small in comparison to the number of set elements. Our development of the Pólya theory depends heavily on the Burnside method of counting equivalence classes.

Definition Let G be a permutation group on a set S. Two elements s and t of S are *equivalent* if and only if there is a g in G such that $g(s) = t$. For each s in S, let H_s be the set $\{h \mid h \in G \land h(s) = s\}$. That is, H_s is the set of permutations that leave s fixed.

Equivalence as defined here is an equivalence relation. You can verify relatively easily that it is symmetric, reflexive, and transitive, because of the group axioms of inverses, identity, and closure, respectively (see Ex. 3.1.1).

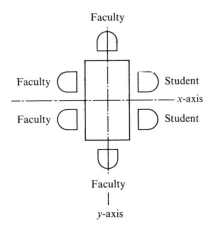

FIG. 3.1.2   One Way of Seating Two Students and Four Faculty Members at a Rectangular Table

As an example, consider the number of ways that we can seat four faculty members and two students around the rectangular table shown in Fig. 3.1.2. Two seatings are equivalent if they differ only by reflections about the axes shown in the figure. Thus the group in this example has four elements: the identity, reflection about the x-axis, reflection about the y-axis, and reflection about both axes. Under the definition of equivalence, the four seatings shown in Fig. 3.1.3 are equivalent. For each of these seatings, $H_s = \{e\}$. For the seating shown in Fig. 3.1.2, $H_s = \{e, \text{reflection about the } x\text{-axis}\}$.

Theorem   For all s in S, the set H_s forms a subgroup of G, and the index of H_s in G is the number of elements equivalent to s.

Proof: If h is in H_s, then h^{-1} is in H_s, because

$$s = e(s) = h^{-1}(h(s)) = h^{-1}(s).$$

Therefore, if h_1 and h_2 are in H_s, then $h_1 \cdot h_2^{-1}$ is in H_s, because

$$h_1 \cdot (h_2^{-1}(s)) = h_1(s) = s.$$

Hence H_s is a subgroup.

Each left coset of G relative to H_s is associated with a distinct element that is equivalent to s. To see this, let g_1 and g_2 be in the same left coset. Then $g_1 = g_2 \cdot h$ for some h in H_s, and

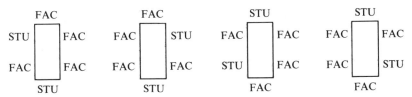

FIG. 3.1.3   Four Equivalent Seatings

117

$$g_1(s) = g_2 \cdot h(s) = g_2(h(s)) = g_2(s).$$

Hence all elements of the same coset map s onto the same element. Similarly, if g_1 and g_2 in G map s onto the same element, then they are in the same left coset. That is, if $g_1(s) = g_2(s)$, then $g_2^{-1} \cdot g_1(s) = s$, so $g_2^{-1} \cdot g_1$ is in H_s, and g_2 and g_1 are in the same left coset. Finally, every element equivalent to s is associated with some coset because, if $t = g(s)$, then t is associated with the coset containing g. This proves the theorem, because the index of H in G is equal to the number of left cosets. $\square$

In Fig. 3.1.3, we illustrate an equivalence class of four seatings. Note that $|H_s| = 1$ for each seating in this class, so that $|G|/|H_s| = 4$, in agreement with the preceding theorem. The seating illustrated in Fig. 3.1.2 has $|H_s| = 2$. Thus there are $|G|/|H_s| = 2$ seatings in its equivalence class. What is the other seating?

The preceding theorem says that we can find the size of an equivalence class containing an element by using the formula

$$|C(s)| = |G|/|H_s|,$$

where $C(s)$ is the set of elements equivalent to s. We use this result to prove Burnside's theorem. The intent of Burnside's theorem is to show that it is possible to enumerate equivalence classes by summing over group elements instead of summing over elements of the set. The following definition will be useful in stating Burnside's theorem.

Definition For each g in G, the notation I_g represents the number of elements in S that are fixed by g. That is, I_g is the cardinality of the set $\{s | g(s) = s\}$.

Theorem (Burnside, Frobenius). The number of equivalence classes of elements acted upon by a group G is given by the expression

$$\frac{1}{|G|} \cdot \sum_{g \in G} I_g.$$

Proof: Let $[s]$ be the equivalence class containing s, and let $[G:H_s]$ be the index of H_s in G. Then the number of classes is equal to

$$\sum_{[s] \subseteq S} 1 = \frac{1}{|G|} \cdot \sum_{[s] \subseteq S} |H_s| \cdot [G:H_s]. \qquad (3.1.1)$$

Using the previous theorem, we find that the right side of Equation 3.1.1 is equal to

$$\frac{1}{|G|} \cdot \sum_{[s] \subseteq S} \sum_{t \in [s]} |H_s| = \frac{1}{|G|} \cdot \sum_{[s] \subseteq S} |H_s| \cdot \sum_{t \in [s]} 1. \qquad (3.1.2)$$

Now Equation 3.1.2 counts every pair g, s where g is in G and s in S such that $g(s) = s$, and the summation is over s in S. If we sum the same pairs by summing over g in G, then we obtain the summation described in the theorem. To see this, let $\delta_{g,s} = 1$ if $g(s) = s$, and otherwise let $\delta_{g,s} = 0$. Then we have

$$|H_s| = \sum_{g \in G} \delta_{g,s},$$

$$I_g = \sum_{s \in S} \delta_{g,s},$$

and

$$\frac{1}{|G|} \cdot \sum_{s \in S} |H_s| = \frac{1}{|G|} \cdot \sum_{s \in S} \sum_{g \in G} \delta_{g,s} = \frac{1}{|G|} \cdot \sum_{g \in G} \sum_{s \in S} \delta_{g,s} = \frac{1}{|G|} \cdot \sum_{g \in G} I_g. \quad \square$$

To apply Burnside's theorem to the problem illustrated in Fig. 3.1.2, we find I_g for each of the four group elements. The values of I_g for each of the group operations are the following: $2^6 = 64$ for the identity transformation; $2^4 = 16$ for reflection about the y-axis; $2^3 = 8$ for reflection about the x-axis; and $2^3 = 8$ for reflection about both axes. Using Burnside's theorem, we find that there are

$$(64 + 16 + 8 + 8)/4 = 96/4 = 24$$

different ways of seating the faculty members and students at the table.

As another example of Burnside's theorem, consider a situation in which circular tables, each seating five persons, are to be used at a university function attended by administration, faculty, and students. For a single table, how many different ways are there to seat members of the three sets of people? In this case, we will call two seatings different if they are not cyclic permutations of each other (Fig. 3.1.4). The group G is the cyclic group of order 5. The various seatings constitute the set S, which has 3^5 elements. The identity element in G leaves all 3^5 seatings fixed. Each of the other group operations leaves three seatings fixed, namely the seatings that contain five members from the same set of people. Therefore, the number of different seatings is

$$(3^5 + 3 + 3 + 3 + 3)/5 = 255/5 = 51.$$

Burnside's theorem sometimes is less useful than it might seem from these examples. It requires a summation over a group and, when the order of the group is large, the summation over G actually may be more difficult to perform than a direct summation.

In practice, the summation can be simplified. Recall from Ex. 2.4.8 that, for g and h in G, we say that g is the *conjugate* of h if there is an

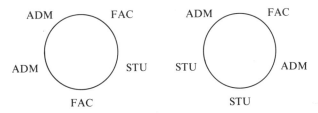

FIG. 3.1.4 Two Different Seatings at a Five-Place Circular Table

x in G such that $g = x \cdot h \cdot x^{-1}$. Conjugacy is an equivalence relation that partitions G into disjoint sets called *conjugacy classes* (Ex. 3.1.2). It so happens that, if g and h are conjugates, then each leaves the same number of items invariant.

Lemma Let g and h be conjugates in G, acting on a set S. Then $I_g = I_h$.

Proof: There is a one-to-one correspondence between elements invariant under g and elements invariant under h (see Ex. 3.1.6). □

With the aid of the preceding lemma, it is trivial to show that the summation in Burnside's theorem reduces to a summation over conjugacy classes and that the size of each conjugacy class appears as a multiplier of I_g. If G is an abelian group, then each element is in a conjugacy class by itself. This follows because, if $g = x \cdot h \cdot x^{-1}$ in an abelian group, then $g = h \cdot x \cdot x^{-1} = h$. Thus the enumeration over conjugacy classes is useful only for nonabelian groups.

As an example of enumeration over conjugacy classes, we return to the problem of seating at circular tables, this time including both cyclic permutations and reflections in the symmetry group. Thus we wish to find the number of different ways of seating administration, faculty, and students around a circular table with five seats under the action of D_5, the dihedral group of degree 5. The four conjugacy classes for this group are the following:

(1)(2)(3)(4)(5);
(1 2 3 4 5), (1 5 4 3 2);
(1 3 5 2 4), (1 4 2 5 3); and
(1)(2 5)(4 3), (2)(1 3)(4 5), (3)(2 4)(1 5), (4)(3 5)(1 2), (5)(1 4)(2 3).

The identity leaves all 3^5 seatings invariant. Each of the two conjugacy classes with 5-cycle permutations leaves just three seatings invariant. Each of the reflections in the fourth conjugacy class leaves $3^3 = 27$ seatings invariant. Thus the number of different seatings is

$$(3^5 + 2 \cdot 3 + 2 \cdot 3 + 5 \cdot 3^3)/10 = 39.$$

Exercises

3.1.1 Let G be a group acting on a set S, and let s and t in S be equivalent if there is a g in G for which $g(s) = t$. Show that we have specified an equivalence relation.

3.1.2 Show that the conjugacy relation in a group is an equivalence relation.

3.1.3 Find the conjugacy equivalence classes for S_3, C_4, C_6, and D_4.

3.1.4 Prove that each conjugacy equivalence class in an abelian group has only one element.

3.1.5 Let g and h be two elements of a permutation group G. They are said to have the *same cycle structure* if they have the same number of cycles of each length in their cycle representations. Thus $(1\,2\,3)(4\,5)$ and $(1\,3\,5)(2\,4)$ have the same cycle structure. Show that two elements are conjugates in S_n if and only if they have the same cycle structure.

3.1.6   Let G be a group of permutations of a set S. Show that, if g and h are conjugate elements of G, then $I_g = I_h$.

3.1.7   Find representatives of each of the equivalence classes for the problem illustrated in Fig. 3.1.2.

3.1.8   Verify by direct count that there are 51 different ways to seat students, faculty, and administration in five seats at circular tables under the action of C_5.

3.1.9   How many different ways can students and faculty be seated at a circular table seating six, where the symmetry group is C_6? Verify your result by direct count.

3.1.10   Repeat Ex. 3.1.9 using the symmetry group D_6.

3.2 Inventories of Functions

In many instances, the Burnside theorem is not sufficiently powerful to permit us to enumerate equivalence classes, even when we take into account the conjugacy classes. The difficulty stems from the problem of computing I_g for a large-order group. As we shall see, the Redfield-Pólya theorem not only provides the tool that is necessary to make this calculation, but it also tells us some information about each of the classes that we cannot obtain with Burnside's theorem. In this section we develop the notion of inventories of functions, a central concept in the Pólya theory.

Enumerations of the type we describe here usually are performed over sets whose elements are functions. In particular, let D be the domain of the functions and let R be the range; the set S is the set R^D of functions from D into R.

There are $|R|$ elements in R, and we shall assign a weight to each such element. The crucial aspect of the weight assignment is that weight may be either a numerical value or a symbol. We will be able to perform different counts by assigning different weights—either numeric or nonnumeric—to each element in the range. To make this idea more precise, we introduce the notion of inventory.

Definition   Let the weight $w(r)$ be assigned to each r in R. The *enumerator*, or *inventory*, of the set R is defined by the following expression:

$$\text{inventory of } R = \sum_{r \in R} w(r).$$

An example will illustrate the reason for the descriptive term "inventory." Let R be a set that contains two dogs, four cats, and a horse. To the dogs we assign the weights d_1 and d_2. Similarly, the other animals receive weights c_1, c_2, c_3, c_4, and h. In this case, the inventory of R is

$$d_1 + d_2 + c_1 + c_2 + c_3 + c_4 + h,$$

and each animal is uniquely identifiable by the dummy symbol that has been selected for its weight. To the owner of the menagerie, the individual distinctions may be quite great, and he may regard this inventory as the appropriate one. To his neighbor, however, the distinctions among individual animals may be quite meaningless. The neighbor assigns the weights d to each dog, c to each cat, and h to the horse. He then calculates the inventory of R to be

$$4c + 2d + h.$$

Notice that the coefficient of each dummy variable gives the number of elements of that weight in R. To the county license agent, the significant feature of an animal may be the cost of its license—in this case, say 2 dollars for a dog or cat and 5 dollars for a horse. The license agent's inventory then is

$$6 \cdot 2 + 1 \cdot 5 = 17.$$

There is a natural extension of the notion of inventory of the range to the notion of inventory of a function. Let $f: D \to R$ be a typical function in the set S. Then $f(d)$ is an element in R for each d in D. Hence we can define $W(f)$, the weight of f, to be

$$W(f) = \prod_{d \in D} w(f(d)).$$

TABLE 3.2.1   Three Functions with Range $X = \{0, 1\}$ and Domain X^2

x_1	x_2	$f_1(x_1, x_2)$	$f_2(x_1, x_2)$	$f_3(x_1, x_2)$
0	0	0	0	1
0	1	1	0	1
1	0	1	1	0
1	1	0	0	1

Consider the functions f_1, f_2, and f_3 (Table 3.2.1) with domain X^2 and range X, where $X = \{0, 1\}$. Let us define the $w(d)$ by $w(0) = a$ and $w(1) = b$. Then the weights of the functions are $W(f_1) = a^2 b^2$, $W(f_2) = a^3 b$, and $W(f_3) = ab^3$.

Definition   The *weight of a set of functions* is the sum of the individual function weights.

The weight of $\{f_1, f_2, f_3\}$ is $a^2 b^2 + a^3 b + ab^3$.

The preceding definitions enable us to apply the rules of ordinary algebra to the symbolic weights, thus enabling us to perform enumerations. We use two different operators, addition and multiplication, to keep the symbols grouped so that we can manipulate them properly. Observe that the expression for the weight of $\{f_1, f_2, f_3\}$ is easily broken into the weights of the individual functions.

To see how we apply algebra, consider the expression for the weight of R^D, the set of all functions from D to R. By definition, the weight of this set is

$$W(R^D) = \sum_{f \in R^D} \prod_{d \in D} w(f(d)). \qquad (3.2.1)$$

We can rewrite Equation 3.2.1 as

$$W(R^D) = \left(\sum_{r \in R} w(r) \right)^{|D|} \qquad (3.2.2)$$

We can verify the equality of Equations 3.2.1 and 3.2.2 by a simple combinatorial argument. Consider the case for which $|D| = |R| = 2$, and let the weight of the range be $(a + b)$. There are four functions from D to R in this case, because each of the two domain elements can be mapped into the range in two possible ways. Thus the weights of the four functions are aa, ab, ba, and bb, where the first and second factors in each of these four products correspond to the weights of the images of the first and second domain· elements, respectively.

Now consider what happens when the products in Equation 3.2.2 are expanded. In this case we obtain

$$(W(R))^{|D|} = (a + b)(a + b) = aa + ab + ba + bb = a^2 + 2ab + b^2.$$

The third expression in this equality has four terms, each corresponding to one of the functions from D to R. In fact, we obtain one term for each way of selecting an a or b from the first $(a + b)$ factor in combination with each way of selecting an a or b from the second $(a + b)$ factor. If we associate the first domain element with the first $(a + b)$ factor and the second domain element with the second such factor, then expansion of the expression $(a + b)(a + b)$ into a sum of products yields one term for each way that we can map the two domain elements onto R.

In the more general case with $|D|$ and $|R|$ greater than 2, expansion of Eq. 3.2.2 into the form of Eq. 3.2.1 yields a summation of product terms, where each product term comes from the selection of one range element for each of the $|D|$ sums in Eq. 3.2.2. Thus each product term is in one-to-one correspondence with a function from D to R, just as in the case for $|D| = |R| = 2$. Hence Eqs. 3.2.1 and 3.2.2 enumerate the same weights.

Example: Let $D = \{0, 1, 2, 3\}$ and $R = \{0, 1\}$. Let $w(0) = r_1$ and $w(1) = r_2$. Using Eq. 3.2.2, we find that

$$W(R^D) = (r_1 + r_2)^4 = r_1^4 + 4r_1^3 r_2 + 6r_1^2 r_2^2 + 4r_1 r_2^3 + r_2^4.$$

Recall from Chapter One that there are 16 functions from $\{0, 1\}^2$ into

{0, 1}. The domain D in this example is $\{0, 1\}^2$. By summing the coefficients of the terms, we find that the inventory of R^D confirms that there are indeed 16 functions from D to R. The coefficient of each term gives the number of functions that have weights equal to the term. Hence the term $4r_1^3 r_2$ indicates that there are four functions that each have three 0's and one 1 in their tables. Note also that the coefficients in this case are binomial coefficients.

We can generalize this notion further. Let $D_1, D_2, \ldots, D_k$ be any partition of D into k disjoint subsets. Then the inventory of the functions that are constant on each of the k subsets of D is given by the expression

$$\prod_{i=1}^{k} \left(\sum_{r \in R} w(r)^{|D_i|} \right). \tag{3.2.3}$$

Here we say that a function is constant on a subset if every element of the subset is mapped onto the same range element. For example, let the set D_1 contain three elements, d_1, d_2, and d_3. If a function is constant on these three elements, then the selection of the range element for d_1 automatically determines the image of d_2 and d_3 because they must map onto the same element. Hence the selection of a range element of weight r as the image of d_1 gives rise to a factor of r^3 in the expression for the weight of the function. After multiplying out the parenthesized terms of Eq. 3.2.3 therefore, we have (as in the case of Eq. 3.2.2) one term for each assignment of a range element to a subset of the partition.

Example: Let D be $\{0, 1, 2, 3\}$, and let R be $\{0, 1\}$. Let $w(0) = r_1$ and $w(1) = r_2$. Now we partition D into the sets $\{1, 2\}$, $\{0\}$, and $\{3\}$, and we proceed to find the inventory of functions that are constant on each of the three subsets. Using Eq. 3.2.3, we find that this inventory is given by

$$(r_1^2 + r_2^2)(r_1 + r_2)(r_1 + r_2) = r_1^4 + 2r_1^3 r_2 + 2r_1^2 r_2^2 + 2r_1 r_2^3 + r_2^4.$$

The coefficient of each term tells how many different functions with that prescribed weight also are functions that assign the same range value to elements 1 and 2 of the domain.

Next we define what we mean by equivalent functions, and we show how this definition relates to inventories.

Definition Let G be a group of permutations of the set D. Let R^D be the set of functions with domain D and range R. Then two functions f_1 and f_2 in R^D are *equivalent* if there exists a permutation π in G such that $f_1(d) = f_2(\pi(d))$ for all d in D.

As an example, consider the functions from $D = \{0, 1\}$ into $R = \{-2, +2\}$. Let G be the set of permutations $\{(0)(1), (0\ 1)\}$. Let the functions f_1 and f_2 be defined by

$$f_1(0) = f_2(1) = 2 \quad \text{and} \quad f_1(1) = f_2(0) = -2.$$

The functions f_1 and f_2 are equivalent.

A very important aspect of the notion of inventory is that equivalent functions have the same inventory. To prove this, observe that

$$\prod_{d \in D} w(f_1(d)) = \prod_{d \in D} w(f_2(\pi(d))) = \prod_{d \in D} w(f_2(d)),$$

where the last equality follows because the same factors appear in the products on both sides of the equality. The permutation π merely changes the order in which the factors appear. If we call an equivalence class a *pattern*, then we define the *weight of a pattern* to be the inventory of any function in the pattern. However, equivalence classes need not be determined uniquely by their weights; it may be possible for two distinct classes to have the same weight.

Example: Consider the set of functions from $Y = \{0, 1, 2, 3\}$ into $X = \{0, 1\}$. Let G be a permutation group that acts on Y and has the set of permutations $\{e, (0)(1\,2)(3)\}$. Table 3.2.2 shows the sixteen functions in the set, with braces indicating the equivalence classes induced by G. If in X we have $w(0) = a$ and $w(1) = b$, then the weights of the patterns are those shown in the righthand column of the table. Clearly, equivalent functions in this set have the same weight, but the weight does not determine uniquely an equivalence class.

TABLE 3.2.2 The Set of Functions from $Y = \{0, 1, 2, 3\}$ into $X = \{0, 1\}$ and the Patterns Induced by $G = \{e, (0)(1\ 2)(3)\}$

f	$f(0)$	$f(1)$	$f(2)$	$f(3)$	Weight of Pattern
f_1	0	0	0	0	} a^4
f_2	0	0	0	1	} a^3b
f_3	0	0	1	0	a^3b
f_4	0	1	0	0	
f_5	1	0	0	0	} a^3b
f_6	0	0	1	1	a^2b^2
f_7	0	1	0	1	
f_8	0	1	1	0	} a^2b^2
f_9	1	0	0	1	} a^2b^2
f_{10}	1	0	1	0	a^2b^2
f_{11}	1	1	0	0	
f_{12}	0	1	1	1	} ab^3
f_{13}	1	0	1	1	ab^3
f_{14}	1	1	0	1	
f_{15}	1	1	1	0	} ab^3
f_{16}	1	1	1	1	} b^4

The notion of the weight of a pattern can be extended to the notion of the *weight of a set of patterns*, or the *pattern inventory*. The weight of a set of patterns is the sum of the weights of patterns in the set. In the set of Table 3.2.2, there are twelve distinct patterns, and the pattern inventory is $a^4 + 3a^3b + 4a^2b^2 + 3ab^3 + b^4$.

Exercises

3.2.1 Consider the functions from $\{0, 1, 2, 3, 4\}$ onto $\{0, 1\}$. Find the weight of this set of functions when $w(0) = a$ and $w(1) = b$.

3.2.2 We say that a function is constant on domain elements $x_1, x_2, \ldots, x_n$ if the function maps all of the x_i's onto the same range element. For the functions defined in Ex. 3.2.1, find the weight of the subset of functions that are constant on $\{0, 1, 2\}$ and also are constant on $\{3, 4\}$.

3.2.3 Repeat Ex. 3.2.2 for the subset of functions that are constant on $\{0, 1, 2, 3\}$.

3.2.4 Compute, by exhaustive enumeration, the pattern inventory of the functions from $\{0, 1, 2, 3\}$ onto $\{0, 1\}$ under the permutation group $G = \{(0)(1)(2)(3), (0)(1\,3\,2), (0)(1\,2\,3)\}$.

3.2.5 Repeat Ex. 3.2.4 for the permutation group

$$H = \{(0)(1)(2)(3), (0\,1)(2\,3), (0\,2)(1\,3), (0\,3)(1\,2)\}.$$

3.2.6 Repeat Ex. 3.2.4 for the permutation group generated by using groups G (Ex. 3.2.4) and H (Ex. 3.2.5) as generators.

3.3 Cycle Index Polynomials

The central result of the Pólya theory is a relation between pattern inventories and a polynomial that represents the behavior of the permutation group acting on the set of functions. Before presenting the central theorem, we develop the notion of cycle index polynomials.

Let g be any element of a permutation group G that acts on the set D, and let S be a set of functions from D into the set R. Because g is a permutation, we can represent g in conventional cycle notation. For example, a typical permutation on a domain D with five elements is $g = (d_1 d_2)(d_3 d_4 d_5)$.

Because g permutes D, it also induces a permutation of the functions from D to R. To see this, consider the function $f : D \to R$ as defined in the second column of Table 3.3.1. When we apply a permutation g to the domain D and then apply the function f, we obtain the function $f' : D \to R$, where $f'(d_i) = f(g(d_i))$. When $g = (d_1 d_2)(d_3 d_4 d_5)$, the function f' is that shown in the third column of Table 3.3.1. In this case, we say that the permutation $g = (d_1 d_2)(d_3 d_4 d_5)$ on the domain D maps the function f onto the function f'. It is easy to show that g maps S onto itself. Thus g, when it is viewed as acting on the functions from D to R, is a permutation of the functions.

TABLE 3.3.1 A Function and Its Permutation by
$g = (d_1 d_2)(d_3 d_4 d_5)$

d_i	$f(d_i)$	$f'(d_i)$
d_1	a	b
d_2	b	a
d_3	a	b
d_4	b	c
d_5	c	a

In order to use a Burnside type of counting argument in the Pólya theory, we must be able to determine precisely which functions from D to R are invariant under the action of a particular g in G. That is, we must compute I_g for each g. In the preceding example, we wish to find the functions that are mapped back onto themselves under the action of $g = (d_1 d_2)(d_3 d_4 d_5)$. A little thought shows that d_1 and d_2 must be mapped onto the same range element by a function invariant under g. In particular, if the function $f(d_i)$ maps d_1 and d_2 onto different range elements, then $f(d_1) \neq f(g(d_1)) = f(d_2)$. However, if d_1 and d_2 are mapped onto the same range element r, then

$$f(d_1) = r = f(d_2) = f(g(d_1)).$$

Extending this argument to the cycle $(d_3 d_4 d_5)$ of length 3, we see that a function invariant under this cycle must assign each of the domain elements d_3, d_4, and d_5 to the same range element. In general, a function is invariant under a permutation g on its domain if and only if the function is constant on the domain elements in each cycle of g. Consequently, the cycle structure of a permutation g contains information sufficient for the calculation of I_g. This observation motivates the following definition.

Definition   Let $x_1, x_2, \ldots, x_m$ be indeterminate symbols. For each element g in the group G acting on the set S, we form the product

$$p(g) = x_1^{b_1} x_2^{b_2} \ldots x_m^{b_m},$$

where b_i is the number of cycles of length i induced by g. The *cycle index polynomial* $Z_G(x_1, x_2, \ldots, x_m)$ then is defined to be

$$Z_G(x_1, x_2, \ldots, x_m) = \frac{1}{|G|} \sum_{g \in G} p(g).$$

Example: The action of the symmetric group S_4 acting on four letters is described by the cycle index polynomial

$$Z_{S_4}(x_1, x_2, x_3, x_4) = \frac{1}{24}(x_1^4 + 6x_1^2 x_2 + 3x_2^2 + 8x_1 x_3 + 6x_4).$$

Some confusion may arise in calculating the cycle index polynomial of groups acting on functions. For example, consider functions from

$\{0, 1\}^2$ into $\{0, 1\}$. Functions of this form usually are written as functions of two arguments, say a and b, so that a typical function is written as $f(a, b)$. It seems obvious that the appropriate permutation group to consider is the group of permutations of a and b, namely the group S_2 of order 2. Thus the unwary might calculate the cycle index polynomial as

$$Z_{S_2} = \frac{1}{2} (x_1^2 + x_2),$$

to reflect the fact that the permutations in S_2 may be written as $(a)(b)$ and $(a\,b)$.

However, there are four elements, not two elements, in the domain of the functions in question. Yet S_2 is a group that permutes only two elements, because its degree by definition is 2. Apparently then, S_2 is not the appropriate group to consider after all. The problem here is that the cycle index polynomial depends on the *representation* of the group, because the polynomial contains explicit information about the objects moved by each permutation. The group that we should consider must be a group of degree 4, because this is the cardinality of the domain. We show in the following paragraphs that the group is isomorphic to S_2, but that it has a different representation.

x_1	x_2	$f(x_1, x_2)$	$f(x_2, x_1)$
0	0	f_0	f_0
0	1	f_1	f_2
1	0	f_2	f_1
1	1	f_3	f_3

FIG. 3.3.1 The Function $f(x_1, x_2)$ and Its Domain $\{0, 1\}^2$

First consider a table (Fig. 3.3.1) with its rows labeled by all the ordered pairs in $\{0, 1\}^2$. Let $f(x_1, x_2)$ be a function defined on $\{0, 1\}^2$, and observe how $f(x_1, x_2)$ is related to the function $f(x_2, x_1)$. Permuting the arguments of f is equivalent to permuting the *columns* in the listing of ordered pairs. However, the domain of f is the set of *rows* of this listing. Moreover, each permutation of columns of this listing has the same effect as some permutation of the rows of the listing. In the example, interchanging the columns of the listing has the same effect as interchanging the two center rows of the listing while leaving the top and bottom rows invariant. Note that the group acting upon the domain of f is the group of row permutations, not the group of column permutations. However, the two groups are isomorphic.

In Fig. 3.3.1, the column-permutation group S_2 has the permutation representation $\{(1)(2), (1\,2)\}$, where we denote the columns by 1 and 2, respectively. The row permutation has the representation

$\{(0)(1)(2)(3), (0)(1\,2)(3)\}$, where the rows are denoted as $0, 1, 2$, and 3, respectively. Observe that, in general, to each column permutation of a domain listing there corresponds a row permutation, and the map from column permutations to row permutations is one-to-one. More-over, if g_1 and g_2 are two column permutations with corresponding row permutations h_1 and h_2, the row permutation corresponding to $g_1 \circ g_2$ is $h_1 \circ h_2$. The properties described in the previous two sentences indicate that the row permutations are a one-to-one homomorphic image of the column permutations, and thus the row-permutation group is isomorphic to the column-permutation group. In Fig. 3.3.1, the isomorphism maps $(1)(2)$ onto $(0)(1)(2)(3)$ and maps $(1\,2)$ onto $(0)(1\,2)(3)$. Note that the group of degree 4 acting on the rows is iso-morphic to S_2. We shall refer to the row permutation group as S_2', where the prime notation by convention distinguishes the group acting on the domain of a function from the group acting on the arguments of the function.

More generally, when we consider the group S_n acting on functions from D^n into R, we must calculate the cycle index polynomial for the group S_n', where the degree of the representation of S_n' is $|D|^n$ rather than n.

For the example of Fig. 3.3.1, the cycle index polynomial for the group S_2' acting on the domain of the function is

$$Z_{S_2'}(x_1, x_2) = \frac{1}{2}(x_1^4 + x_1^2 x_2).$$

In the next section, we show that the pattern inventory can be computed from a cycle index polynomial.

Exercises

3.3.1 Consider the set of functions from $\{0, 1\}^3$ into $\{0, 1\}$. Let S_3 act on the arguments of the functions, and find a permutation representation for S_3' on the domain of the function.

3.3.2 Find the degree of the group S_n' that acts on the domain of the functions from $\{0, 1\}^n$ into $\{0, 1\}$, where S_n acts on the arguments of the functions. What is the order of S_n'?

3.3.3 Let S_n' be the group defined in Ex. 3.3.2. For what values of n is the degree of S_n' equal to n?

3.3.4 Consider the set of functions from $\{0, 1\}^3$ into $\{0, 1\}$. Let C_3 act upon the arguments of these functions, and find the permutation repre-sentation of C_3'.

3.3.5 Consider the set of functions from $\{0, 1\}^p$ into $\{0, 1\}$, where p is prime. Let C_p act upon the arguments of these functions, and find the permu-tation representation of C_p'.

3.3.6 Let g be a permutation of a set D, and consider the effect of g on the functions from D into a set R. Show that this effect is a permutation of the functions from D into R. Hint: Associate a mapping P_g with each g, where P_g maps the functions in R^D into themselves. Show that P_g is a permutation.

3.3.7 Prove that a function from D to R is invariant under a permutation of D if and only if the function is constant on each of the cycles of the permutation.

3.4 The Redfield-Pólya Theorem

Our problem is to enumerate the equivalence classes of functions from D to R under the action of a group G acting on the domain D of the set of functions. We can solve this problem (and more) if we are able to calculate the pattern inventory of the set of functions under the action of G. Recall that the pattern inventory of the set is the sum of the weights of the patterns in the set, and that the weight of a pattern is the inventory of any function in the corresponding equivalence class. We now are ready to state the central result of this chapter, the *Redfield-Pólya Theorem*.

Theorem   The pattern inventory of the set of functions from D into R is the cycle index polynomial

$$Z_G\left(\sum_{r\in R} w(r), \sum_{r\in R} w(r)^2, \ldots, \sum_{r\in R} w(r)^m\right).$$

Proof: The proof of this theorem follows an argument similar to the proof of Burnside's theorem. We want to show that each pattern is counted exactly once when we substitute the power sums for the indeterminate symbols of the cycle index polynomial. Here we use an informal argument to outline the full proof.

Consider what happens when x_i in the cycle index polynomial is replaced by the power sum $\Sigma w(r)^i$, where the sum is taken over all r in R. Because we replace each indeterminate x_i by the sum of the ith powers of range weights, each term t in Z_G becomes a product of power sums. But each such product is the pattern inventory of the functions left invariant by the group element whose cycle structure is described by the term t. In other words, the expansion of the product of power sums is the pattern inventory of the functions left invariant by the corresponding group element. Thus, Z_G is a sum over all group elements, and all invariant functions for each group element are inventoried. Consequently, each function is inventoried once for each group element that leaves it invariant.

Consequently, if we let $[s]$ represent the pattern class containing the function s, then each s in $[s]$ is inventoried $|H_s|$ times. Now we note that $1/|G|$ is a multiplier of Z_G, so that each function $s \in [s]$ contributes $|H_s|/|G| = 1/|[s]|$ to the overall sum. This contribution is made for each $s \in [s]$ so that each pattern contributes $|[s]|/|[s]| = 1$ to the final sum. $\square$

Example: We find the pattern inventory of the set of functions from $\{0, 1\}^2$ into $\{0, 1\}$ under the action of S_2 acting on the arguments of the functions. In Section 3.3 we mentioned that the cycle index polynomial for this case is

$$Z_{S_2}(x_1, x_2) = \frac{1}{2} (x_1^4 + x_1^2 x_2).$$

Let $w(0) = a$ and $w(1) = b$. Using the Redfield-Pólya theorem, we find that the pattern inventory is equal to

$$Z_{S_2}((a + b), (a^2 + b^2)) = \frac{1}{2} ((a + b)^4 + (a + b)^2 (a^2 + b^2))$$

$$= \frac{1}{2} (2a^4 + 6a^3 b + 8a^2 b^2 + 6ab^3 + 2b^4)$$

$$= a^4 + 3a^3 b + 4a^2 b^2 + 3ab^3 + b^4.$$

Note that, when we replace x_1^4 by $(a + b)^4$, the expansion of the latter product is $a^4 + 4a^3 b + 6a^2 b^2 + 4ab^3 + b^4$, which is the pattern inventory of every function left invariant by the identity in the group. Similarly, the term $x_1^2 x_2$ expands into $a^4 + 2a^3 b + 2a^2 b^2 + 2ab^3 + b^4$, which is the inventory of the functions that are invariant under the second group element. Obviously every function should be inventoried in the first inventory, and we see by inspection that all functions are properly accounted. The second inventory accounts for the functions shown in Fig. 3.4.1, and inspection reveals that this inventory is correct as well. From the pattern inventory, we obtain an enumeration of equivalence classes of the functions. The term a^4 indicates that there is

x_1	x_2	f_1	f_2	f_3	f_4	f_5	f_6	f_7	f_8
0	0	0	0	1	0	1	0	1	1
0	1	0	0	0	1	0	1	1	1
1	0	0	0	0	1	0	1	1	1
1	1	0	1	0	0	1	1	0	1

FIG. 3.4.1   Functions Inventoried by the Term $x_1^2 x_2$

one class containing a function with four 0's in its map. The term $3a^3 b$ indicates that there are three classes having one 1 and three 0's in each of their maps. The other terms are interpreted similarly. Fig. 3.4.2 shows maps of the functions in the first four classes.

Corollary The number of equivalence classes of functions from D to R under the action of G on D is equal to

$$Z_G(|R|, |R|, \ldots, |R|).$$

x_1	x_2	f_1	f_3	f_9	f_{10}	f_2
0	0	0	1	0	0	0
0	1	0	0	1	0	0
1	0	0	0	0	1	0
1	1	0	0	0	0	1
Pattern Weights		a^4	a^3b	a^3b		a^3b
Pattern Inventory		a^4		$3a^3b$		

FIG. 3.4.2   Maps of Functions in Four Equivalence Classes

Proof: For any cycle, a function is invariant on that cycle if and only if it maps all the domain elements of the cycle onto the same range element. Hence there are $|R|$ different assignments to range elements for each cycle that leave the function invariant. Now consider a term of the form $x_1^{b_1} x_2^{b_2} \ldots x_m^{b_m}$ in the cycle index polynomial. When each x_i is replaced by $|R|$ to enumerate the $|R|$ assignments of that cycle, each term in Z_G becomes a power of $|R|$, where the power is the number of cycles in that term. This power of $|R|$ is exactly equal to I_g, the number of functions left invariant by the corresponding group operator. Therefore we have

$$Z_G(|R|, |R|, \ldots, |R|) = \frac{1}{|G|} \cdot \sum_{g \in G} I_g.$$

which, by Burnside's theorem, is the number of equivalence classes.   □

For the preceding example, the sum of the coefficients of the pattern inventory is equal to the number of equivalence classes, so we know that there are 12 equivalence classes in the example. This result is in agreement with the formula

$$Z_{S_2'}(2, 2) = \frac{1}{2}(2^4 + 2^3) = 24/2 = 12.$$

Example: In how many different ways can we use two different colors to paint the faces of a cube? In order to answer this question, we must investigate the group of rigid motions of the cube. This group has order 24, and its permutations are of five types:

(a)   the identity, which gives rise to the term x_1^6 in the cycle index polynomial;

(b)   90° rotations about the three axes that connect centers of opposite faces, with the six such rotations giving rise to six terms of the form $x_1^2 x_4$;

(c)   180° rotations about the axes that connect centers of opposite faces, with the three such rotations each inducing a term of the form $x_1^2 x_2^2$;

(d) 180° rotations about the six axes that connect midpoints of parallel, diagonally opposite edges, with the six such rotations each inducing a term of the form x_2^3; and

(e) 120° rotations about the four axes that connect diagonally opposite vertices, with the eight such rotations each inducing a term of the form x_3^2.

From this group of motions we have

$$Z_G(x_1, x_2, x_3, x_4) = \frac{1}{24}(x_1^6 + 3x_1^2 x_2^2 + 6x_1^2 x_4 + 6x_2^3 + 8x_3^2).$$

Let a and b be the weights of the two colors. From the Redfield-Pólya theorem, we find the pattern inventory to be

$Z_G((a + b), (a^2 + b^2), (a^3 + b^3), (a^4 + a^4))$

$= \dfrac{1}{24}((a + b)^6 + 3(a + b)^2(a^2 + b^2)^2 + 6(a + b)^2(a^4 + b^4)$

$\quad + 6(a^2 + b^2)^3 + 8(a^3 + b^3)^2)$

$= a^6 + a^5 b + 2a^4 b^2 + 2a^3 b^3 + 2a^2 b^4 + ab^5 + b^6.$

The first term indicates that there is precisely one way of coloring the cube with all faces painted color a; the second term indicates that there is only one way to paint the cube with a single face painted color b; the third term indicates that there are two ways to paint the cube with two faces colored b; and so on. The sum of the coefficients is 10, the total number of different ways of coloring the cube.

If we did not wish to have the pattern inventory, we could compute the number of different colorings more directly by using the corollary to the Redfield-Pólya theorem:

$$Z_G(2, 2, 2, 2) = \frac{1}{24}(2^6 + 3 \cdot 2^4 + 6 \cdot 2^3 + 6 \cdot 2^3 + 8 \cdot 2^2)$$

$$= 240/24$$

$$= 10.$$

Exercises For each of Exercises 3.4.1 through 3.4.8, develop the cycle index polynomial and calculate the pattern inventory.

3.4.1 Consider the set of functions from $\{0, 1\}^2$ into $\{0, 1\}$ under complementation of function arguments. Let $\bar{0} = 1$ and $\bar{1} = 0$. Then $f(x_1, x_2)$ is equivalent to $f(x_1, \bar{x}_2)$, $f(\bar{x}_1, x_2)$, and $f(\bar{x}_1, \bar{x}_2)$. Verify the pattern inventory by exhibiting the equivalence classes.

3.4.2 Consider the number of different ways of coloring the edges of regular polygons of prime order with two colors, when the group is the set of rotations of the polygon. Exhibit the classes for pentagons and septagons.

3.4.3 Repeat Ex. 3.4.2 for the group that includes both rotations and reflections of the polygon.

3.4.4 Consider the number of different ways of coloring the nodes of a regular hexagon with two colors, under the group of rotations of the hexagon. Exhibit representatives of each class.

3.4.5 Consider the set of functions from $\{0,1\}^2$ into $\{0,1\}$ under the group of permutations and complementations of function arguments. Exhibit the equivalence classes.

3.4.6 Consider the set of functions from $\{0,1\}^3$ into $\{0,1\}$ with the group S_3 acting on function arguments.

3.4.7 Consider the number of different ways of labeling the vertices of a cube with the letters $\{0,1\}$ under the group of rotations of the cube.

3.4.8 Consider the number of different ways of labeling the edges of a cube with two letters under the group of rigid rotations of the cube.

3.4.9 Find the number of equivalence classes on the functions from $\{0,1\}^n$ into $\{0,1\}$ under the group of complementations of function arguments.

3.4.10 Let G be a group acting on a set D, and let H be a group acting on a set R. Let two functions f_1 and f_2 from D to R be equivalent if there exist g in G and h in H such that $f_1(g(d)) = h(f_2(d))$ for all d in D.

(a) Show that this definition of equivalence is an equivalence relation.

(b) Show that the number of equivalence classes under the action of the composite group in this case is

$$Z_{G,H}(x_1, x_2, \ldots, x_m) = \frac{1}{|G||H|} \sum_{g \in G} \sum_{h \in H} I_{g,h},$$

where $I_{g,h}$ is the number of functions f that satisfy $f(g(d)) = h(f(d))$.

3.5 Historical Background and Applications

The development in this chapter roughly parallels George Pólya's development in his classic paper (Pólya 1937). The key enumeration theorem was first proved by J. Howard Redfield (1927) in a paper that anticipated Pólya's results, but Redfield's work was not widely understood. Redfield seems to have vanished just before his paper was published, and this paper stands as his solitary contribution to mathematics. By the time Pólya's paper appeared, the mathematical community had forgotten Redfield's work, and Pólya himself was unaware of it. Redfield's contribution did not receive due recognition until more than two decades after the appearance of Pólya's paper.

The expositions of Pólya theory by de Bruijn (1964), Liu (1968), and Berge (1971) may prove illuminating for the reader.

The importance of the Pólya theory depends upon the ability to construct a cycle index polynomial for specific groups. Although this is always possible in principle, groups are encountered in practice that are too large for direct manipulation. However, in many cases of interest, it is possible to construct the cycle index polynomial for the group from the cycle index polynomials of the subgroups. For example, if $G = H \times K$ where H and K are groups, then it is advantageous to compute Z_G from Z_H and Z_K, because both H and K are much smaller groups than is G. Harrison (1963) has shown that a formula does exist for performing this operation, although some mathematicians may not consider it to be a "closed" formula. The formula involves a summation over partitions of an integer, a very common occurrence in combinatorial analysis. Although the formula may lack a certain elegance, it does provide an algorithm for computing cycle index polynomials with the aid of a computer.

The direct product certainly is not the only structure that is found in groups, and Harrison showed how to construct cycle index polynomials for several different important classes of group structures. De Bruijn (1959) has investigated the problem of counting equivalence classes when a group G acts on the domain D of a set of functions and a group H acts simultaneously on the range R.

The enumeration of equivalence classes of switching functions has strongly influenced some of the research in finding minimal-cost networks for the realization of switching functions. We mention at the beginning of this chapter the project undertaken by the Harvard Computation Laboratory in 1950, involving the enumeration of the equivalence classes of 4-variable switching functions where the variables can be permuted or complemented. This computer program revealed that there are 402 equivalence classes of these functions, and that this number can be reduced to 222 equivalence classes by including function complementation with the group operations. A correct mathematical analysis of this problem was provided a few years later by Slepian (1953). His technique used Burnside's theorem and the theory of group characters (the latter following the lead of Frobenius, 1896). Later, Harrison (1963) applied his techniques for computing cycle index polynomials to the problem and re-enumerated the classes. The fact that 65,536 functions can be characterized by as few as 222 functions has led to the publication of several tables of minimal-cost realizations of switching functions for several different hardware technologies. It would be virtually impossible to create such tables without the notion of equivalence of functions. One such table of the most economical switching networks for each equivalence class of the 4-variable functions is given by Harrison (1965).

To extend the practical utility of equivalence classes, several researchers have investigated the switching functions under the actions of various groups. In general, the larger groups induce fewer equivalence classes, but the problem of finding a representative of each equivalence class becomes very difficult as the classes become large. Ninomiya (1961) made one of the early important contributions to this area by introducing a hierarchy of four equivalence relations and showing how each switching function can be identified with an equivalence class in each of the four relations. Essentially, the identification algorithm involves taking the Fourier transform of the switching function. Lechner (1971) provides an excellent summary of the work in this area, including substantial contributions of his own. Stone and Jackson (1969) tabulate the Ninomiya equivalence classes. Harrison (1971) enumerates equivalence classes of the switching functions for a large class of different groups.

FOUR

Applications of
Group Theory
to Computer
Design

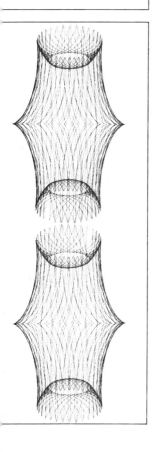

FOUR | Applications of Group Theory to Computer Design

In this chapter we examine three applications of group theory to computer design. Two of these applications concern computer arithmetic; the third deals with the design of computer memories using recent device technologies. Both of the first two applications concern the problem of performing addition in minimal time. The first application is the derivation of good lower bounds on the time required to perform addition; the second application is the design of fast adders. The third application deals with the embedding of permutation networks in memory systems. This problem arises for dynamic memories, in which stored data are in constant circulation. In such memories, data-access time depends on the time required for data to circulate to a read station. If we use conventional interconnection schemes, the circulation time can be quite large for large memories. However, we can find permutation interconnections that yield small access time, even for very large memories.

4.1 How Fast Can We Add? One of the challenges of computer design is the design of high-speed arithmetic circuits, particularly high-speed adders. Because digital computers inherently are confined to arithmetic systems with a finite number of elements, computer arithmetic usually is arithmetic modulo n for some large value of n. As the design of high-speed computers has developed from art to science, so has the design of the high-speed adders that are central modules of such computers. Winograd (1965) investigated the problem of adder design from a theoretical point of view in order to determine the ultimate limits on the speed of addition circuits. His results enable us to measure the effectiveness of high-speed adder designs, as well as to design adders that approach the theoretical lower bound. Group theory plays a central role in Winograd's work because the structure $\langle \mathbf{Z}_n, +_n \rangle$ is a group—in fact, is a cyclic group for all $n \geq 1$. In this section we present a derivation of his theoretical lower bound.

Before investigating the properties of groups and how they relate to the time for addition, let us first model the devices that are used in adders. All of the devices that we consider here have r or fewer inputs for some fixed value of $r \geq 2$, and each input is *binary* (two-valued).

For convenience we assume that the two possible values are 0 and 1. This assumption is consistent with the design of computers, in which the values are represented by physical variables such as voltage or current. Fig. 4.1.1 shows a typical device. In terms of the notation developed in earlier chapters, each device performs some function from $(Z_2)^r$ into Z_2, and the function differs from device to device.

r Inputs

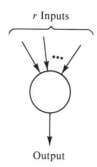

Output

FIG. 4.1.1 A Typical Device with *r* Inputs and One Output

To construct an adder, we assume that the integers in Z_n are encoded as binary m-tuples in a one-to-one fashion. That is, to each integer x in Z_n, there corresponds a unique m-tuple

$$h(x) = \langle h_1(x), h_2(x), \ldots, h_m(x) \rangle$$

such that each $h_i(x)$ is in Z_2 and such that the function $h : Z_n \to (Z_2)^m$ is an injection. An adder is a network of devices that computes a function $\oplus$ from $(Z_2)^m \times (Z_2)^m$ into $(Z_2)^m$, with the property that $h(x) \oplus h(y) = h(x +_n y)$ for each x and y in Z_n. Hence addition of binary m-tuples is an isomorphism of addition modulo n. Fig. 4.1.2 illustrates one way that we might construct an adder from devices of the type shown in Fig. 4.1.1. Each device has its input connected to the outputs of other devices or to components of the m-tuple inputs. There are m network outputs, one for each component in the m-tuple representation of the sum. We assume throughout the remainder of this section that m (the dimension of the representation of the integers in Z_n) is larger than r (the maximal number of inputs to each device). This assumption is quite realistic because, in high-speed computers, m may be as low as 24 or as high as 64, but r seldom exceeds 4.

To model the computation time of a network, we assume that each device takes one unit time to compute its output. Thus a network such as the one diagrammed in Fig. 4.1.3(a), in which there are three levels of devices, takes three units of time to produce its outputs. We also assume a special property of our networks. The network shown in Fig. 4.1.3(b) is functionally identical to the one in Fig. 4.1.3(a), but the networks differ in their internal structures. Specifically, each device output in Fig. 4.1.3(b) is connected to a single device input, whereas

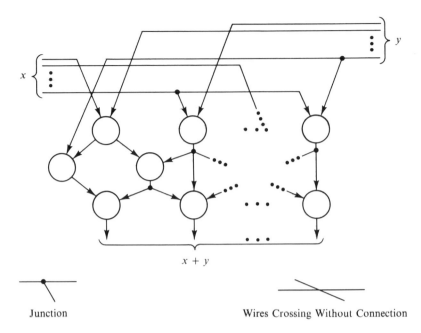

Junction   Wires Crossing Without Connection

FIG. 4.1.2 *r*-**Input 1-Output Devices Combined in a Network for Adding Two Numbers Encoded as Binary** *m*-**tuples**

the shaded device in Fig. 4.1.3(*a*) drives more than one other device. Given a network like that of Fig. 4.1.3(*a*), we can always change it into one like that of Fig. 4.1.3(*b*) by replicating one or more devices, as shown by the shaded devices in Fig. 4.1.3(*b*). Although replication increases the number of devices, it does not change the computation time. However, our analyses become somewhat simpler if we assume that no device drives more than one other device, so we limit our attention here to networks of the type shown in Fig. 4.1.3(*b*). Note that the adder network of Fig. 4.1.2 includes devices whose output drives two or more other devices. Therefore we consider a modified design for such an adder network that computes an *m*-tuple representation of the sums of two integers. Our adder network is composed of *m* networks, each of which computes a function of Z_2^{2m} into Z_2, where the $2m$ arises because each module has 2 *m*-tuple inputs. These *m* networks share inputs but are otherwise disjoint. Thus the time required to do addition is the maximum of the times required to compute each of the *m* functions. Fig. 4.1.4 shows our mathematical model of such a network, with each rectangle representing one of the *m* disjoint networks.

Before turning to the design of high-speed adders, we must consider the ways in which we construct *m*-ary functions from devices with *r* or fewer inputs. For example, how can we construct functions of 10 variables with devices that can depend on no more than 3 variables?

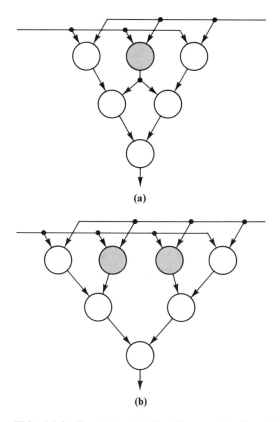

(a)

(b)

FIG. 4.1.3   Two Networks That Compute the Same Function

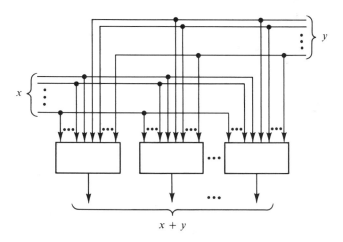

FIG. 4.1.4   An Adder Composed of *m* Disjoint Networks

The precise answer to this question varies with the specific function under consideration, but we can find some lower bounds on the computation time for such a network. Generally speaking, we always can construct a network of r-input devices to perform an m-ary function by building a treelike network such as that shown in Fig. 4.1.5. Note that we may have to connect some of the input variables to more than one device input. After the following definition, we prove a lemma that establishes a lower bound on the computation time for networks of the type shown in Fig. 4.1.5.

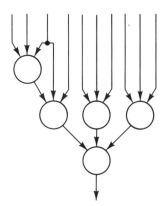

FIG. 4.1.5   A Network of 3-Input Devices for Computing a Function of 10 Inputs

Definition   For any real number x, the *ceiling of x*, denoted by $\lceil x \rceil$, is the smallest integer not less than x, and the *floor of x*, denoted by $\lfloor x \rfloor$, is the largest integer not greater than x.

Lemma   The time required to compute an m-ary function with r-input devices is at least $\lceil \log_r m \rceil$.

Proof: A network with t levels cannot have more than r^t distinct inputs, because there are at most r network inputs at the bottom level, and each level has no more than r times as many inputs as the maximal number of inputs of the level beneath it. Hence, if a network computes an m-ary function, then it must have at least t levels, where $m \leq r^t$, or $\lceil \log_r m \rceil \leq t$.   □

Let us use this lemma in a trivial way to obtain a lower bound for computation time for addition in which we use conventional binary notation. The *binary representation* of an integer x is the unique binary m-tuple $\langle x_{m-1}, x_{m-2}, \ldots, x_0 \rangle$ with the property that

$$x = \sum_{i=0}^{m-1} x_i 2^i.$$

In this representation, each of the m adder-network functions computes one of the m components of the binary representation. Because the leading digit of the sum of two numbers depends on all digits of both the addends, the network to compute x_{m-1} has at least $2m$ inputs, and this computation therefore requires at least $\lceil \log_r 2m \rceil = \lceil \log_r 2 \lceil \log_2 n \rceil \rceil$ time units, where n is the modulus of the arithmetic. To see that the leading digit of the sum does indeed depend on all addend digits, observe how it depends on the least significant digits through carry propagation in this example:

$$011111_2 + 000001_2 = 100000_2.$$

If we wish to find faster adders, we must use number representations other than the usual binary representation. The computation-time lemma is a powerful tool in the search for a faster adder, because it tells us that we must seek a representation in which none of the m adder functions depends on very many of the input variables. Group theory comes into play later when we show that, for every representation, at least one of the network functions for addition modulo n depends on $2\lceil \log_2 t(n) \rceil$ inputs, where t is a function of n that we will define. Because $t(n)$ is less than n for some n, some representations may lead to faster addition circuits than those for conventional binary arithmetic. We discuss one such representation in Section 4.2, but for the remainder of this section we continue with the derivation of Winograd's lower bound.

In the remainder of this section, we assume that each integer x in $\mathbf{Z}_n$ is represented by a binary m-tuple $\langle x_1, x_2, \ldots, x_m \rangle$ and we use the notation $(x)_i$ to denote the ith component of the representation of x. Without loss of generality, we assume that 0 in $\mathbf{Z}_n$ is represented by the m-tuple $\langle 0, 0, \ldots, 0 \rangle$.

Definition For each nonzero x in $\mathbf{Z}_n$, $x \cdot \mathbf{Z}_n$ is the set of all multiples of x taken modulo n.

For example, $2 \cdot \mathbf{Z}_5 = \{0, 2, 4, 1, 3\} = \mathbf{Z}_5$.

Lemma For each nonzero x in $\mathbf{Z}_n$, $x \cdot \mathbf{Z}_n$ is a subgroup of $\langle \mathbf{Z}_n, +_n \rangle$.
Proof: It is sufficient to show that the set is closed under addition modulo n. This easy proof is left as an exercise (Ex. 4.1.2). □

In order to discover the smallest number of inputs that may influence a particular output of an adder network, we investigate what happens when a network output is independent of some components of an m-tuple representation. In this discussion, the group operation for $\mathbf{Z}_n$ is addition modulo n if not stated explicitly.

Lemma Let the representations of x and y in $\mathbf{Z}_n$ differ only in components that do not influence the ith output of an adder network for this representation. Then every element z in the subgroup $(x - y) \cdot \mathbf{Z}_n$ satisfies $(z)_i = 0$.
Proof: Because x and y are indistinguishable to the ith output, we have $(x +_n z)_i = (y +_n z)_i$ for each z in $\mathbf{Z}_n$. Then

$$((x - y) +_n z)_i = (x +_n (-y + z))_i = (y +_n (-y + z))_i = (z)_i$$

for all z in $\mathbf{Z}_n$. Thus the ith component of z is equal to the ith component of $(x - y) +_n z$ in this representation. In particular, for $z = 0$ we obtain $(x - y)_i = 0$. Similarly, for $z = x - y$ we obtain $(2(x - y))_i = 0$, and generally $(k(x - y))_i = 0$ for every k in $\mathbf{Z}_n$. $\qquad\square$

Thus, whenever we find elements x and y that are indistinguishable to a particular output, we know that the subgroup generated by $x - y$ must be constant on the ith coordinate in the chosen representation. By limiting the size of the subgroup, we can bound the number of elements that do not influence an output. We need one more definition before proving the central result of this discussion.

Definition   An element g of a group G is said to be a *ubiquitous element* of G if $g \neq e$ and if g is in every nontrivial subgroup of G.

Theorem   (Winograd 1965). Let n be such that $\langle \mathbf{Z}_n, +_n \rangle$ has a ubiquitous element. Then, for every representation of $\mathbf{Z}_n$, at least one output of a modulo n adder depends on no less than $2\lceil \log_2 n \rceil$ distinct inputs.

Proof: Let u be the ubiquitous element of $\mathbf{Z}_n$. Because u is not 0, the group identity, let i be an index for which $(u)_i \neq 0$. We show that the ith output depends on at least $\lceil \log_2 n \rceil$ components of each addend.

Assume that the ith adder output depends on fewer than $\lceil \log_2 n \rceil$ components of the integer x in the sum $x +_n y$. Because each input can assume only 2 states, the ith output can distinguish among at most $2^{\lceil \log_2 n \rceil} - 1 < n$ integers in $\mathbf{Z}_n$ as the x operand. Hence there are two integers x and x' such that $(x +_n y)_i = (x' +_n y)_i$. Then, by the preceding lemma, $(x - x') \cdot \mathbf{Z}_n$ is a subgroup for which the ith component of each element is 0. But, because u is ubiquitous, u is in $(x - x') \cdot \mathbf{Z}_n$, and $(u)_i = 0$. This contradiction proves the theorem. $\qquad\square$

We now have a lower bound on the time required for addition in $\mathbf{Z}_n$, for some n.

Corollary   If $\mathbf{Z}_n$ has a ubiquitous element, then the time needed for addition must be at least $\lceil \log_r (2\lceil \log_2 n \rceil) \rceil$ time units.

Corollary   If $\mathbf{Z}_n$ does not have a ubiquitous element, but H is a subgroup that does, then addition in $\mathbf{Z}_n$ must take at least $\lceil \log_r (2\lceil \log_2 |H| \rceil) \rceil$ time units.

Proof: The bound given is the bound for computation in H. The time for computation in Z_n must be at least as great as the computation time for H (see Ex. 4.1.3). $\qquad\square$

Definition   Let $t(n)$ be the order of the largest subgroup of $\mathbf{Z}_n$ that has a ubiquitous element.

Now we can rephrase the preceding corollary in terms of $t(n)$.

Corollary   Addition in $\mathbf{Z}_n$ must take at least $\lceil \log_r (2\lceil \log_2 t(n) \rceil) \rceil$ time units.

Fortunately, we can compute the function $t(n)$ very easily.

Lemma   If $\mathbf{Z}_n$ is isomorphic to $\mathbf{Z}_{n_1} \times \mathbf{Z}_{n_2}$, then $t(n)$ is the maximum of $t(n_1)$ and $t(n_2)$.

Proof: In this case, $\mathbf{Z}_n$ does not have a ubiquitous element, because the subgroups isomorphic to $\mathbf{Z}_{n_1}$ and $\mathbf{Z}_{n_2}$ have only the identity in common. However, computation time in $\mathbf{Z}_n$ is at least as great as either of the computation times in $\mathbf{Z}_{n_1}$ and $\mathbf{Z}_{n_2}$.   ☐

Lemma   If $n = p^i$ where p is prime, then $\mathbf{Z}_n$ has a ubiquitous element, and $t(n) = n = p^i$.

Proof: $\langle \mathbf{Z}_n, +_n \rangle$ is a cyclic group for all n, and the element 1 is a generator. By making use of the fact that every subgroup of a cyclic group also is cyclic, it is not difficult to show that every nontrivial subgroup includes the element p^{i-1} (Ex. 4.1.6). Hence p^{i-1} is ubiquitous and, by Winograd's theorem, $t(n) = n$.   ☐

The following theorem establishes the lower bound on computation time for all n.

Theorem   Let n have the prime factorization $n = p_1^{i_1} p_2^{i_2} \ldots p_k^{i_k}$, where the primes are distinct primes. Then $t(n)$ is the largest of the prime power factors of n (that is, the largest p^i).

Proof: Because $\mathbf{Z}_n$ is cyclic and of order n, it has the direct product decomposition

$$\mathbf{Z}_n \cong \mathbf{Z}_{p_1^{i_1}} \times \mathbf{Z}_{p_2^{i_2}} \times \cdots \times \mathbf{Z}_{p_k^{i_k}}.$$

Our previous results indicate that each of the factor groups has a ubiquitous element and that no larger subgroups of $\mathbf{Z}_n$ have a ubiquitous element. Because $t(n)$ is equal to the order of the largest subgroup that has a ubiquitous element, the theorem is proved.   ☐

Winograd's theorem extends to lower bounds on multiplication time. To derive the bound, we represent each integer by an m-tuple where each component designates a power of a prime number, and where each integer is equal to the product of the prime powers. Thus, if the components in a 3-tuple represent powers of 2, 3, and 5, respectively, then $\langle 2, 1, 0 \rangle$ represents $2^2 \cdot 3 = 12$ and $\langle 1, 1, 2 \rangle$ represents $2 \cdot 3 \cdot 5^2 = 150$. In this representation, multiplication is done by componentwise addition. Thus the bounds for addition time can be applied to bounds on multiplication. Full details are given by Winograd (1967).

Although we have discussed Winograd's lower bound in the context of addition modulo n, his results apply to computation times for group operations in all groups. Spira (1969) has sharpened Winograd's results by raising his lower bound and by constructing circuits that perform any group composition in a time no more than one unit greater than the theoretical lower bound. For cyclic groups, Spira's bound is in agreement with Winograd's bound as described here. The lower bound for addition modulo 2^n was first obtained by Ofman (1962).

The practical implications of Winograd's results are affected by some peripheral considerations. First, an unconventional representation is required to reach the lower bound in computation time, and such a representation may have significant impact on the speed and implementation cost of other common computer operations. In particular, given two integers x and y in $\mathbf{Z}_n$, we often wish to determine if $x \geq y$, and if the sum $x + y$ exceeds the largest representable number. Winograd (1967) showed that each of these functions requires at least $\lceil \log_r(2\lceil \log_2 n \rceil) \rceil$ time units to compute (Ex. 4.1.4). In conventional binary notation, both functions can be computed easily by making minor modifications to an adder. However, in other representations, these functions require extensive, relatively slow, and costly networks. In fact, representations that tend to be fast for addition tend to be slow and ill-suited for magnitude comparison and overflow detection. Hence, when we view the problem from a broader perspective, the advantages of conventional binary representation may offset the disadvantage of longer computation time indicated by Winograd's lower bound.

Another aspect of Winograd's problem has been investigated more thoroughly by Avizienis (1961). He pursued a question involving number representations in which an integer in $\mathbf{Z}_n$ may be represented by several different m-tuples rather than by a unique binary m-tuple as in Winograd's representation. Avizienis has pursued this investigation from a practical point of view, and he has demonstrated fast arithmetic circuits that can be implemented quite reasonably (Avizienis 1964). His circuits violate Winograd's assumption that the representation is a one-to-one correspondence, and indeed these circuits can perform addition more rapidly than is predicted by Winograd's lower bound.

Exercises

4.1.1 Evaluate Winograd's lower bound for devices with three or fewer inputs and for addition in $\mathbf{Z}_n$ when
   (a) $n = 2^{32}$;
   (b) $n = 2^{36}$;
   (c) $n = 2^{48}$;
   (d) $n = 2^{32} - 1 = 2{,}147{,}483{,}647$ [a prime number];
   (e) $n = 2^{24} - 1 = 3^2 \cdot 5 \cdot 7 \cdot 13 \cdot 17 \cdot 241$.

4.1.2 Show that $x \cdot \mathbf{Z}_n$ is a subgroup of $\mathbf{Z}_n$ with respect to addition modulo n for all x in $\mathbf{Z}_n$.

4.1.3 In the proofs of the corollaries to Winograd's theorem, we assume that addition in $\mathbf{Z}_n$ cannot be performed more rapidly than addition in any of its subgroups. Why is this so?

4.1.4 (Winograd 1967). For x and y in $\mathbf{Z}_n$, let $f(x, y)$ be the *magnitude comparison* function. That is,

$$f(x, y) = \begin{cases} 1 \text{ if } x \geq y; \text{ or} \\ 0 \text{ if } x < y. \end{cases}$$

Show that $f(x, y)$ requires at least $\lceil \log_r (2 \lceil \log_2 n \rceil) \rceil$ time units to compute.

4.1.5 We have been assuming that each input or output line can take on either of two possible values. Suppose that these lines can assume d different values, where $d > 2$. Derive a lower bound on arithmetic computation time for this case.

4.1.6 Let g be a generator of a cyclic group of order p^n where p is a prime. Show that every nontrivial subgroup of this group contains $g^{p^{n-1}}$.

2 The Residue Number System

In the preceding section, we indicate that integers in $\mathbf{Z}_n$ may have binary representations for which addition can be performed somewhat more rapidly than can addition in conventional binary representation. In this section, we briefly investigate one such representation.

Because $\langle \mathbf{Z}_n, +_n \rangle$ is a cyclic group, our previous results on the structure of cyclic groups show that this group is isomorphic to the group

$$\mathbf{Z}_{n_1} \times \mathbf{Z}_{n_2} \times \cdots \times \mathbf{Z}_{n_m},$$

where each of the factors is a cyclic group whose order is a power of a prime, the primes are distinct, and

$$n = \prod_{i=1}^{m} n_i.$$

This result suggests that we can perform arithmetic in $\mathbf{Z}_n$ by doing arithmetic in $\mathbf{Z}_{n_i}$ for each n_i in the prime decomposition of n.

Svoboda and Valach (1955) and Garner (1959) showed that it is indeed possible to make use of the decomposition structure of $\mathbf{Z}_n$ to construct fast adders. Moreover, there is a relatively straightforward method for converting integers from conventional binary representation into the unconventional representation and back again.

Before showing how to construct arithmetic systems from the decomposition structure of $\mathbf{Z}_n$, let us verify that such systems will indeed be suitable for fast arithmetic. Speaking intuitively, we know that the time required for arithmetic in conventional representation is due largely to the carry-propagation time. If we can perform addition in

Z_n with a representation that uses addition in Z_k (where k is small compared to n), then the maximal carry-propagation time in the Z_k representation is much smaller than the carry-propagation time for Z_n using conventional representation. Hence, if n has many different prime factors, then each of the groups in the direct-product decomposition of Z_n has a relatively small carry-propagation time as compared to that for addition in Z_n using conventional representation. Fig. 4.2.1 illustrates the way that addition in Z_n is performed by doing addition in each of the groups Z_{n_i}. This diagram makes it clear that the addition time in the composite network will be precisely the time that it takes to do addition in the largest of the subgroups of Z_n in its direct-product decomposition.

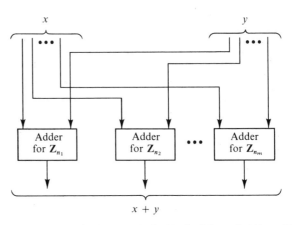

FIG. 4.2.1   An Adder for Integers in Z_n Derived from the Direct-Product Decomposition of Z_n

Now we examine the construction of a representation of Z_n that makes use of its direct-product decomposition.

Definition   Let $n = p_1^{x_1} p_2^{x_2} \ldots p_m^{x_m}$, where the p_i's are distinct primes. Then, if k is an integer in Z_n, its *residue* (or *modular*) *representation* is the m-tuple $\langle k \bmod p_1^{x_1}, k \bmod p_2^{x_2}, \ldots, k \bmod p_m^{x_m} \rangle$.

Definition   Let Z_n* be the direct-product decomposition of Z_n. That is,

$$Z_n* = Z_{n_1} \times Z_{n_2} \times \cdots \times Z_{n_m},$$

where $n_i = p_i^{x_i}$ in the prime factorization of n.

Theorem   (*Chinese Remainder Theorem*). The function $h: Z_n \rightarrow Z_n*$ that carries each x in Z_n into its residue representation in Z_n* is a bijection.

Proof: First note that Z_n* has $\Pi n_i = n$ elements (where the product is taken over all i). We can use the pigeonhole principle to prove the lemma by establishing that the residue representation is an injection. Now suppose that x and y in Z_n have identical residue representations.

Then $x \equiv y \bmod n_i$ for each i, so that $x - y$ is a multiple of n_i. Then $x - y$ is a multiple of the least common multiple of $n_1, n_2, \ldots,$ and n_m. Because the n_i's are relatively prime, their least common multiple is equal to n. Hence $x - y$ is a multiple of n, and both x and y cannot be in $\mathbf{Z}_n$. This is a contradiction, which proves that h is an injection. Consequently h is a bijection. $\square$

Having shown that h carries $\mathbf{Z}_n$ bijectively onto $\mathbf{Z}_n*$, we next show that arithmetic in $\mathbf{Z}_n$ can be done by doing arithmetic in each of the subgroups $\mathbf{Z}_{n_i}$.

Definition For each x and y in $\mathbf{Z}_n$, let $\oplus$ be the binary operation defined on $\mathbf{Z}_n*$ such that

$$h(x) \oplus h(y) = \langle (x \bmod n_1) +_{n_1} (y \bmod n_1),$$
$$(x \bmod n_2) +_{n_2} (y \bmod n_2), \ldots,$$
$$(x \bmod n_m) +_{n_m} (y \bmod n_m) \rangle.$$

Note that $\oplus$ is addition of residue representations, where the sum is obtained by componentwise addition and where addition in the ith component is modulo n_i. The following theorem shows that $\langle \mathbf{Z}_n*, \oplus \rangle$ is isomorphic to $\langle \mathbf{Z}_n, +_n \rangle$.

Theorem The function $h : \mathbf{Z}_n \to \mathbf{Z}_n*$ that carries each x in $\mathbf{Z}_n$ into its residue representation is an isomorphism of $\langle \mathbf{Z}_n, +_n \rangle$ onto $\langle \mathbf{Z}_n*, \oplus \rangle$.

Proof: We have shown previously that h is a bijection, so it is sufficient here to show that it is a homomorphism. The homomorphism property follows because $h(x +_n y) = h(x) \oplus h(y)$ for all x and y in $\mathbf{Z}_n$. In particular, the ith component of $h(x +_n y)$ is $(x + y) \bmod n_i$, and the ith component of $h(x) \oplus h(y)$ is $(x \bmod n_i) +_{n_i} (y \bmod n_i)$. These clearly are equal. $\square$

As an example of residue arithmetic, consider the residue representation of $\mathbf{Z}_{30}$. Because $30 = 2 \cdot 3 \cdot 5$, we construct the residue representation by finding the residues modulo 2, 3, and 5 of each of the integers in $\mathbf{Z}_{30}$, as shown in Table 4.2.1. Now observe that

$$h(x +_{30} y) = h(x) \oplus h(y).$$

For example, if $x = 7$ and $y = 13$, then $x +_{30} y = 20$. The residue representations of 7 and 13 are $\langle 1, 1, 2 \rangle$ and $\langle 1, 1, 3 \rangle$, respectively. Then $h(x) \oplus h(y) = \langle 0, 2, 0 \rangle$, which is the residue representation of 20, as it should be.

In order to use the residue representation in computers, we must have convenient ways to translate from conventional representation to residue representation and back again. The method of computing residue representation from conventional representation is clear, for to do this we need only carry out the standard arithmetic operation of computing $x \bmod n_i$. But suppose that we have the residue representation of x and wish to obtain its conventional representation. As the next theorem shows, this transformation also is easy to perform.

Theorem   Let x be an integer in $\mathbf{Z}_n$, and let $(x)_i$ be x mod n_i, or the ith coordinate in the residue representation of x. Let w_i, $1 \leq i \leq m$, be the m integers in $\mathbf{Z}_n$ that have residue representations with a 1 in the ith coordinate and have 0's in all other coordinates. Then

$$x = \sum_{i=1}^{m} (x)_i w_i,$$

where the summation is done modulo n.

Proof: We prove the theorem by showing that $\Sigma(x)_i w_i$ is an integer in $\mathbf{Z}_n$ that has the residue representation $\langle x_1, x_2, \ldots, x_n \rangle$. Because the summation is done modulo n, the summation must be an integer in $\mathbf{Z}_n$. Because w_i is a multiple of all the moduli except n_i, we see that $(\Sigma(x)_i w_i) \bmod n_j$ excludes contributions from all terms that are multiples of n_j, and therefore the sum is congruent to $(x)_j w_j$ modulo n_j. Because $w_j \bmod n_j = 1$,

$$(\Sigma(x)_i w_i) \bmod n_j = (x)_j = x \bmod n_j$$

for all j. Thus the summation has the desired residue representation.   □

Turning to Table 4.2.1 for an example, we observe that

$$w_1 = \langle 1, 0, 0 \rangle = 15, \quad w_2 = \langle 0, 1, 0 \rangle = 10, \quad \text{and} \quad w_3 = \langle 0, 0, 1 \rangle = 6.$$

The integer corresponding to $\langle 1, 2, 3 \rangle$ is the integer $(15 \cdot 1 + 10 \cdot 2 + 6 \cdot 3)$ mod $30 = 23$. Similarly, the integer whose residue representation is $\langle 0, 1, 4 \rangle$ is the integer $(15 \cdot 0 + 10 \cdot 1 + 6 \cdot 4)$ mod $30 = 4$. In each case, the transformation gives a result that agrees with the table.

The preceding theorem establishes the fact that we can quite easily

TABLE 4.2.1   Residue Representations of the Integers in $\mathbf{Z}_{30}$

x	Residues Modulo			x	Residues Modulo			x	Residues Modulo		
	2	3	5		2	3	5		2	3	5
0	0	0	0	10	0	1	0	20	0	2	0
1	1	1	1	11	1	2	1	21	1	0	1
2	0	2	2	12	0	0	2	22	0	1	2
3	1	0	3	13	1	1	3	23	1	2	3
4	0	1	4	14	0	2	4	24	0	0	4
5	1	2	0	15	1	0	0	25	1	1	0
6	0	0	1	16	0	1	1	26	0	2	1
7	1	1	2	17	1	2	2	27	1	0	2
8	0	2	3	18	0	0	3	28	0	1	3
9	1	0	4	19	1	1	4	29	1	2	4

transform integers in $\mathbf{Z}_n$ from conventional representation to residue representation and back again, using a standard repertoire of computer instructions. However, the process can be excessively time consuming if performed frequently. Designers of high-speed computers have been very interested in the advantages of the residue representation. Unfortunately, the problems of magnitude comparison and overflow detection are severe, thus nullifying some of the advantages of the representation. However, residue representation is highly satisfactory for certain computations that do not require frequent magnitude comparisons or overflow checks. Borosh and Fraenkel (1966) used this representation in attacking the problem of finding a complete set of independent solutions to 111 linear equations in 120 unknowns. They solved the equation sets separately, using arithmetic modulo n_i for prime moduli n_i, and were able to compute exact solutions substantially more rapidly than approximate solutions could be computed with the more usual floating-point techniques.

Knuth (1969) gives a brief but thorough survey of residue arithmetic. The book by Szabo and Tanaka (1967) is an excellent reference for a more detailed discussion.

Exercises

4.2.1 Construct the residue representation for $\mathbf{Z}_{21}$ and determine the weights w_i for converting from residue representation to conventional representation.

4.2.2 We define the binary operation $\odot$ on $\mathbf{Z}_n*$ to be

$$x \odot y = \langle (x \bmod n_1) \cdot_{n_1} (y \bmod n_1),$$
$$(x \bmod n_2) \cdot_{n_2} (y \bmod n_2), \ldots,$$
$$(x \bmod n_m) \cdot_{n_m} (y \bmod n_m) \rangle.$$

Show that $\langle \mathbf{Z}_n*, \odot \rangle$ is isomorphic to $\langle \mathbf{Z}_n, \cdot_n \rangle$. That is, we can use the residue representation to do both addition and multiplication by performing addition and multiplication modulo n_i on each of the components.

4.2.3 Let $f(x, y)$ be the magnitude-comparison function on $\mathbf{Z}_n$. That is, $f(x, y) = 1$ if $x \geq y$, and otherwise $f(x, y) = 0$. Use Table 4.2.1 to show that $f(x, y)$ depends on all of the residues in the residue representation of x and y when $n = 30$. Prove that f depends on all of the residues for all n.

3 Permutation Interconnections for Dynamic Memories

Some computer memories require that stored data be in constant circulation. In each such memory also there is a restriction on interconnections between the memory and the other system modules. The memory has a read station through which each datum must pass if it is requested by another module. Therefore, data access in such a system requires **(a)** that the position of each item in the memory be known at every moment to the access-control mechanism, and **(b)** that the access-control mechanism be able to determine the shortest route from any memory position to the read station.

Dynamic memories have been constructed from technologies as diverse as integrated circuits, magnetic bubbles, and charge-coupled capacitors. Most dynamic memories are constructed as cyclic shift-registers, so that data are shifted cyclically at unit-time intervals.

Memory-access control is very simple in such cyclic memories. Suppose that the memory contains n items. Then the controller can keep track of the positions of words in memory by using a modulo-n counter. When item 0 is at the read station, the counter contains the number 0. Each time a shift occurs, the counter increases by 1 modulo n. If item i is at the read station before a shift, then it is replaced by item $i + 1$ after the shift, where the index of the item is increased by 1 modulo n. Therefore the counter always contains the index of the item at the read station. If the controller receives a request for item i, it waits until the number i appears in the counter and then copies the item at the read station onto an output line. Fig. 4.3.1 illustrates the access mechanism for this type of memory.

The problem with a cyclically organized memory is that the access time increases linearly as the size of the memory increases. Obviously, the longest possible access time will result when a request is received for an item that has just been shifted out of the read station. If there are

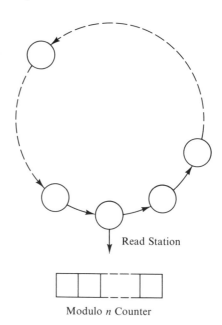

Read Station

Modulo n Counter

FIG. 4.3.1 A Model of a Cyclic Dynamic Memory

n items in memory, this worst-case access time is $n - 1$ time units. If requests are distributed uniformly, the average access time is $n/2$ units. In the remainder of this section we explore ways to decrease the access time by including more interconnections in the memory. The interconnections we discuss here make it possible to achieve the theoretical lower bound on access time, a bound that is proportional to $\log_2 n$ rather than to n. Hence, for n of the order of 10^6, the worst-case access time is reduced from 500,000 time units to only 20 time units.

To begin our discussion, we develop a general model of a dynamic memory and derive a lower bound on access time. Three assumptions characterize our model:

(i) there is a single location in memory designated as a read station (the read station also can function as a write station for storing new information into memory);

(ii) each location in memory holds one item, and it can transfer the item to one of r other locations, each identified by an integer i such that $1 \leq i \leq r$;

(iii) at each unit-time interval, the access-control circuit broadcasts an integer i, $1 \leq i \leq r$, to all memory locations, and each location then transfers its stored datum to its ith neighboring location.

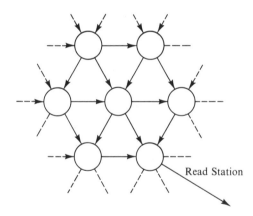

FIG. 4.3.2   A Model of a Dynamic Memory with $r = 3$ Interconnections

Fig. 4.3.2 shows a model of such a memory. The memory is said to have r interconnections, because each cell can transfer its datum to one of r other cells. Each datum or item stored in a cell is called a word.

For $r = 1$, the cyclic memory discussed earlier is the only memory with the proposed structure and with the property that all items in the memory are accessible. Next we show that access time for memories with $r \geq 2$ can grow as $\log_r n$ rather than linearly as n.

Lemma The worst-case access time for a memory with n words and r interconnections cannot be less than T, where T is the least integer that satisfies the inequality

$$(r^{T-1} - 1)/(r - 1) \geq n.$$

Proof: One word (that at the read station when the request is received) is accessible in 0 time units. In 1 time unit, the access-control circuit can bring to the read station any of the r words that are stored in cells directly connected to the read station. In 2 time units, the access-control circuit can bring to the read station any of the r^2 words stored in cells directly connected to the r cells just mentioned. Of course, depending on the specific structure of the network, some of these cells may not be distinct. For example, the contents of a particular cell might be accessible by one path in 1 time unit, by two other paths in 2 time units, etc. However, we know that, after T units of time, no more than $1 + r + r^2 + \cdots + r^T$ words are accessible. To ensure that all n items are accessible, this sum must be at least as large as n. Using the well-known relation for geometric sums,

$$\sum_{i=0}^{\infty} r^i = 1/(1 - r),$$

we obtain

$$1 + r + r^2 + \cdots + r^T = (r^{T+1} - 1)/(r - 1) \geq n. \qquad (4.3.1) \quad \square$$

Solving Eq. 4.3.1 for r^{T+1}, we obtain

$$r^{T+1} \geq (r - 1)n + 1. \qquad (4.3.2)$$

For large n, we can neglect the 1 on the righthand side of Eq. 4.3.2. Then taking logarithms, we obtain the inequality

$$T \geq \lceil \log_r (r - 1) + \log_r n \rceil - 1,$$

or $$T \geq \lceil \log_r n \rceil.$$

For sufficiently large n, therefore, the lower bound on access time is $\lceil \log_r n \rceil$.

Next we show a pair of permutations that together achieve the minimal access time for $r = 2$ and $n = 2^m$ for some integer m. Note that substitution of $n = 2^m$ and $r = 2$ into Eq. 4.3.2 yields the inequality

$$2^{T+1} \geq 2^m + 1,$$

or $$T \geq \lceil \log_2 (2^m + 1) \rceil - 1 = m + 1 - 1 = \log_2 n.$$

From this model of dynamic memories, we see that the interconnection pattern in memory must be r distinct permutations. If the ith interconnection pattern were not a permutation of memory, then two different memory locations would each have the same ith neighbor and consequently, when the ith interconnection is activated, their common neighbor would receive two items to store. By assumption,

each location can store only one item, so a nonpermutation interconnection is not permissible.

The problem at hand is to find two generators for a permutation
group that acts upon the memory locations. Let the locations be indexed
$0, 1, \ldots, n - 1$, and let location 0 be the read station in memory.
Any given location must be connected to location 0 by some permutation
in the group that we construct. Thus the permutation group must
contain a cycle $(i\ 0 \ldots)$ for every i such that $1 \le i \le n - 1$. To hold
the worst-case access time to the lower bound computed earlier, we
impose the condition that every location be able to reach location 0
in no more than $\log_2 n$ shifts. In other words, we require that every
permutation containing the cycle $(i\ 0 \ldots)$ be expressible as a product
of no more than $\log_2 n$ generator factors.

We use two permutations called the *perfect shuffle* and the *exchange-
shuffle*. Next we describe these permutations and then show that a
dynamic memory with such permutations has the desired properties.

Fig. 4.3.3 shows the perfect-shuffle permutation pattern. The pattern
takes its name from the fact that it is analogous to the shuffling of a
deck of playing cards. The memory cells on the left of the figure are
shuffled by first dividing the cells into two groups with a "cut" in the
middle of the memory and then alternating between the two groups in
connecting them to the cells on the right, just as the two halves of a
deck of playing cards are interlaced when undergoing a perfect shuffle.
The first cell on the right receives the datum from the first cell of the

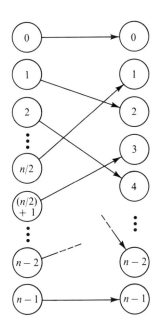

FIG. 4.3.3 The Perfect Shuffle of *n* Memory Cells

first group on the left; the second cell receives the datum from the first cell of the second group on the left; the third cell receives the datum from the second cell of the first group; and so on. In a dynamic memory, the two columns of cells shown in Fig. 4.3.3 are actually one and the same column.

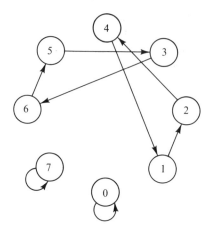

FIG. 4.3.4   A Dynamic Memory with the Perfect-Shuffle Interconnection Pattern and $n = 8$

Now suppose that we fill the n memory locations with integers $0, 1, \ldots, n - 1$ as shown in Fig. 4.3.4 for $n = 8$. (Recall that $n = 2^m$.) What happens to the integers after a perfect shuffle? It is not difficult to show that the following properties hold.

Property 1.   A perfect shuffle connects cell i to cell $p(i)$, where $p(i)$ is defined by

$$p(x) = \begin{cases} 2x & \text{for } 0 \le x \le (n/2) - 1; \text{ or} \\ 2x - n + 1 & \text{for } n/2 \le x \le n - 1. \end{cases}$$

Property 1′.   If the binary representation of i is

$$i = i_0 \cdot 2^0 + i_1 \cdot 2 + \cdots + i_{m-1} \cdot 2^{m-1},$$

then $p(i)$ has the binary representation

$$p(i) = i_{m-1} \cdot 2^0 + i_0 \cdot 2 + i_1 \cdot 2^2 + \cdots + i_{m-2} \cdot 2^{m-1}.$$

Property 2.   Consider the $n/2$ adjacent pairs of cells in this memory, as shown in Fig. 4.3.5. When the binary representations of the integers $0, 1, \ldots, n - 1$ are placed in memory in ascending order, the two integers in any adjacent pair differ only in the coefficient of 2^0 in their binary expansions. After one shuffle, paired integers differ only in the coefficient of 2^{m-1} in their binary representations. After j shuffles, where $1 \le j \le m$, the paired integers differ only in the coefficient of 2^{m-j} in their binary representations. After any number of shuffles,

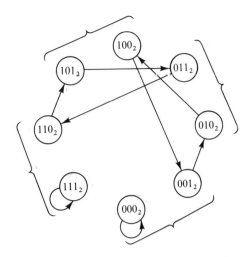

FIG. 4.3.5   The Memory of Fig. 4.3.4 with Cells Grouped in Adjacent Pairs and with Integers in Binary Representations

paired integers differ by precisely one coefficient in their binary expansions.

Fig. 4.3.6 shows the second interconnection pattern, the exchange-shuffle. As the second part of the figure shows, the pattern is the composition of an exchange of adjacent pairs of cells followed by a perfect shuffle. Fig. 4.3.7 shows an 8-cell memory with both interconnection patterns. The perfect shuffle is indicated by solid lines and the exchange-shuffle by dashed lines.

Fig. 4.3.7 also shows the control registers for data access. We need three registers, each with $\log_2 n$ bits. The *A-register* (or *address register*) identifies the item to be accessed. In this model, a unique address is associated with each item, and retrieval is by address. The address of an item remains constant, even though the item visits many different locations in the memory. The *S-register* contains the address of the item that is currently in the read station (location 0) of the memory. The *C-register* is a circulating shift-register that contains a single 1 bit. The 1 bit cycles through the register, moving one stage at each time unit and shifting end-around when it reaches the final register stage.

The memory-access mechanism is based upon the following notions.

Consider the addresses of items in adjacent pairs of locations. Property 2 guarantees that, after any number of perfect shuffles, the addresses in any pair differ by precisely one coefficient in their binary expansions, and this coefficient is the same for all pairs. We call this coefficient the *pivot bit*. Because the exchange permutation does not alter this property, the exchange-shuffle does not alter it either. Therefore, at any time, the addresses of items in adjacent pairs of locations have binary expansions that differ only in their pivot-bit positions. After each shuffle

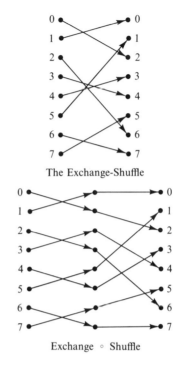

The Exchange-Shuffle

Exchange ∘ Shuffle

FIG. 4.3.6   The Exchange-Shuffle Permutation with $n = 8$

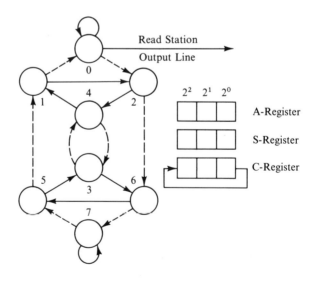

FIG. 4.3.7   The Perfect Shuffle and Exchange-Shuffle in a Dynamic Memory with $n = 8$

or exchange-shuffle, the pivot bit shifts to the next coefficient of less significance; from the coefficient of 2^0, it cycles back to the most significant position (the coefficient of 2^{m-1}). We use the 1 bit in the C-register to keep track of this pivot-bit position. This bit is cycled in the appropriate direction after each perfect shuffle or exchange-shuffle.

The memory-access mechanism requires that we modify the contents of the S- and C-registers after each permutation so that the S-register continues to identify the address of the item in the read station and the C-register continues to identify the pivot bit. We have already shown that the pivot bit should be cycled after a perfect shuffle. Because a perfect shuffle does not alter the contents of the read station, the S-register remains unchanged after a perfect shuffle. We have yet to determine how to update the registers after an exchange-shuffle.

Now let us consider the exchange-shuffle as a combination of two distinct shifts—an exchange shift followed by a perfect shuffle—as shown in Fig. 4.3.6. We already know how the S- and C-registers must change for the perfect shuffle, so we need only examine the effects of the exchange permutation. Because an exchange modifies the contents of the read station, we must update the S-register after an exchange. The pivot-bit position is not affected by the exchange shift, so no change is made to the C-register after the exchange. After the exchange, the S-register should have the address of the item that formerly was in location 1. However, we know that this address differs only in the pivot-bit position from the address currently in the S-register (the address of the item formerly in location 0). Hence we can update the S-register after an exchange by simply complementing the bit in the pivot position, which is indicated by the 1 bit in the C-register.

Combining the exchange shift and the perfect shuffle, we see that after an exchange-shuffle we must first complement the pivot bit of the S-register (that is, the bit in the position corresponding to the position of the 1 bit in the C-register) and then cycle the C-register.

In order to access an item, we must place it in location 0. We know that the item is in that location when its address appears in the S-register. If the desired address already appears in the S-register, we can access the item immediately. If the address is not in the S-register, we can place it there by complementing appropriate bits in the S-register. Each exchange-shuffle complements the pivot bit of the S-register and then cycles the pivot bit; each perfect shuffle cycles the pivot bit without altering the S-register. Thus we can force any address we wish into the S-register by a sequence of no more than $\log_2 n$ perfect shuffles and exchange-shuffles. The following algorithm summarizes the procedure.

Step 1. Load into the A-register the address of the item to be accessed.

Step 2. If the A-register and the S-register are identical, go to step 6. If not, continue.

Step 3. If the A-register and the S-register agree in the pivot-bit position (the position of the 1 bit in the C-register), go to step 4; otherwise go to step 5.

Step 4. Permute the memory with a perfect shuffle. Cycle the C-register to indicate the new pivot-bit position. Return to step 2.

Step 5. Permute the memory with an exchange-shuffle. Complement the bit in the pivot-bit position of the S-register. (It now agrees with the corresponding bit of the A-register.) Cycle the C-register to indicate the new pivot-bit position. Return to step 2.

Step 6. The desired item is in location 0. Copy it onto the output line. Terminate.

Table 4.3.1 shows two examples of the access algorithm. In each case, the integer i is stored as the item with address i, so that the permutations of the addresses can be followed through the algorithm.

It is clear that this access algorithm requires no more than $\log_2 n$ shifts, because the S-register contains $\log_2 n$ bits and the processing of each bit takes exactly one shift. In some cases, the desired item will be accessed in fewer than $\log_2 n$ shifts, but this does not appear to have a significant effect on the average access time. If we assume that the requested addresses for access are independent and identically distributed with a uniform distribution, we know that half of the accesses will require $m = \log_2 n$ shifts, because in half of the cases the mth pivot bit of the address will disagree with the mth pivot bit of the S-register. By similar reasoning, one quarter of the accesses will be

TABLE 4.3.1 Examples of Memory Access in a Perfect-Shuffle Shift-Register Memory

Word	Example I: Memory Contents after				Example II: Memory Contents after		
	Step 1	Step 5	Step 4	Step 5	Step 1	Step 5	Step 5
A-Register	101	101	101	101	011	011	011
S-Register	000	100	100	101	101	001	011
C-Register	100	010	001	100	100	010	001
Location 0	0	4	4	5	5	1	3
1	4	6	5	1	1	3	2
2	1	0	6	4	4	5	1
3	5	2	7	0	0	7	0
4	2	5	0	7	7	0	7
5	6	7	1	3	3	2	6
6	3	1	2	6	6	4	5
7	7	3	3	2	2	6	4

accomplished with $m - 1$ shifts, one eighth of the accesses with $m - 2$ shifts, etc. Thus the average access time, $\overline{T}$, is given by

$$\overline{T} = \sum_{i=0}^{m} 2^{-(i+1)}(m - i) = m \sum_{i=0}^{m} 2^{-(i+1)} - \sum_{i=0}^{m} i \cdot 2^{-(i+1)}$$

$$= m(1 - 2^{-(m+1)}) - [1 - 2^{-m}(1 + m/2)]$$

$$= m - 1 + 2^{-m}$$

$$\cong m - 1 = (\log_2 n) - 1.$$

If we access the addresses in numerical order, we obtain an average access time of $\log_2 n$, because successive addresses then always differ in the least significant bit.

Although no dynamic memory of the type we have described has been constructed for commercial use, the properties of such a memory clearly indicate its utility for dynamic memory technologies. The properties of the perfect-shuffle permutation make it useful in many other applications, particularly in parallel computation. Pease (1968) describes its use in fast Fourier transforms; Batcher (1968) applies it to sorting networks; and Stone (1971) discusses both of these applications and others. The material in this section is drawn from Stone (1972).

Exercises

4.3.1 Prove that properties 1, 1′, and 2 hold as claimed for the perfect shuffle.

4.3.2 Consider the problem of transposing a 2^n by 2^n matrix M in a memory of size 2^{2n}. The matrix is said to be stored in *row major order* if the matrix elements are stored so that $[M]_{i,j}$ precedes $[M]_{i',j'}$ if $i < i'$ or if $i = i'$ and $j < j'$. The matrix is said to be stored in *column major order* if the elements are stored so that $[M]_{i,j}$ precedes $[M]_{i',j'}$ if $j < j'$ or if $j = j'$ and $i < i'$. Show that n perfect shuffles will transpose a matrix from either major order to the other. How can we transpose a 2^m by 2^n matrix with perfect shuffles?

4.3.3 (Ullman). Consider a dynamic memory (with $n = 2^m$ cells) that has a perfect-shuffle interconnection and an interconnection that connects location i to location $i + 1$ for $1 \le i \le n - 2$ and connects location $n - 1$ to location 1. (Location 0 is disconnected from the rest of the memory.) Let the read station be location 1. Show that every word in the memory other than that in location 0 is accessible in no more than $2 \log_2 n$ permutation shifts. Show that it is possible to access a sequence of words at ascending addresses $i, i + 1, i + 2, \ldots, i + r$ so that the first word is accessible in no more than $2 \log_2 n$ time units, and thereafter the next word is available in one time unit.

FIVE Group Codes

FIVE | Group Codes

In this chapter we explore another application of group theory to computer design. Specifically, we consider one method for increasing the reliability of computer systems by protecting them against component failures. The idea here is to incorporate redundancy in the encoding of data, so that we can make consistency checks on the representation of each datum. Component failures often manifest themselves by causing spurious changes to data passing through a faulty module. If the spurious changes cause the consistency checks to be violated, we can detect the failures. Some encodings are sufficiently redundant to allow correction of highly probable errors.

In order to design data representations that are suitable for error-detection and error-correction schemes, we must find representations for which the most probable errors introduce changes that can be detected by the consistency checks. Group theory has been a significant help in the solution of this problem. In fact, the most frequently applied method uses an encoding in which code words are a subgroup of a group. Data corrupted by errors due to component failures are likely to fall in cosets relative to the subgroup.

In this chapter we investigate the area of coding theory as it has been developed by Richard W. Hamming (1950) and David Slepian (1956). Hamming invented the first single-error correcting codes, and his codes still are among the more widely used codes. Slepian placed coding theory on a firm group-theoretical foundation and opened the way for the significant advances that followed. In the final section of this chapter, we discuss codes that are particularly useful for checking addition. Such codes depend strongly on the structure of the cyclic group $\langle \mathbf{Z}_n, +_n \rangle$ and on properties of group homomorphisms.

5.1 An Error Model for Computer Systems

In this section we describe briefly the ways that component failures in computer systems manifest themselves through errors in data. The failures that we consider here are those that cause incorrect computations rather than entirely preventing the operation of the computer system. Failures in arithmetic circuitry and in memory systems are typical examples of this type.

In computers, all data, instructions, and control signals are coded as binary n-tuples of varying dimension. Memories normally are constructed so that stored data are represented as binary n-tuples where n

is the same for each datum. In the remainder of this section we assume that our codes are binary n-tuples for some fixed n, and that n can be chosen to be consistent with the constraints of a computer design.

Fig. 5.1.1 outlines our model of component failures. In part (a) of the figure, n data lines form n independent networks. One faulty module (indicated by shading) can influence only a single data line. Suppose that the n data lines represent the n components of a binary n-tuple, and that the correct function of the networks is to convert each component x_i into the component x'_i. In such a situation, failure of a single module can at worst introduce the error of complementing one of the n-tuple components in the output. This type of behavior is characteristic of certain types of failures in memories and in data buses.

Part (b) of the figure shows a model of a different type of network. Here the n signals pass through a single network in which failure of a single component can affect many lines. The shaded area represents the portion of the network that is affected by a single faulty device. In this example, failure of the device can lead to complementation of three of the output components. Failures of this type are characterized by errors in a number of signals, all in close proximity to one another. Failures in arithmetic circuits, shifters, and similar types of circuits tend to approximate the behavior of this model.

The model shown in Fig. 5.1.1(a) is called the *independent error model*. The second model covers several different situations that differ in the nature of the error dependencies. We consider specifically here the error dependencies due to failures in adders. In this case we use the

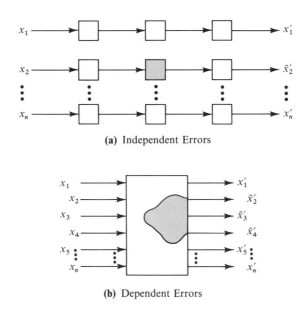

(a) Independent Errors

(b) Dependent Errors

FIG. 5.1.1 Component Failures and the Errors They Influence on Signal Lines

arithmetic error model for data errors. For other situations, particularly for communication channels, the appropriate model is called the *burst error* model. We do not treat the generalized burst error model in this chapter.

In the following sections, we use some of the elementary statistical properties of the models to help us in the design of good codes.

5.2 Parity Check Codes for Independent Errors

In this section we investigate group codes to protect against independent errors. Our development begins with a simple example, to show the properties of such codes from an intuitive point of view. Then we take a more rigorous look at the encoding and decoding procedures for such codes.

5.2.1 A Simple Error-Correcting Code

First let us explore, through examples, the use of redundancy to detect and correct errors. Suppose that we wish to guard against errors on a single binary line. Fig. 5.2.1(*a*) shows such a line connected to a typical module. One way to guard against errors is to use three lines instead of one, as shown in Fig. 5.2.1(*b*). In this model, we encode the signal by sending a copy of the signal on each of the three lines. Thus we encode the signal 0 as the 3-tuple $\langle 0, 0, 0 \rangle$, and we encode the signal 1 as $\langle 1, 1, 1 \rangle$. These two 3-tuples are the only legitimate words in this code.

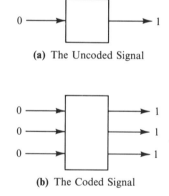

(a) The Uncoded Signal

(b) The Coded Signal

FIG. 5.2.1   Redundancy Coding for Error Protection on a Signal Line

If no errors occur along the input lines, the module receives identical copies of the signal on each of its inputs. Suppose that there is a component failure, causing an error to occur on one of the signal lines. (Using the independent error model, we assume that any single component failure influences at most one signal line.) Let us assume that the correct input signal for the module was 0, but that the failure along one line causes the code word $\langle 0, 0, 1 \rangle$ to reach the module. Because this is not a legitimate code word, module can detect that an error has occurred.

Not only can the module detect a single error in any of the three signal lines, but it also can detect any combination of double errors. The module will be fooled only if errors occur independently on all three lines. In such a case, if the actual code word is $\langle 0, 0, 0 \rangle$, then the module receives $\langle 1, 1, 1 \rangle$ and accepts this as a legitimate word. Because this code enables the module to detect the presence of any combination of one or two errors, we call this a *double-error detecting* code.

Now suppose that we wish not only to detect the errors, but also to correct errors if we can find a correction procedure that has a good statistical foundation. What should we do if the module receives the signal $\langle 1, 0, 0 \rangle$? We know that an error has occurred, but it might have been a single error in the first position (if the actual code word were $\langle 0, 0, 0 \rangle$) or a double error in the other positions (if the actual code word were $\langle 1, 1, 1 \rangle$). The well-known *maximum-likelihood estimation* method tells us that we will minimize the probability of a mistake in our decoding decision if we always assume that the most probable error occurred (see Cramér 1945). In our model we assume that a single error is more likely than two errors, so we decode $\langle 1, 0, 0 \rangle$ as if it were $\langle 0, 0, 0 \rangle$ rather than $\langle 1, 1, 1 \rangle$.

Fig. 5.2.2 provides a graphic view of this situation. The eight nodes of the cube correspond to the eight possible 3-tuples. An edge connects two nodes if and only if the corresponding triples differ by a single error. For example, $\langle 1, 0, 1 \rangle$ is connected to $\langle 1, 1, 1 \rangle$ because an error in the second position will convert either triple into the other. In this graph, the triples $\langle 1, 0, 0 \rangle$, $\langle 0, 1, 0 \rangle$ and $\langle 0, 0, 1 \rangle$ are each connected to the code word $\langle 0, 0, 0 \rangle$ by a path of length 1, but the shortest path from any of these triples to the code word $\langle 1, 1, 1 \rangle$ is a path of length 2. Therefore, it is more probable that these triples should be decoded as $\langle 0, 0, 0 \rangle$ than as $\langle 1, 1, 1 \rangle$. Similarly, the triples $\langle 0, 1, 1 \rangle$, $\langle 1, 0, 1 \rangle$, and $\langle 1, 1, 0 \rangle$ should each be decoded as $\langle 1, 1, 1 \rangle$ because they are "nearer" that code word in the graph.

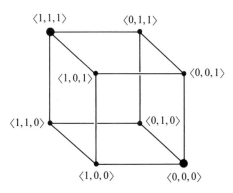

FIG. 5.2.2   Graph Showing Code Words $\langle 0, 0, 0 \rangle$ and $\langle 1, 1, 1 \rangle$ with Other 3-tuples

A decoding procedure for an error-detecting or error-correcting code is simply a rule that states the interpretation to be given for every possible received n-tuple, or vector. For any received vector, two actions are possible:

(i) we can decode the received vector as a code word, possibly not the same n-tuple as the received vector; or

(ii) we can signal that an error has been detected and then terminate the decoding process without decoding the received vector.

If the decoder takes choice **(i)** and decodes the received vector, it assumes that the most probable error occurred and always decodes the received n-tuple as the nearest legitimate code word, where "nearest" is defined in terms of path lengths as discussed earlier. A decoding decision of type **(i)** always is subject to error because even the most improbable errors can occur occasionally, and any errors other than the most probable ones will fool the decoder.

What are the probabilities of correct and incorrect decoding for a given decoding rule? We note that there are three possible outcomes of a single decoding decision:

(i) Correct decoding. The decoder correctly decodes the received n-tuple as a specific code word.

(ii) Incorrect decoding. The decoder decodes the received n-tuple as a specific code word, but picks the wrong code word.

(iii) Error detected without decoding. The decoder notes that the received n-tuple is corrupted by one or more errors, but makes no attempt to decode the n-tuple into a code word.

For any given code, we have some flexibility in choosing decoding rules. We can minimize the probability of error—the probability of outcome **(ii)**—by using a decoding rule that never attempts error correction, thus eliminating all possibility of incorrect error-correction. (Note, however, that this does not totally eliminate the possibility of incorrect decoding, because there remains the possibility of errors that convert one code word into another legitimate code word.) Table 5.2.1 illustrates a decoding rule of this type for the code graphed in Fig. 5.2.2. In this table, we assume that the correct code word is $\langle 0,0,0 \rangle$, and we indicate the decoding decision for each of the possible received 3-tuples.

For this error-detection decoding rule, the only received 3-tuples that are decoded are the code words $\langle 0,0,0 \rangle$ and $\langle 1,1,1 \rangle$. Therefore, if the correct code word is $\langle 0,0,0 \rangle$, this decoding rule makes a correct decision on the received 3-tuple $\langle 0,0,0 \rangle$ and an incorrect decision on the received 3-tuple $\langle 1,1,1 \rangle$. All other 3-tuples are not decoded. Hence, for this decoding rule, the probability of correct decoding is the probability that no errors occur, and the probability of incorrect decoding is the probability that three errors occur. The probability that the decoder will halt the computation with an error report is equal to the probability that one or two errors will occur.

TABLE 5.2.1 Outcomes of Decoding Decisions Under Two Different Decoding Rules When the Correct Code Word is $\langle 0, 0, 0 \rangle$

Received n-tuple	Outcome of Decision Under	
	Error-Detecting Rule	Error-Correcting Rule
$\langle 0,0,0 \rangle$	Correct	Correct
$\langle 0,0,1 \rangle$	Not Decoded	Correct
$\langle 0,1,0 \rangle$	Not Decoded	Correct
$\langle 1,0,0 \rangle$	Not Decoded	Correct
$\langle 0,1,1 \rangle$	Not Decoded	Incorrect
$\langle 1,0,1 \rangle$	Not Decoded	Incorrect
$\langle 1,1,0 \rangle$	Not Decoded	Incorrect
$\langle 1,1,1 \rangle$	Incorrect	Incorrect

The table also illustrates an error-correcting decoding rule under which single errors are corrected. Under this rule, the decoder decodes every received 3-tuple as the nearest code word. The probability of a correct decision is higher under this rule, because the decoder makes the correct decision if it receives any of the 3-tuples $\langle 0,0,0 \rangle$, $\langle 1,0,0 \rangle$, $\langle 0,1,0 \rangle$, or $\langle 0,0,1 \rangle$. However, the probability of an incorrect decision also is higher for this rule, because more received 3-tuples are decoded incorrectly when error correction is attempted. Thus error correction, when compared to error detection, increases both the probability of correct decision and the probability of incorrect decision, while it decreases the probability of an error report without decoding. Error detection, on the other hand, minimizes the chance of mistakes in decoding, but does so at the expense of indecision.

5.2.2 Code Generation by Parity Checks

In this section we describe the single-error correcting codes developed by Hamming (1950) and discuss the general idea of parity checks. The discussion in the preceding subsection provides some intuitive justification for the more formal approach that we take here. We begin by introducing the notion of Hamming distance.

Definition

Let x and y be binary n-tuples. The *Hamming distance* between x and y, denoted as $H(x, y)$, is the number of coordinates (components) in which they differ.

For example, the Hamming distance between $\langle 1,0,1 \rangle$ and $\langle 1,1,0 \rangle$ is 2. We notice immediately that the Hamming distance between two n-tuples is equal to the number of independent single errors needed to change one n-tuple into the other. In our discussion, we make use of the following properties of Hamming distance:

(i) $H(x, y) \geq 0$ for all x and y;
(ii) $H(x, y) = 0$ if and only if $x = y$;
(iii) $H(x, y) = H(y, x)$ for all x and y;
(iv) $H(x, y) \leq H(x, y) + H(y, z)$ for all x, y, and z.

Verification of these properties is left to the reader (Ex. 5.2.1). When we attempt to determine the error-detecting and error-correcting capabilities of a code, we find that the notion of Hamming distance plays a central role. In the next lemma we establish this fact for error detection.

Definition   The *minimum distance* of an n-coordinate code (a code whose words are n-tuples) is the minimum of the Hamming distances between all possible pairs of n-tuples in that code.

Lemma   A code C can detect all combinations of d or fewer errors if and only if its minimum distance is at least $d + 1$.

Proof: We fail to detect a combination of errors if and only if that combination changes some code word x into some other code word y. If the minimum distance is at least $d + 1$, at least $d + 1$ errors must occur in order to change a code vector x into any other code vector y, and therefore we can detect all combinations of d or fewer errors. $\square$

For the code shown in Fig. 5.2.2, the minimum distance obviously is 3. According to the preceding lemma, this code can detect any combination of two or fewer errors, a result that is in agreement with our previous observations.

The error-correction capability of a code also is a function of the minimum code distance. In order to derive the relationship, we use the following rule to decode n-tuples.

Maximum-likelihood decoding rule. To decode an n-tuple, decode it as the code word that is nearest to it in terms of Hamming distance. Ties are broken arbitrarily.

We justified this rule earlier by mentioning that a decoding rule that selects the most probable code word must assume that the most probable error has occurred (in this case, that the fewest single errors have occurred).

Lemma   A code C can correct every combination of t or fewer errors if and only if its minimum distance is at least $2t + 1$.

Proof: First we assume that the minimum code distance is $2t + 1$, and we show that the code can correct t errors. Let x be a code word, and let x' be a received n-tuple that is corrupted by not more than t errors. Then, if a decoding rule correctly decodes x' as x, we know that

x' must be nearer to the code word x than to any other code word y. From the properties of the Hamming distance, we have

$$H(x, y) \leq H(x, x') + H(x', y),$$

or
$$H(x, y) - H(x, x') \leq H(x', y).$$

But $H(x, y) \geq 2t + 1$, and $H(x, x') \leq t$, so that $H(x', y) \geq t + 1$. Therefore, every code word y is farther from x' than is x, and x' can be correctly decoded.

To prove the converse half of the theorem, assume that the minimum distance of the code is less than $2t + 1$. Then we show that the code cannot correct all combinations of t or fewer errors. In particular, because the minimum distance is $2t$ or less, we can find two code words x and y that satisfy $H(x, y) \leq 2t$. Then there exists some n-tuple x' that satisfies $H(x, x') = t$ and $H(x', y) \leq t$. To find such an x', observe the coordinate positions in which x and y differ, and then let x' differ from x in precisely t of these positions. In this case, if x' is a received n-tuple for the code word x, a decoder cannot correctly decode x' with certainty, because the code word y is at least as near to x' as x is. □

The problem of designing a good code is essentially the problem of finding a subset of n-tuples that are far from each other. There have been several approaches to solving this problem, but group-theoretic approaches and their algebraic generalizations stand out as the most powerful among them. Next we show how group theory enables us to design single-error correcting codes.

Definition   A *group code* is a code for which the n-tuple code words form a group with respect to the operation $\oplus$, where

$$x \oplus y = \langle x_1 +_2 y_1, x_2 +_2 y_2, \ldots, x_n +_2 y_n \rangle.$$

In this definition, the operation $\oplus$ is componentwise addition modulo 2. Hence $\langle 1, 1, 0 \rangle \oplus \langle 0, 1, 1 \rangle = \langle 1, 0, 1 \rangle$.

The code shown in Fig. 5.2.2 happens to be a group code with $\langle 0, 0, 0 \rangle$ as an identity. The error-correcting capability of a group code is intimately related to the number of 1's among the coordinates of each code word. We establish this relation after the following definition.

Definition   The *weight* of a code word x, denoted as $W(x)$, is the number of its coordinates that are 1's.

The 4-tuples $\langle 1, 1, 1, 1 \rangle$, $\langle 1, 1, 0, 0 \rangle$, and $\langle 1, 0, 0, 0 \rangle$ have weights of 4, 2, and 1, respectively. If we denote the n-tuple $\langle 0, 0, 0, \ldots, 0 \rangle$ simply as 0, then $W(x) = H(x, 0)$. Also, we can express the distance between two n-tuples as the weight of their sum:

$$H(x, y) = H(x \oplus y, 0) = W(x \oplus y).$$

Lemma   The minimum distance of a group code is equal to the minimum weight of its nonzero code words.

   Proof: Let x and y be code words such that $H(x, y)$ is equal to the minimum distance for the code. Then $x \oplus y$ is a code word (by the closure axiom), and $W(x \oplus y) = H(x, y)$. Hence there is a code word whose weight equals the minimum distance.

   To complete the proof, we also must show that, if z has minimum weight, then there are two code words separated by the distance $W(z)$. The 0 n-tuple is a code word, because $x \oplus x = 0$ for any x. Thus $H(z, 0) = W(z)$, and the proof is complete.   □

   Table 5.2.2 shows an example of a group code. Observe that the minimum weight of this code is 3, so that this code can correct all single errors and can detect all combinations of one or two errors. This particular code was constructed by using the matrix

$$H = \begin{bmatrix} 1 & 1 & 0 & 1 & 1 & 0 & 0 \\ 1 & 0 & 1 & 1 & 0 & 1 & 0 \\ 0 & 1 & 1 & 1 & 0 & 0 & 1 \end{bmatrix}.$$

The code is the set of all n-tuples x that satisfy the equation $x \cdot H^t = 0$, where H^t denotes the matrix transpose of H, and where the matrix multiplication is done with arithmetic modulo 2. We show next that a code defined by such matrix equations always is a group code, and that the minimum weight of the code can be computed from the H matrix. Although we do not prove it here, it also is true that every group code can be defined by such a matrix equation.

TABLE 5.2.2   A Group Code with Sixteen Binary 7-tuples

$$\langle 0, 0, 0, 0, 0, 0, 0 \rangle$$
$$\langle 0, 0, 0, 1, 1, 1, 1 \rangle$$
$$\langle 0, 0, 1, 0, 0, 1, 1 \rangle$$
$$\langle 0, 0, 1, 1, 1, 0, 0 \rangle$$
$$\langle 0, 1, 0, 0, 1, 0, 1 \rangle$$
$$\langle 0, 1, 0, 1, 0, 1, 0 \rangle$$
$$\langle 0, 1, 1, 0, 1, 1, 0 \rangle$$
$$\langle 0, 1, 1, 1, 0, 0, 1 \rangle$$
$$\langle 1, 0, 0, 0, 1, 1, 0 \rangle$$
$$\langle 1, 0, 0, 1, 0, 0, 1 \rangle$$
$$\langle 1, 0, 1, 0, 1, 0, 1 \rangle$$
$$\langle 1, 0, 1, 1, 0, 1, 0 \rangle$$
$$\langle 1, 1, 0, 0, 0, 1, 1 \rangle$$
$$\langle 1, 1, 0, 1, 1, 0, 0 \rangle$$
$$\langle 1, 1, 1, 0, 0, 0, 0 \rangle$$
$$\langle 1, 1, 1, 1, 1, 1, 1 \rangle$$

Definition Let H be an $r \times n$ binary matrix. Then the set of binary n-tuples x that satisfy $x \cdot H^t = 0$ is called the *null space* of H.

Theorem The null space of an $r \times n$ binary matrix is a group under $\oplus$, componentwise addition modulo 2.

Proof: Associativity holds for the operation $\oplus$ on n-tuples. Closure holds because, if $x \cdot H^t = 0$ and $y \cdot H^t = 0$, then it also is true that

$$(x \oplus y) \cdot H^t = (x \cdot H^t) \oplus (y \cdot H^t) = 0.$$

The 0 n-tuple is in the null space because $0 \cdot H^t = 0$, and thus there is an identity in the null space. The axiom of inverses holds because $x \oplus x = 0$ for every x, and thus every n-tuple is self-inverse. $\qquad\square$

Inspection of the code in Table 5.2.2 and of the matrix from which we generated the code reveals some other interesting properties that also hold for group codes in general. Observe that, when we compute $x \cdot H^t$ with arithmetic modulo 2, we compute three different sums of selected coordinates of a code word. The row $[1\,1\,0\,1\,1\,0\,0]$ of H requires that the sum modulo 2 of the first, second, fourth, and fifth coordinates be zero for every code word. In other words, there must be an even number of 1's in the coordinate set $\{1, 2, 4, 5\}$ of every code word. Similarly, the other two rows of the matrix require that there be even numbers of 1's in the coordinate sets $\{1, 3, 4, 6\}$ and $\{2, 3, 4, 7\}$ of each word. The specification of the code as the null space of the H matrix is equivalent to the specification of selected subsets of coordinates that must contain even numbers of 1's. Each row of H is therefore called a *parity check*, and the H matrix is called a *parity-check matrix*.

Next we face the question of determining the minimum weight of a group code from properties of the H matrix that defines the code. The next theorem answers this question.

Theorem Let $c_1, c_2, \ldots, c_d$ be d distinct columns of the H matrix. Then the r-tuple sum $c_1 \oplus c_2 \oplus \ldots \oplus c_d$ is 0 if and only if the null space of H has a code word of weight d.

Proof: Let the sum of the d columns be 0. Then let x be the n-tuple with 1's in these coordinates and 0's elsewhere. We have $x \cdot H^t = 0$, so x is in the code defined by H, and x has weight d. Conversely, if x is in the code and has weight d, the d columns of H in the positions of the 1's of x must sum to 0. $\qquad\square$

The preceding theorem shows that the minimum weight of a code word is equal to the minimum number of columns of H that sum to 0. To construct a code of minimum weight 3 or more (a code with single-error correction capability), we must have no column of H all 0's and must have no pair of columns that sum to 0. If we fail to satisfy the former condition, the code generated by H has code words of weight 1; if we fail to satisfy the latter condition, the code has code words of weight 2. Note that $(x \oplus y = 0) \Leftrightarrow (x = y)$, so that two columns sum to 0 if and only if they are identical. Hamming (1950) made the preceding observations and thereby established the following simple condition for the construction of single-error correcting codes.

Theorem H is a parity-check matrix for a code of minimum weight at least 3 if and only if no column of H is all 0's and no two columns are identical.

Proof: The theorem follows from previous observations. □

The H matrix for the code listed in Table 5.2.2 clearly satisfies the conditions of the theorem, and therefore the code is a single-error correcting code.

Our final observation about parity-check matrices and the codes they generate has to do with the size of the code that we obtain in a null space.

Theorem Let H be an $r \times n$ binary parity-check matrix of the form

$$[P|I_r],$$

where I_r is an $r \times r$ identity matrix, and P is an arbitrary $r \times (n - r)$ matrix. Then the code defined by H has 2^{n-r} code words.

Proof: We can pick the first $n - r$ coordinates of a code word arbitrarily, and from these and the parity checks we can determine uniquely all of the remaining r coordinates by a procedure that we describe in the following paragraphs. Thus there is at least one code word for each of the 2^{n-r} ways of picking the first coordinates. There are no other code words, because any such code word x must be identical to some code word y in the first $n - r$ positions and therefore must differ from y in at least one of the last r positions. But we choose the last r coordinates of y in the only possible way that satisfies the parity checks. Therefore, x must violate one parity check for each coordinate in which it differs from y, and thus x cannot be a code word unless it is equivalent to y. □

The preceding theorem not only tells us how many code words we obtain from a given matrix, but it also provides the basis for an encoding procedure. If H has the form required by the preceding theorem, we call it a *canonical parity-check matrix*. Let $k = n - r$, where n and r are the dimensions of H. Then the following procedure maps the set of all possible binary k-tuples into the code of n-tuples defined as the null space of H. The first k coordinates of the code word are set equal to the coordinates of the k-tuple. The remaining r coordinates are computed to satisfy the parity checks and thus ensure that the resulting n-tuple is in the null space of H.

Encoding Procedure. Given a k-tuple $x = \langle x_1, x_2, \ldots, x_k \rangle$, we compute the coordinates of the corresponding n-tuple code word $y = \langle y_1, y_2, \ldots, y_n \rangle$ by

(i) setting y_i equal to x_i for $1 \le i \le k$; and

(ii) computing y_{k+i} for $1 \le i \le r$ as the modulo-2 sum

$$y_{k+i} = \sum_{j=1}^{k} x_j [H]_{i,j}.$$

As an example of this encoding procedure, consider the matrix

$$H = \begin{bmatrix} 1 & 1 & 0 & 1 & 1 & 0 & 0 \\ 1 & 0 & 1 & 1 & 0 & 1 & 0 \\ 0 & 1 & 1 & 1 & 0 & 0 & 1 \end{bmatrix}$$

that we discussed previously. Using the procedure, we see that the code word for the 4-tuple $\langle 1, 0, 1, 0 \rangle$ will be $\langle 1, 0, 1, 0, y_5, y_6, y_7 \rangle$, where

$$y_5 = x_1 +_2 x_2 +_2 x_4 = 1 +_2 0 +_2 0 = 1,$$
$$y_6 = x_1 +_2 x_3 +_2 x_4 = 1 +_2 1 +_2 0 = 0, \text{ and}$$
$$y_7 = x_2 +_2 x_3 +_2 x_4 = 0 +_2 1 +_2 0 = 1.$$

Thus the code word is $\langle 1, 0, 1, 0, 1, 0, 1 \rangle$, which we find as the eleventh code word in Table 5.2.2. You can verify that the encoding procedure constructs that entire table of code words from the H matrix and the set of all possible binary 4-tuples.

From the encoding procedure, it is clear that the first k coordinates of a code word are identical to the k-tuple that it encodes. Hence we call these positions the *information positions* of the code. The final r positions are called the *check positions*. The encoding procedure guarantees that each code word is in the null space of the parity-check matrix, because the components of a code word in the check positions are selected to satisfy each of the parity checks defined by H. In error-free conditions, the decoding of such codes is trivial because the first k coordinates represent the encoded k-tuple.

Suppose that we are constrained to use r check positions in a code, and that we wish to protect as many information positions as possible against single errors. What is the maximum number of information positions that we can have? Recall that the canonical parity-check matrix is made up of the two submatrices P and I_r. With r check positions, H has r rows, because I_r must be an $r \times r$ matrix. Because the number of distinct binary r-tuples is 2^r, there cannot be more than 2^r columns in H if no two columns are to be the same. This satisfies one of the conditions for H to be the parity-check matrix of a single-error correcting code; the other condition is that H have no column with all 0's. Hence H cannot have more than $2^r - 1$ columns. Because r of these columns are check positions, there cannot be more than $2^r - 1 - r$ information positions. This establishes the bound that $k \le 2^r - 1 - r$ for a single-error correcting code with r check positions and k information positions. The bound can be reached if the submatrix P includes among its columns all binary r-tuples with two or more 1's.

Definition　Let H be a canonical parity-check matrix of dimension $r \times (2^r - 1)$, and let its submatrix P have as columns all r-tuples of weight 2 or more. Then H generates a code called the *Hamming single-error correcting code of length n*, where $n = 2^r - 1$.

From the definition we see that H generates a code of minimum weight at least 3, so the code is single-error correcting. In fact, the minimum weight is exactly 3, because we can find many sets of three columns of H that sum to zero.

5.2.3 Decoding Group Codes

For codes of any sort, the problem of decoding reduces to the problem of finding the code word that is nearest to the n-tuple being decoded. Although this problem has not been solved in general, it is tractable for some group codes, particularly for Hamming's single-error correcting codes. In this section we apply group theory to the decoding problem.

For this model, we assume that the true code word is the n-tuple x, but that we observe the n-tuple x', which is x after it has been corrupted by errors. Let ε, the *error n-tuple*, be the n-tuple that satisfies

$$x' = x \oplus \varepsilon.$$

Then ε has 1's in coordinates where x' has errors. Our problem is to determine ε, for then we can compute $x = x' \oplus \varepsilon$. We show that the problem of finding ε reduces to the problem of finding the coset to which x' belongs.

Let C be the group code, with individual words c_i. For each c_i, let us find the error vector ε_i that satisfies $x' = c_i \oplus \varepsilon_i$. Our general decoding principle indicates that the ε_i of smallest weight is associated with the c_i into which we should decode x'. Because $\varepsilon_i = c_i \oplus x'$, the error vectors ε_i form the set $E = C \oplus x'$. Because C is a subgroup of the group of binary n-tuples under the operation $\oplus$, we know that $C \oplus x'$ is a coset of the group of n-tuples with respect to C. Hence we wish to find ε, the n-tuple of least weight in the coset that contains x'. This n-tuple is called the *coset leader* for that coset.

In principle, we can enumerate all of the cosets of the group code and then identify the coset leader for each coset by examining weights. Then the following algorithm provides a valid decoding procedure.

Decoding Procedure for Group Codes:
　(i)　determine the coset to which the observed n-tuple x' belongs;
　(ii)　find the coset leader ε for that coset; and
　(iii)　decode x' as the n-tuple $x = x' \oplus \varepsilon$.

In order to reduce the probability of incorrect error-correcting, we may choose not to use the full error-correcting capability of the code.

We may choose to decode x' only if the most probable error involves t or fewer single errors (for any t such that the minimum distance of the code is at least $2t + 1$). In this case, we modify the third step of the decoding procedure as follows:

> **(iii)** if ε has weight t or less, then decode x' as the n-tuple $x = x' \oplus \varepsilon$; otherwise, signal that an error has been detected and terminate.

Some of these steps are difficult to perform for large group codes, particularly for multiple-error correcting codes. Fortunately, the first step is trivial for any group code, as we discover after the next definition.

Definition For any observed n-tuple x', the *syndrome* of x' is the r-tuple $x' \cdot H'$, where r is the number of parity-check digits.

Theorem Two n-tuples are in the same coset if and only if they have the same syndrome.

Proof: We show that the syndrome of $c \oplus x'$ is the same as that of x' for all code words c. This follows because

$$(c \oplus x') \cdot H^t = (c \cdot H^t) \oplus (x' \cdot H^t) = x' \cdot H^t$$

for all code words c. To complete the proof, we must show that x and x' are in the same coset if they have the same syndrome. If

$$x \cdot H^t = x' \cdot H^t,$$

then $(x \oplus x') \cdot H^t = 0$, and $x \oplus x'$ is a code word. Then $x = c \oplus x'$ for some code word c, and x and x' are in the same coset. $\square$

Given the syndrome, the next problem in decoding is to find the coset leader. When the number of check digits, r, is small, we can do this by consulting a table of coset leaders that is prepared by exhaustive enumeration. Because the number of table entries grows as 2^r, such tables are impractical for large r. Single-error correcting codes have been implemented in computer systems for k as large as 64 in which $r = 8$ and $n = r + k = 72$. The table size required here is $2^8 = 256$, which is not unreasonably large. In this case, each table entry need hold only an integer in the range $1 \leq i \leq 72$ to indicate the error position, and this integer can be encoded in seven bits.

Special classes of codes have been developed in recent years through the sophisticated use of modern algebra. However, with some notable exceptions, most computer systems use Hamming codes or codes derived from Hamming codes. Most memory systems use a simple parity-check scheme for error detection without error correction, although some use a more complex Hamming code with $n = 72$, $k = 64$, and $r = 8$. Auxiliary storage devices (particularly those used for archival storage) have used more complex codes (see Oldham et al. 1968). Srinivasan (1971) describes some clever coding techniques for computer memories. Peterson and Weldon (1972) have written a thorough and excellent text on the subject of coding theory, and

Berlekamp (1968) has written a particularly good advanced monograph on the subject.

In this section we discussed certain notions as they apply to coding theory, but these notions have also been put to other uses. For example, the design of core-memory access switches is equivalent in some sense to the design of good error-correcting codes. This equivalence was noted by Chien (1959) and is discussed at some length by Minnick and Haynes (1962).

Exercises

5.2.1 Prove that the Hamming distance has the following properties:
   (a) $H(x, y) \geq 0$ for all x and y;
   (b) $H(x, y) = 0$ if and only if $x = y$;
   (c) $H(x, y) = H(y, x)$ for all x and y;
   (d) $H(x, z) \leq H(x, y) + H(y, z)$ for all x, y, and z.

5.2.2 Prove that a code can correct all combinations of t or fewer errors and can detect all combinations of $t + 1$ to d errors (where $t \leq d$) if and only if it has minimum distance at least $t + d + 1$.

5.2.3 Let x and y be binary n-tuples, and let H be a binary matrix with n columns. Prove that $x \cdot H^t \oplus y \cdot H^t = (x \oplus y) \cdot H^t$, where the matrix multiplication is done with modulo-2 arithmetic.

5.2.4 Let H be an $r \times (2^r - 1)$ matrix for which the ith column is the binary representation of the integer i.
   (a) Prove that the null space of H is a single-error correcting code.
   (b) Prove that, if a received n-tuple has a nonzero syndrome, then its syndrome is the binary representation of the single coordinate in which it differs from a code word.

5.2.5 Let H be an $r \times (2^r - 1)$ parity-check matrix for a Hamming code. Let H' be created from H by appending a row of all 1's. Show that the null space of H' is a group code with minimum distance 4.

5.2.6 Let the null space of an $r \times n$ canonical parity-check matrix be a group code that satisfies the following conditions: **(i)** for each coordinate there is some code word with a 1 in that position; and **(ii)** for each pair of coordinates there is some code word that has different values in those two positions.
   (a) Prove that the set of code words with a 0 in the ith coordinate is a subgroup of the code.
   (b) Prove that the average weight of a code word is $n/2$. HINT: The cosets of the subgroup of part **(a)** are of equal size.
   (c) Prove that the average square of the weight of a code word is $n(n + 1)/4$. HINT: Find a formula for the average square weight that involves a double summation, and evaluate the summation by reversing the order of summation.

5.3 Arithmetic Codes

In this section we investigate the construction of codes that protect data from errors introduced by faults in an adder. Following the approach in the previous section, we assume that code words are a subset of the set of binary n-tuples.

Our basic problem is to find a code that is preserved by addition, so that addition of the code words for two integers x and y yields the code word for their sum. In algebraic notation, if $c(x)$ is the code word for x, then $c(x) + c(y) = c(x + y)$. Actually, we need not insist that addition of code words be ordinary addition. Any binary operation $\odot$ on code words will do if $c(x) \odot c(y) = c(x + y)$.

Because digital computers inherently deal with finite sets, we assume that x and y are in $\mathbf{Z}_n$ for some value of n, and that the operation $\odot$ satisfies $c(x) \odot c(y) = c(x +_n y)$. No matter how we phrase the problem, the constraints require that addition of code words be a homomorphic image of addition of integers. Because the mapping between code words and the integers they represent is one-to-one, addition of code words must be isomorphic to addition of integers.

We can satisfy the constraints in one way if we represent each integer x in $\mathbf{Z}_n$ by the integer $A \cdot x$, where A is a fixed constant greater than 1. If code words are added modulo $A \cdot n$, then we have

$$c(x) +_{A \cdot n} c(y) = A \cdot x +_{A \cdot n} A \cdot y = A \cdot (x +_n y) = c(x +_n y),$$

which is precisely the relationship required for our code. Therefore this code allows us the convenience of performing addition modulo n by doing addition modulo $A \cdot n$ on code words (rather than decoding, adding, and encoding again). We can construct our arithmetic networks for coded integers as standard arithmetic networks.

We next determine a way of choosing A so that the code has good error-protection qualities. To do so, we introduce the notion of arithmetic distance, which is analogous in this context to Hamming distance in the context of independent errors. We begin with a definition of weight in this context.

Definition

The *arithmetic weight* of an integer x is the minimum number of nonzero coefficients in any expansion of x of the form

$$x = a_m \cdot 2^m + a_{m-1} \cdot 2^{m-1} + \cdots + a_1 \cdot 2 + a_0,$$

where the coefficients a_i may assume the values $+1$, -1, or 0.

The expansion of x differs from the usual binary expansion in that we permit negative as well as positive coefficients. Thus the integer 29 has the unique binary expansion $1 \cdot 16 + 1 \cdot 8 + 1 \cdot 4 + 1 \cdot 1$, but it also has the expansions $1 \cdot 32 - 1 \cdot 2 - 1 \cdot 1$ and $1 \cdot 32 - 1 \cdot 4 + 1 \cdot 1$ when we permit both positive and negative coefficients. Thus the weight of 29 in the Hamming-distance sense is 4, whereas its arithmetic weight is 3.

We define weight in this manner because the most probable errors in adders are errors that have small arithmetic weights, just as the most probable errors in the independent error model have small Hamming weights.

Definition   The arithmetic distance between two integers x and y, denoted as $D(x, y)$, is the arithmetic weight of $x - y$.

The four properties given for Hamming distance hold for arithmetic distance as well. That is,

   (i) $D(x, y) \geq 0$ for all x and y;
   (ii) $D(x, y) = 0$ if and only if $x = y$;
   (iii) $D(x, y) = D(y, x)$ for all x and y;
   (iv) $D(x, z) \leq D(x, y) + D(y, z)$ for all x, y, and z.

Arithmetic distance is the appropriate concept for checking addition. To see this, consider the addition of some binary n-tuples.

$$011111_2 + 000001_2 = 100000_2$$

is the binary form of $31_{10} + 1_{10} = 32_{10}$. Suppose that the binary adder fails to generate a carry from the least significant position. In this case, we obtain the erroneous result

$$011111_2 + 000001_2 = 011110_2.$$

Although only a single error has occurred, the correct answer and the erroneous answer differ in five out of six positions. Similarly, we can find cases where five errors produce an erroneous sum that differs in only two positions from the correct answer. Suppose that we add

$$000000_2 + 000001_2 = 000001_2,$$

and that each of the first five stages fail so that 000000_2 is treated as 011111_2. The faulty adder produces the sum

$$011111_2 + 000001_2 = 100000_2.$$

Thus we see that very improbable errors can produce results that are nearer to the correct result in the Hamming sense than are some results produced by very probable errors. Clearly, the Hamming distance is an inappropriate measure of error probability in the context of addition.

On the other hand, the arithmetic distance is appropriate in this context, because probable errors produce results that are near to the correct results in the arithmetic sense. The examples just given show. why this is true. Suppose that a failure causes a carry to change incorrectly from 1 to 0 or from 0 to 1 at the ith stage in the adder. Then the sum differs from the true sum by $\pm 2^i$, with the sign depending on the direction of the change. Similarly, if a failure causes the adder to misread one of the components in the binary representation of an input, the error in the sum is $\pm 2^i$. Thus a single error in an adder is likely to change the arithmetic weight of the sum by 1.

Under reasonable sets of assumptions, the errors introduced by two component failures are of the form $\pm 2^i \pm 2^j$, changing the arithmetic weight by 2. More generally, m component failures produce an error of arithmetic weight m. Therefore we can immediately apply the results derived in the preceding section for codes in the independent error model. In the remainder of this section, we use the terms weight and distance in the arithmetic sense, not in the Hamming sense.

Theorem An arithmetic code can detect any combination of d or fewer errors if and only if the minimum distance of the code is at least $d + 1$. It can correct any combination of t or fewer errors if and only if the minimum distance is at least $2t + 1$.

Proof: The proofs for these statements are similar to the corresponding proofs given earlier for independent-error codes. □

Returning to the $A \cdot n$ codes, observe that the multiples of A in the group $\langle \mathbf{Z}_{A \cdot n}, +_{A \cdot n} \rangle$ form a subgroup. Again this is similar to the case for group codes. We can use the group properties of $A \cdot n$ codes to establish some relations between the minimum distance of an $A \cdot n$ code and the values of A and n. The next theorem also is the counterpart of a theorem already developed for independent-error codes.

Theorem An $A \cdot n$ code has minimum distance d if and only if the minimum weight of a nonzero code word is d.

Proof: Suppose that a code has minimum distance d, and suppose that $A \cdot x$ and $A \cdot y$ are two code words distance d apart. Then the code word $[A \cdot x +_{A \cdot n} (-A \cdot y)] = A \cdot (x - y)$ has weight d. Conversely, if $A \cdot x$ is a nonzero code word of weight d, then it is distance d from the code word $A \cdot 0 = 0$. □

Now we can determine how to design codes that can detect single errors.

Theorem An $A \cdot n$ code has distance 2 or more if and only if A is not a power of 2.

Proof: If A is a power of 2, then $A \cdot 1 = A$ is in the code, and the code therefore has a code word of weight 1. Conversely, if the code has a code word $A \cdot x$ of weight 1, then $A \cdot x = 2^i$ for some i, and A must be a power of 2. □

From the preceding theorem, we see that it is very simple to construct $A \cdot n$ codes with minimum distance 2; we merely pick A so that it is not a power of 2. If we want a single-error correcting code, we must have a code with minimum distance at least 3. The next theorem shows how to construct such codes.

Theorem An $A \cdot n$ code has minimum distance 3 if and only if, for all positive powers of 2 in the range $0 \leq 2^i < A \cdot n$, the quantities $2^i \bmod A$ and $(-2^i) \bmod A$ are distinct and nonzero, with the possible exception of 2^i and 2^j that satisfy $2^i \equiv -2^j \bmod A$ and $|2^i + 2^j| \geq A \cdot n$.

Proof: We first examine codes with minimum distance 1 or 2 and verify that the conditions given in the theorem are not satisfied. If an $A \cdot n$ code has minimum distance 1, there is a code word $A \cdot x$ such that x is in $\mathbf{Z}_n$ and $A \cdot x = 2^i$. Then $2^i \bmod A = 0$, and the conditions are not satisfied. If an $A \cdot n$ code has minimum distance 2, there exists a code word $A \cdot x$ such that x is in $\mathbf{Z}_n$, $A \cdot x = 2^i \pm 2^j$, and $A \cdot x < A \cdot n$. Hence, $2^i \equiv \pm 2^j \bmod A$, for the appropriate choice of sign, and the conditions again are not met.

Conversely, if the conditions are not met for a particular $A \cdot n$ code, then the code has minimum distance less than 3. If $2^i \equiv 0 \bmod A$ for some i, then 2^i is a code word of weight 1, and the minimum distance is 1. If $|2^i \pm 2^j| < A \cdot n$ and $2^i \equiv \pm 2^j \bmod A$, then $2^i \pm 2^j$ is a code word of weight 2, and the code has minimum distance 2 or less.

Note that the integer $2^i + 2^j$ is not a code word if $|2^i + 2^j| > A \cdot n$. Hence we do not require that the residues of 2^i and -2^j modulo A be distinct if $2^i + 2^j > A \cdot n$; the code can have minimum distance 3 even if they are not distinct. $\qquad\square$

The preceding theorem provides a tool that we can use in searching efficiently for good single-error correcting arithmetic codes. Several researchers have used number-theoretic approaches to aid the search for good $A \cdot n$ codes; details are given in their papers (Peterson 1961; Massey 1964; Rao and Garcia 1971). Table 5.3.1 gives some values of A and n for single-error correcting codes. The values given in the table have been reported by Brown (1960) and Peterson (1961), who obtained these values from properties of integers of the forms $2^i - 1$ and $2^i + 1$.

TABLE 5.3.1 Single-Error Correcting Arithmetic Codes

A	n	$\lceil \log_2 n \rceil$	$A \cdot n$
11	3	2	$2^5 + 1$
13	5	3	$2^6 + 1$
19	27	5	$2^9 + 1$
23	89	7	$2^{11} - 1$
29	565	10	$2^{14} + 1$
37	7,085	13	$2^{18} + 1$
47	178,481	18	$2^{23} - 1$
53	1,266,205	21	$2^{26} + 1$
59	9,099,507	24	$2^{29} + 1$
61	17,602,325	25	$2^{30} + 1$
67	128,107,979	27	$2^{33} + 1$
71	483,939,977	29	$2^{35} - 1$
79	6,958,934,353	33	$2^{39} - 1$
83	26,494,256,091	35	$2^{41} + 1$

The value of n given with each A is the largest value for which the $A \cdot n$ code has minimum distance 3 or more.

Research in the area of $A \cdot n$ codes was stimulated largely by the work of Diamond (1955) and Brown (1960). Massey (1964) and Massey and Garcia (1972) give particularly good discussions of the area in tutorial surveys. Chien and Hong (1972) show codes for correcting errors in multiplier networks.

Exercises

5.3.1 Prove that every $A \cdot n$ code with minimum distance 2 has a code word of the form $2^i + 1$ or $2^i - 1$.   HINT: $2^k + 2^j = 2^j \cdot (2^{k-j} + 1)$.

5.3.2 (Brown 1960). Let A be an integer that is not a power of 2, let r be the least integer for which A divides either $2^r + 1$ or $2^r - 1$, and let $n = 2^r + 1$ or $n = 2^r - 1$, whichever is a multiple of A. Show that every $A \cdot m$ code for $m \leq n$ has minimum distance at least 3.

5.3.3 Let n be an integer such that $n \leq 2^m$. Recall that the arithmetic weight of n is defined in terms of the fewest number of distinct powers of 2 that can be added or subtracted together to yield n. Show that the arithmetic weight of n need never involve powers of the form 2^r for $r > m$.

5.3.4 Prove that an arithmetic code can correct any combination of t or fewer errors if and only if the minimum arithmetic distance between two code words is at least $2t + 1$.

SIX Semigroups

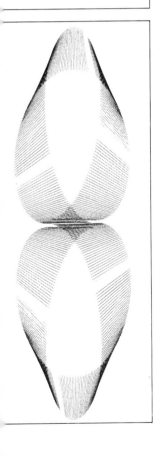

The theory of groups contains a wealth of results that follow from the group axioms. We recall that an essential characteristic of groups is the unique solvability of equations within the group. This property characterizes addition modulo n, which is one of the reasons the theory of groups can be applied to the construction of adders in computers. A second important property of groups is the relation between groups and sets of permutations. This property also proved important in our application of Pólya theory to computer design.

In this chapter we investigate semigroups, algebraic structures that satisfy only two or three of the four group axioms. We cannot guarantee the unique solvability of equations in such structures, nor can we associate each semigroup with a set of permutations as we did with each group. Rather, we find that we can associate each semigroup with a set of transformations. The crucial difference is that a permutation is a bijection of a set onto itself, whereas a transformation is any mapping of a set into itself. The theory of semigroups tends to have fewer major results than does group theory, and the results tend to be more general. Nevertheless, semigroup theory includes several important notions, many of which cannot be described by group theory. As we show in this chapter, these notions have applications to various aspects of computers.

In particular, a sequential machine is associated intimately with a set of transformations of a set of states. We find that a sequential machine computes operations in a semigroup. Thus some of the concepts we discuss in this chapter have applications in Chapter Seven, where we discuss sequential machines.

Semigroups also have important applications relating to computer arithmetic. We have seen that the group axioms do not hold for all computer arithmetic systems. On most computers, integer multiplication is a semigroup operation, and the theory of semigroups does have application to the implementation of multiplication for computers.

This chapter is a brief survey of some of the fundamental notions of semigroups, providing a context for applications considered in following chapters. Although the later applications explicitly use few of the theorems and lemmas derived in this chapter, the central ideas of the proofs given here are essential to the later material. Of particular importance in this chapter is the material on semigroup homomorphism, because this material is applied to the sequential-machine decomposition theory in Chapter Seven.

.1 Semigroups and Monoids The material in this section is developed from the semigroup axioms. We examine several properties of semigroups that are analogous to particular properties of groups. First we determine the various kinds of distinguished elements that may exist in a semigroup. These elements are identities, idempotents, and zeros as discussed in Chapter One. Then we examine the representations of a semigroup, and we discover that every semigroup has a transformation representation, just as every group has a permutation representation. Next we observe that semigroups can be defined by generator relations much as groups can be. Finally, we investigate some infinite semigroups that have generators but no generator relations.

Definition A *semigroup* is an algebraic structure $\langle S, \cdot \rangle$ for which the composition $\cdot$ is closed and associative. If $\cdot$ also is commutative, we say that $\langle S, \cdot \rangle$ is a *commutative semigroup*.

Definition A *monoid* is a semigroup with an identity element.

Among the semigroups mentioned in earlier chapters are $\langle \mathbf{N}, \cdot \rangle$, $\langle \mathscr{P}(X), \cup \rangle$, and $\langle \mathbf{N} - \{0\}, + \rangle$. The first two semigroups are monoids, because they have identities 1 and $\varnothing$, respectively. The third semigroup has no identity, so it is not a monoid. However, $\langle \mathbf{N}, + \rangle$ is a monoid.

We discuss some essential properties of semigroups in Section 1.7. For example, we prove there that, if an element of an associative structure is invertible, then it has a unique inverse. In this section we explore several other properties of elements of semigroups and relate them to the special situation when the structure is a group.

Definition An element e of a structure $\langle S, \cdot \rangle$ is called a *left identity* if $e \cdot s = s$ for all s in S. Similarly, the element e is called a *right identity* if $s \cdot e = s$ for all s in S. An element that is both a left and a right identity is a *two-sided identity*, or simply an *identity*.

A semigroup may have several left identities or it may have several right identities, but it cannot have both distinct left and right identities.

Theorem If a semigroup has a left identity e_1 and a right identity e_2, then they are the same element, and this element is the unique identity of the semigroup.

Proof: By definition, $e_1 \cdot e_2 = e_2$ because e_1 is a left identity, and $e_1 \cdot e_2 = e_1$ because e_2 is a right identity. Therefore $e_1 = e_2$. This element then is a two-sided identity. From results of Section 1.7, it must be the unique identity in S. $\quad\square$

Several familiar algebraic structures have one-sided identities. Consider, for example, $\langle \mathbf{Z} - \{0\}, / \rangle$ and note that 1 is a right identity but not a left identity for this structure. Note also that this structure is not associative, and therefore is not a semigroup.

Because a monoid has a two-sided identity, it can have no other one-sided or two-sided identity. Obviously, this result holds for groups as well, because every group is a monoid.

In Chapter One we define the element called a zero; here we introduce the one-sided zero.

Definition An element z of a semigroup S is said to be a *left zero* if $z \cdot s = z$ for all s in S. Similarly, a *right zero* is an element for which $s \cdot z = z$ for all s in S. An element that is both a left and a right zero is a two-sided zero, or simply a zero.

Semigroups may contain zeros, but no zero can exist in a group of order 2 or more. The existence of a zero in any nontrivial algebraic structure guarantees that the axiom of inverses cannot be valid in that structure. This follows because, if z is a zero, then $x \cdot z = y \cdot z = z$ for all x and y, so that z cannot have an inverse.

Just as with one-sided identities, a semigroup may have several left zeros or several right zeros, but it can never have both a distinct left and a right zero.

Theorem If a semigroup S has both a left zero and a right zero, then the left zero equals the right zero and is the unique zero in the semigroup.

Proof: Let z_1 be a left zero and z_2 be a right zero. Then $z_1 = z_1 \cdot z_2$ because z_1 is a left zero, and $z_1 \cdot z_2 = z_2$ because z_2 is a right zero. Therefore $z_1 = z_2$. By similar reasoning, if z_1 and z_2 are two-sided zeros, then they are equal. $\square$

Fig. 6.1.1 is the composition table for a three-element semigroup that contains only one-sided zeros; the semigroup in this example consists of three right zeros. The table obviously is closed, so we need only show that it is associative in order to verify that it is a semigroup table. In this case, associativity is rather trivial to prove (Ex. 6.1.1).

Inspection of the table reveals that an element is a right zero if and only if the column labeled by that element has all entries equal to that element. Similarly, all of the entries in a row labeled by a left zero are equal to that label. These observations provide another proof of the preceding theorem, for if a row and a column each have constant entries,

$\cdot$	x	y	z
x	x	y	z
y	x	y	z
z	x	y	z

FIG. 6.1.1 Composition Table for a 3-Element Semigroup with Three Right Zeros

then they must have the same entries because they have a common entry at their intersection.

Among the familiar semigroups with zeros are $\langle \mathbf{Z}, \cdot \rangle$, $\langle \mathscr{P}(X), \cup \rangle$, and $\langle \mathscr{P}(X), \cap \rangle$ (see Ex. 6.1.2).

One of the key facts about groups is that every group can be represented as a set of permutations. In many cases, this representation is very compact. Semigroups also can be represented as sets of mappings, but in most cases the mapping representation of a semigroup is not a permutation. Recall that a permutation representation is said to be of degree n if the representation involves n distinct letters. We achieve a very compact representation when the degree is much less than the order of the group, as is the case for the degree-n representation of the order-$n!$ group S_n. Similarly, we can represent a semigroup by a set of functions, and this representation in some cases is very compact.

We generally speak of a map from a set into itself as a *transformation* of the set. In the special case that the transformation is a bijection, the transformation is a permutation of the set. A semigroup always is representable as a set of transformations in the same way that a group always is representable as a set of permutations.

Theorem For every semigroup S there is some set X, such that S can be represented as a set of transformations of X.

Proof: We construct a transformation semigroup that is isomorphic to S. With each s in S we associate the transformation that corresponds to the row labeled s in the semigroup composition table. That is, $F_s(x) = s \cdot x$. Then $F_s(F_t(x)) = s \cdot t \cdot x = F_{s \cdot t}(x)$ for all s and t in S. Hence the map $f: S \to \{F_s\}$, such that $f(s) = F_s$, is a homomorphism of S. This homomorphism is an isomorphism if and only if all rows of the composition table are distinct, for only in this case is the homomorphism a bijection. Hence, if S contains an identity, the rows are distinct, and f is an isomorphism. If S does not contain an identity, then let e be an identity not in S and construct the table for $S \cup \{e\}$ from the table for S. In this table, all rows are distinct, and hence all transformations in the set $\{F_s | s \in S\}$ are distinct. The function f is an isomorphism on this set of transformations. □

If S has order n, then the representations for S constructed in the previous proof have degree n if S has an identity or degree $n + 1$ if S has no identity.

Fig. 6.1.2(a) shows the composition table of a 4-element semigroup, $\langle \{s, t, u, v\}, \cdot \rangle$. We obtain a degree-4 representation of this semigroup by transposing the table. Fig. 6.1.2(b) shows the image of S under F_x for each x in S, thus defining four maps, each corresponding to one of the rows in the semigroup table. Fig. 6.1.2(c) shows how the four transformations act under function composition. For example,

$$F_s \circ F_t = F_s$$

in this table, because $F_s \circ F_t(x) = F_s(F_t(x)) = F_s(x)$ for all x in S. This transformation representation clearly is isomorphic to $\langle S, \cdot \rangle$. For example, $F_s \circ F_t$ (under left composition) is the map F_s, corresponding to $s \cdot t = s$. This representation sometimes is called the *left regular transformation representation* of the semigroup. The *right regular transformation representation* is the corresponding representation obtained from the right-multiplication maps (the columns of the composition table) using right function composition. We make use of the latter representation when we discuss automata in Chapter Seven.

$\cdot$	s	t	u	v
s	s	s	s	s
t	s	t	u	v
u	v	u	t	s
v	v	v	v	v

(a) Composition Table for Semigroup S

x	$F_s(x)$	$F_t(x)$	$F_u(x)$	$F_v(x)$
s	s	s	v	v
t	s	t	u	v
u	s	u	t	v
v	s	v	s	v

(b) Four Maps Defined from the Set S into S

$\circ$	F_s	F_t	F_u	F_v
F_s	F_s	F_s	F_s	F_s
F_t	F_s	F_t	F_u	F_v
F_u	F_v	F_u	F_t	F_s
F_v	F_v	F_v	F_v	F_v

(c) Composition Table for the Four Maps

FIG. 6.1.2   A Semigroup and Its Left Regular Transformation Representation

The regular representation of a semigroup is no more compact than the semigroup composition table, but you will recall that the regular permutation representation of a group similarly offered no gain in compactness. Fig. 6.1.3 shows a more compact transformation representation of the semigroup we have been discussing. In this degree-2 representation, the functions are transformations of the set $X = \{0, 1\}$. You can verify easily that the composition table for these functions is identical to the table of Fig. 6.1.2(c).

A semigroup, like a group, is specified by a generator set and a set of generator relations. For the semigroup of Fig. 6.1.2, $\{u, v\}$ is a generator set because $t = u^2$ and $s = u \cdot v$. Many sets of generator

x	$F_s(x)$	$F_t(x)$	$F_u(x)$	$F_v(x)$
0	0	0	1	1
1	0	1	0	1

FIG. 6.1.3   A Degree-2 Transformation Representation of the Semigroup in Fig. 6.1.2

relations exist for this semigroup. One such set includes the following relations (where e is a semigroup identity):

$$u^2 = t = e;$$

and

$$v \cdot u = v^2 = v.$$

With these relations we can show that every finite product of u's and v's reduces to an element of the set $\{u^2 = t, u, v, u \cdot v = s\}$.

We can construct semigroup graphs from generator relations in much the same way that we constructed group graphs for groups. In a semigroup graph there is one node for each semigroup element and one edge from each node for each generator. As in a group graph, there is an edge labeled g from node s_1 to node s_2 if $s_1 \cdot g = s_2$. Fig. 6.1.4 shows the graph for the semigroup of Fig. 6.1.2. This graph clearly is not a group graph because, for example, there are two v edges entering node s.

There is a family of infinite semigroups that have special importance to us—the family of free semigroups.

Definition   Let X be an arbitrary finite set. The *free semigroup* generated by X, denoted as X^+, is the set of all nonempty finite sequences of elements in X, under the operation of concatenation.

If x and y are elements in X, then the semigroup composition $x \cdot y$ in X^+ is defined to be xy, the concatenation of x and y. Similarly, if xyz and yxx are two sequences in X^+, then $xyz \cdot yxx$ is their concatenation, $xyzyxx$. Concatenation clearly is associative, and it is closed because X^+ contains all nonempty sequences. Hence this structure is a semigroup. In fact, this is a semigroup generated by X and for which

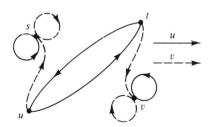

FIG. 6.1.4   Graph of the Semigroup in Fig. 6.1.2

the defining set of relations is empty. This statement completely characterizes the free semigroup. The free semigroup generated by a set with one element is isomorphic to the semigroup generated by $\{1\}$ with the operation $+$. The latter is the semigroup $\langle \mathbf{N} - \{0\}, + \rangle$. It is easy to show that, for each n, all free semigroups with n generators are isomorphic. Note also that X^+ has an infinite number of elements, although each sequence in X^+ is finite.

We encounter the free semigroup in our discussion of sequential machines in Chapter Seven. In that discussion, X is the set of letters in the input alphabet of a machine, and X^+ is the set of all nonempty finite sequences of input letters. Compilers for programming languages can be viewed as programs that accept sequences in X^+ as their inputs, and much of the formal theory behind compilers has developed with this point of view (see Hopcroft and Ullman 1969).

The next theorem shows that every finite semigroup is the homomorphic image of a free semigroup.

Theorem Every semigroup with n generators is a homomorphic image of the free semigroup with n generators.

Proof: Let X be the set of generators of the free semigroup X^+, let Y be the set of generators of a semigroup $\langle S, \circ \rangle$, and let $|X| = |Y|$. Take any bijection from X to Y, say the bijection $f(x_i) = y_i$ for all i. This bijection has a unique extension into a homomorphism from X^+ onto S. In particular, we extend f to operate on the domain X^+ by defining $f(\alpha) = f(x_{i_1}) \circ f(x_{i_2}) \circ \cdots \circ f(x_{i_m})$ for every sequence

$$\alpha = x_{i_1} x_{i_2} \ldots x_{i_m}.$$

With this definition, it follows from the associativity of $\circ$ on S that $f(\alpha\beta) = f(\alpha) \circ f(\beta)$ for every α and β in X^+. That is, f is a homomorphism from X^+ onto S. $\qquad \square$

Fig. 6.1.5 shows how the function f is extended in the proof of the preceding theorem. Initially we have $f: X \to Y$, and from this we construct $f: X^+ \to S$.

As an example of the use of the theorem, consider the semigroup of Fig. 6.1.2 again, $S = \langle \{s, t, u, v\}, \cdot \rangle$ with generators u and v. The theorem says that there is a homomorphism from the free semigroup on two generators into the semigroup S. Let X^+ be the free semigroup generated by $\{u', v'\}$, so that our homomorphism is the function f that carries u' and v' into u and v, respectively. Now consider a typical

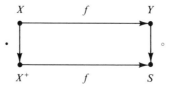

FIG. 6.1.5 **S as a Homomorphic Image of X^+ under the Homomorphism f**

sequence α in X^+. We show that its image under f is one of the four elements of S.

Because $v \cdot u = v$, we have $f(v'u') = f(v')$, and we can eliminate every u' immediately to the right of a v' in α without changing $f(\alpha)$. By repeated applications of this procedure, we can eliminate all occurrences of u' in α except for those that precede the first v' (if there is a v'). From $v^2 = v$ we have $f(v'v') = f(v')$, so we can eliminate all but one v' (if there is one), still leaving $f(\alpha)$ unchanged. Finally, because $u^2 = e$, we can eliminate any pairs of the form $\ldots u'u' \ldots$ without changing $f(\alpha)$. Thus we always can reduce $f(\alpha)$ to one of the four forms $f(u')$, $f(u'u')$, $f(v')$, or $f(u'v')$. Hence it is obvious that the only possible images are $\{u, u^2 = t = e, v, u \cdot v = s\} = S$.

A free semigroup becomes a free monoid when we append an identity to it. In X^+, where the semigroup operation is concatenation, the appropriate identity element is the empty sequence.

Definition   Let X be an arbitrary finite set. The *free monoid* generated by X, denoted as X^*, is the set of all sequences of letters of X, including the empty sequence Λ.

In the definition of a free monoid, we assume that $\Lambda \cdot \alpha = \alpha = \alpha \cdot \Lambda$ for every sequence α, where the operation $\cdot$ is concatenation. An example of a free monoid is $X^* = \{\Lambda, 0, 1, 00, 01, 10, 11, 000, \ldots\}$, where $X = \{0, 1\}$ and $X^+ = \{0, 1, 00, 01, 10, 11, 000, \ldots\}$. A free monoid with one generator is isomorphic to the monoid $\langle \mathbf{N}, + \rangle$ with the generator set $\{1\}$ and the identity 0.

As with free semigroups, the free monoid with n generators is unique to within isomorphism. It also follows that every monoid with n generators is a homomorphic image of the free monoid with n generators.

Summary   Fig. 6.1.6 shows the algebraic structures introduced in this section and summarizes their relations to one another in terms of their defining axioms. A semigroup is an algebraic structure that satisfies the axioms of closure and associativity. If the axiom of identity also holds for the structure, then the semigroup is a monoid. If both the axioms of identity and of inverses hold for the semigroup, then the semigroup is a group.

Because the axiom of inverses need not hold for a semigroup, we cannot guarantee the unique solvability of equations in the semigroup. Thus an equation of form $a \cdot x = b$ may have no solution, one solution, or many solutions in a semigroup, whereas such an equation always has a unique solution in a group.

Every semigroup is isomorphic to some transformation semigroup, in the same sense that every group is isomorphic to a permutation group. The transformation representation of a semigroup, like the permutation representation of a group, is not unique. The degree of a transformation representation is not related to the order of the semigroup; the degree may be less than, equal to, or greater than the order

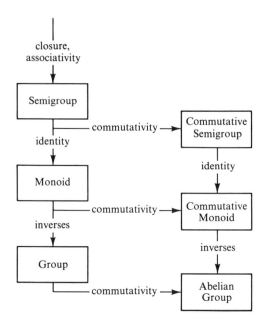

FIG. 6.1.6 Algebraic Structures Derived from Semigroups and Axioms That Hold for Each

of the semigroup. The degree of the regular transformation representation of a semigroup is equal to or one greater than the order of the semigroup. (Note the contrast with the regular permutation representation of a group, which has a degree exactly equal to the order of the group.)

Semigroups can be specified by defining relations on generators, much as groups are specified by such relations. From the defining relations, we can construct a semigroup graph, which is similar to a group graph but does not necessarily satisfy all properties of group graphs.

A free semigroup is a semigroup generated by a set when no defining relations hold for the generators. The free-semigroup operation is concatenation, and the elements of the free semigroup are the finite nonempty sequences of elements of the generator set. The free monoid has all of the elements of the free semigroup and includes the empty sequence.

Exercises

6.1.1 Prove that the composition table of Fig. 6.1.1 is a semigroup table.

6.1.2 What are the zeros in the following semigroups?
(a) $\langle \mathbf{Z}, \cdot \rangle$;
(b) $\langle \mathcal{P}(X), \cup \rangle$;
(c) $\langle \mathcal{P}(X), \cap \rangle$.

6.1.3 Let S be a finite semigroup. Suppose that there exist a, s, and t in S such that s is a left identity and $t \cdot a = s$. Show that every equation of the form $a \cdot x = b$ has a unique solution. Find an example to show that some equations of the form $x \cdot a = b$ might not be solvable under these conditions.

6.1.4 Consider the first quadrant of the complex plane—that is, numbers of the form $a + bi$, where $a \geq 0$ and $b \geq 0$. Show that this quadrant is a monoid under addition with generators $\{1, i\}$ and an identity 0. Is this monoid isomorphic to the free monoid with two generators?

6.1.5 Find a semigroup that is not a group, but in which every element is cancellable.

6.1.6 Show that a semigroup with a left identity is a group if every element has a left inverse with respect to that left identity.

6.1.7 Show that, if S is a semigroup, then it always is possible to append to S a new element e that is an identity, so that $S \cup \{e\}$ is a monoid. Similarly, show that it also is always possible to append a zero to S so that the resulting structure is a semigroup.

6.1.8 Show that a semigroup with two idempotents cannot be a group. Construct a semigroup of order 3 or more in which every element is an idempotent.

6.1.9 An element s of a semigroup is said to be *left-cancellable* if
$$(s \cdot x = s \cdot y) \Rightarrow (x = y)$$
for all x and y in S. Show that, if s and t in S are left-cancellable, then so is $s \cdot t$.

6.1.10 Let S be a semigroup, and let f and g be endomorphisms of S. That is, $f : S \to S$ and $f(s) \cdot f(t) = f(s \cdot t)$. Show that $f \circ g$ and $g \circ f$ also are endomorphisms of S.

6.1.11 Prove that the set of all endomorphisms of a semigroup is itself a semigroup under function composition. Is this semigroup a monoid?

6.1.12 Let X be any finite set, and let $|X| = 2$. Prove that neither X^+ nor X^* is a commutative semigroup. Prove that, if $|X| = 1$, then both X^+ and X^* are commutative semigroups.

6.1.13 Prove that $\langle \mathbf{Z} - \{0\}, \cdot \rangle$ is a monoid. What is the least set of generators for this monoid?

6.2 Sub-semigroups and Submonoids

In this section we continue our investigation of the properties of semi-groups. The notion that corresponds to the subgroup of a group is the subsemigroup of a semigroup. In this section we discuss a particular type of subsemigroup, called an ideal, for which there is no correspond-ing notion in group theory. A subsemigroup is closed under composition of any two elements in the subsemigroup, but an ideal is closed under composition of an ideal element and any semigroup element. We prove here that a semigroup is a group if and only if it has no nontrivial ideals; thus the presence of ideals is a distinguishing feature of a semigroup.

Later in the section we examine homomorphisms of semigroups. We also observe that the endomorphisms of an algebraic structure form a semigroup. Thus semigroups arise naturally in the context of maps that preserve algebraic structures.

We begin with a brief examination of subsemigroups and submonoids.

Definition   Let T be a subset of a set S that forms a semigroup $\langle S, \cdot \rangle$. If T is closed with respect to $\cdot$, then T is a *subsemigroup* of S.

Consider the semigroup $S = \langle \{s, t, u, v\}, \cdot \rangle$, whose composition table appears in Fig. 6.1.2(a). The set $\{u, t\}$ forms a subsemigroup of S. Because t is an idempotent, the singleton set $\{t\}$ also forms a subsemi-group of S. (See Ex. 6.2.1.) The semigroup S is a monoid. The sub-semigroup formed by $\{u, t\}$ also is a monoid, and t is an identity both in S and in $\{u, t\}$. This observation suggests the next definition.

Definition   If S is a monoid and if T is a subsemigroup of S that has the same identity as S, then T is called a *submonoid* of S.

Suppose that some semigroup S has a subsemigroup T, where the identity of T is not the identity of S. According to our definition, we would *not* call T a submonoid of S in such a case. However, some authors use a less restrictive definition, in which T is called a submonoid even though its identity is not the same as that of S.

We turn next to certain subsemigroups that satisfy a very strong closure property.

Definition   A subsemigroup T of a semigroup S is a *left ideal* of S if $S \cdot T \subseteq T$. The subsemigroup T is a *right ideal* of S if $T \cdot S \subseteq T$, and T is called a *two-sided ideal*, or simply an *ideal*, of S if it is both a left and a right ideal of S. If T is a (left, right, or two-sided) ideal of S and also the set T is a proper subset of the set S, then we say that T is a *proper* (left, right, or two-sided) *ideal* of S.

In the semigroup S of Fig. 6.1.2(a), the set $\{s, v\}$ forms a proper two-sided ideal. Obviously, every one- or two-sided ideal satisfies the closure axiom, so every ideal is a subsemigroup.

An example of an ideal is the set of multiples of an integer n in the semigroup $\langle \mathbf{Z}, \cdot \rangle$. If we take any multiple of n and multiply it by any integer, we still have a multiple of n. When we enlarge $\langle \mathbf{Z}, \cdot \rangle$ to $\langle \mathbf{Q}, \cdot \rangle$, the multiples of n no longer form an ideal. In this case, we can multiply kn by $(kn)^{-1}$ to produce a number that is not a multiple of n.

A group has no proper ideals, either one-sided or two-sided, because $g \cdot G = G = G \cdot g$ for every g in a group G as a direct consequence of the axiom of inverses. With the aid of the following definition and lemma, we can prove a stronger statement than the preceding. We wish to show that a semigroup with no proper one-sided ideals is a group.

Definition A semigroup S is said to be *left-simple* if it has no proper left ideals, and S is said to be *right-simple* if it has no proper right ideals.

Lemma If S is a left-simple semigroup, then $S \cdot s = S$ for all s in S. That is, right multiplication by any element of S is a bijection.

Proof: Let S be a left-simple semigroup. Suppose that $S \cdot s = T$, and that T is a proper subset of S. Then, because $S \cdot S \subseteq S$, we have

$$S \cdot T = S \cdot (S \cdot s) = (S \cdot S) \cdot s \subseteq S \cdot s = T.$$

Thus T is a proper left ideal of S, which contradicts the initial assumption that S is left-simple. □

Clearly, a similar lemma holds for right-simple semigroups and left multiplication. We are now ready to prove the theorem we promised.

Theorem A semigroup S is a group if and only if it is both left-simple and right-simple.

Proof: The proof that a group is both left-simple and right-simple follows from statements made earlier. It remains to prove that a semigroup with no proper left or right ideal is a group. To prove this, it is sufficient to prove that there exists an identity in the semigroup and that each element has an inverse.

To find an identity, it is sufficient to find a left identity. If one exists, then similar reasoning shows that there is a right identity, and by a previous theorem the two must be the same two-sided identity. If S is left-simple, then $S \cdot s = S$ for all s in S. Therefore, for a particular s in S, there is an element e such that $e \cdot s = s$. Hence e is a candidate for a left identity. Because S is right-simple, $s \cdot S = S$, for every y in S there is some element t such that $s \cdot t = y$. Now we have

$$e \cdot y = e \cdot (s \cdot t) = (e \cdot s) \cdot t = s \cdot t = y.$$

Hence e is a left identity and consequently, as mentioned previously, is a two-sided identity.

For a finite semigroup, the pigeonhole principle provides a simple proof that every element has an inverse. However, the theorem holds for infinite semigroups as well. Because $x \cdot S = S = S \cdot x$ for all x in S, there must be x_1 and x_2 such that $x \cdot x_1 = e = x_2 \cdot x$. Hence every element has both a left and a right inverse. By a previous theorem these elements must be the same two-sided inverse. □

In some sense, ideals are subsemigroups that act like traps. In any composition of elements of a semigroup (such as $x \cdot y \cdot x \cdot z \cdot y \cdot y$), if one of the elements (say z) is a member of an ideal T, then the product is "trapped" and must lie within T. There is no way to compose the product with other semigroup elements to create a new product that is not within T. We can make similar, but weaker, statements about one-sided ideals. The proof of the previous theorem suggests that the notion of a trap is related to the lack of inverses.

Several familiar algebraic structures have ideals containing zeros. For example, $\{0\}$ is an ideal of $\langle \mathbf{Z}, \cdot \rangle$. Every product of integers that contains 0 as a factor is equal to 0, so we very clearly see the trap at work. Similarly, $\{X\}$ is an ideal in $\langle \mathscr{P}(X), \cup \rangle$, and $\{\varnothing\}$ is an ideal in $\langle \mathscr{P}(X), \cap \rangle$. A zero of any semigroup is an ideal, and it is not difficult to see that any subsemigroup composed of only left zeros is a right ideal. For another example of an ideal, consider the multiples of an integer n in the algebraic structure $\langle \mathbf{Z}, \cdot \rangle$. Let $k \cdot n$ be a multiple of n in $\mathbf{Z}$, and note that $k \cdot n$ multiplied by any other element of $\mathbf{Z}$ is still a multiple of n. Hence the set $\{k \cdot n \mid k \in \mathbf{Z}\}$ forms an ideal in $\langle \mathbf{Z}, \cdot \rangle$.

The ideals and other subsemigroups of a semigroup help us to characterize the structure of the semigroup, just as the subgroups of a group help us to characterize the structure of the group. We have seen that group homomorphisms also help us to understand the structure of a group in the sense that, if G' is a homomorphic image of a group G, then G' exhibits some of the properties of G. The analogous notion for semigroups is the notion of semigroup homomorphism.

Definition A map f from a semigroup $\langle S, \cdot \rangle$ into a semigroup $\langle T, \circ \rangle$ is called a *semigroup homomorphism* if $f(s_1 \cdot s_2) = f(s_1) \circ f(s_2)$ for all s_1 and s_2 in S. If S and T are monoids and if f also maps the identity of S onto the identity of T, then f is called a *monoid homomorphism*.

It is easy to construct examples where the identity of $f(S)$ is not the same as the identity of T, even though $f(S)$ is properly included in T and $f(S)$ is a monoid. According to our definition, f in such a case is *not* a monoid homomorphism from S to T, but some authors do not impose this restriction that a monoid homomorphism must map one identity onto the other.

To clarify the difference between semigroup and monoid homomorphisms, let us look at some examples. Consider the map $f: \mathbf{N} \to \{0, 1\}$ that maps all nonzero integers onto 0 and maps 0 onto 1. This map is a homomorphism of $\langle \mathbf{N}, + \rangle$ onto $\langle \{0, 1\}, \cdot \rangle$. In this case, 0 is the identity of $\langle \mathbf{N}, + \rangle$, and $f(0) = 1$ is the identity of $\langle \{0, 1\}, \cdot \rangle$. The map f is a monoid homomorphism. Now consider the map $g: \mathbf{Z} \to \{0, 1\}$ that maps all nonzero integers onto 1 and maps 0 onto 0. This map is a homomorphism of $\langle \mathbf{Z}, \cdot \rangle$ onto $\langle \{0, 1\}, \cdot \rangle$. Here 1 is the identity of $\langle \mathbf{Z}, \cdot \rangle$, and $g(1) = 1$ is the identity of $\langle \{0, 1\}, \cdot \rangle$, so g also is a monoid homomorphism.

Now consider the semigroup $S = \langle \{e, 0, 1\}, \circ \rangle$ with the composition

$$
\begin{array}{c|ccc}
\circ & e & 0 & 1 \\
\hline
e & e & 0 & 1 \\
0 & 0 & 0 & 0 \\
1 & 1 & 0 & 1 \\
\end{array}
$$

FIG. 6.2.1   Composition Table for the Semigroup $S = \langle \{e, 0, 1\}, \circ \rangle$

table shown in Fig. 6.2.1. Let us use S instead of $\langle \{0, 1\}, \cdot \rangle$ as the range of the two semigroup homomorphisms just described. Thus we have $f' : \mathbf{N} \to S$ and $g' : \mathbf{Z} \to S$, where neither map carries any element of the domain into the element e of S. The maps f' and g' are semigroup homomorphisms, but they are not monoid homomorphisms.

Next we consider the composition of semigroup homomorphisms. Let S, T, and W be three semigroups, and let $f : S \to T$ and $g : T \to W$ be semigroup homomorphisms. Then we can construct a new homomorphism from S into W by composing f and g.

Theorem   Let S, T, and W be semigroups with compositions $\cdot$, $+$, and $*$, respectively. If $f : S \to T$ and $g : T \to W$ are semigroup homomorphisms, then $(f \circ g) : S \to W$ is a semigroup homomorphism.

Proof: For all s_1 and s_2 in S, we have

$$
g \circ f(s_1 \cdot s_2) = g(f(s_1 \cdot s_2)) = g(f(s_1) + f(s_2))
$$
$$
= g(f(s_1)) * g(f(s_2)) = (g \circ f(s_1)) * (g \circ f(s_2)).
$$

Hence $g \circ f$ is a homomorphism.   □

A semigroup endomorphism is a homomorphism of a semigroup into itself. Because the domains and ranges of semigroup endomorphisms are automatically compatible, we see that it is perfectly natural to compose endomorphisms of a semigroup S. This observation suggests the next theorem.

Theorem   The set of all endomorphisms of a semigroup forms a semigroup under function composition.

Proof: The closure property follows from the preceding theorem. Associativity is a consequence of the associativity of function composition.   □

The set of all endomorphisms of a semigroup obviously forms a monoid, because the identity mapping on the semigroup is a homomorphism and is an identity in the semigroup of endomorphisms.

As an example of a monoid of endomorphisms, let us find all of the endomorphisms of the group $\langle \mathbf{Z}_4, +_4 \rangle$, whose composition table is given in Fig. 6.2.2. Because all of the endomorphisms of a group must have normal subgroups of the group as kernels, we can turn our attention immediately to functions that map either $\{0\}$, the normal subgroup $\{0, 2\}$, or the group itself onto the identity. We observe that 0 is an

$+_4$	0	1	2	3
0	0	1	2	3
1	1	2	3	0
2	2	3	0	1
3	3	0	1	2

FIG. 6.2.2   Composition Table for the Group $\langle Z_4, +_4 \rangle$

identity in this group, that 0 and 2 are self-inverse, and that 1 and 3 are inverses of each other. The only endomorphisms for which 0 is the image only of 0 are bijective. Therefore they leave 0 and 2 fixed, and they either swap 1 and 3 or leave 1 and 3 fixed. The only endomorphism that maps 0 and 2 onto 0 must map 1 and 3 onto 2. Lastly, there is the endomorphism that maps every element onto 0. Fig. 6.2.3 shows the four endomorphisms and their composition table.

Next we show that a certain subset of the endomorphisms forms a group.

Theorem   The set of all automorphisms of a semigroup forms a group under function composition.

   Proof: The axioms of associativity and identity follow directly for the set of automorphisms. Next we show that closure holds. From a previous theorem we know that the composition of two automorphisms is an endomorphism. Because automorphisms are bijections, the composition of two automorphisms is a bijection. But a bijective endomorphism is an automorphism, so closure is proved.

   To prove the property of inverses we must show that, if f is an automorphism, then its inverse f^{-1} is an automorphism. The inverse of an automorphism is guaranteed to be a function because an automorphism

g	$f_1(g)$	$f_2(g)$	$f_3(g)$	$f_4(g)$
0	0	0	0	0
1	1	3	2	0
2	2	2	0	0
3	3	1	2	0

$\circ$	f_1	f_2	f_3	f_4
f_1	f_1	f_2	f_3	f_4
f_2	f_2	f_1	f_3	f_4
f_3	f_3	f_3	f_3	f_4
f_4	f_4	f_4	f_4	f_4

FIG. 6.2.3   The Endomorphisms of the Group $\langle Z_4, +_4 \rangle$ and Their Composition Table

is a bijection. Given that $f(x) \cdot f(y) = f(x \cdot y)$ for all x and y and that $f^{-1}(f(x)) = x$ for all x, we have

$$f^{-1}(f(x) \cdot f(y)) = f^{-1}(f(x \cdot y)) = x \cdot y = f^{-1}(f(x)) \cdot f^{-1}(f(y)).$$

Now as x and y range over all elements of the semigroup, then $f(x)$ and $f(y)$ range over all elements of the semigroup. Hence

$$f^{-1}(x \cdot y) = f^{-1}(x) \cdot f^{-1}(y)$$

for all x and y, and f^{-1} is an automorphism as well as the inverse of f. $\square$

In our example of the monoid of endomorphisms of $\langle \mathbf{Z}_4, +_4 \rangle$, functions f_1 and f_2 are both automorphisms. Inspection of the composition table reveals that these two functions form a group isomorphic to S_2, in agreement with the preceding theorem.

Summary   An ideal T of a semigroup S is a subsemigroup that satisfies $S \cdot T \subseteq T$ and $T \cdot S \subseteq T$. A group has no proper ideals, either one-sided or two-sided, and a semigroup that is not a group has at least one proper one-sided or two-sided ideal.

A homomorphism of a semigroup is a structure-preserving map of that semigroup into another semigroup. A semigroup endomorphism is a homomorphism that carries a semigroup into itself. The set of all semigroup homomorphisms is itself a semigroup under function composition, and the set of semigroup automorphisms forms a group. The endomorphisms and automorphisms of a semigroup give the structure of the semigroup, in that they identify the symmetries, the ideals, and the subsemigroups that exist within the semigroup. In Chapter Seven we see how the notion of semigroup homomorphism extends naturally to machine homomorphism. We use this idea to construct economical realizations of sequential machines.

Exercises

6.2.1   Find all of the subsemigroups of the semigroup S whose composition table appears in Fig. 6.1.2(a).

6.2.2   A cyclic semigroup is a semigroup with one generator. Show that every finite cyclic semigroup has an idempotent. HINT: For the generator g, there must be m and n such that $g^{m+n} = g^n$. Show that $g^{km+n} = g^n$ for all k; then look for the idempotent.

6.2.3   Show that every finite semigroup has an idempotent.

6.2.4   Let S be a monoid. Prove that the set of all the invertible elements of S forms a group under the composition of the monoid.

6.2.5   If a semigroup S has a subgroup H, then we define the left cosets of S with respect to H to be sets of the form $s \cdot H$ for s in S. Show that, if the identity in H is an identity in S, then the left cosets partition S.

6.2.6 Construct a transformation representation of the semigroup of endomorphisms of S_3.

6.2.7 Show that a left ideal and a right ideal of a semigroup cannot be disjoint.

6.2.8 The *center* of a semigroup is the set of elements that commute with every element of the semigroup. Show that the center forms a subsemigroup.

6.2.9 Show that the set of noncancellable elements of a commutative semigroup forms an ideal.

6.2.10 Show that, if T_1 and T_2 are ideals in a semigroup $\langle S, \cdot \rangle$, then $\langle T_1 \cdot T_2, \cdot \rangle$ and $\langle T_1 \cup T_2, \cdot \rangle$ are ideals.

6.2.11 Show that, for each element s of a semigroup S, the set of elements that commute with s forms a subsemigroup.

6.2.12 Show that the set of idempotents of a commutative semigroup forms a subsemigroup.

6.2.13 Let T be any subset of elements of a semigroup S. Show that $S \cdot T$ forms a left ideal of S.

6.2.14 An element s of a semigroup S is called *nilpotent* if there is an $i \geq 1$ such that s^i is a zero of S. Prove that the set of nilpotent elements of a commutative semigroup forms an ideal.

6.2.15 Let T be a subset of elements of a semigroup S, and define $T^2 = T \cdot T$, $T^3 = T \cdot T \cdot T$, etc. We say that T is *nilpotent* if $T^i = \{z\}$ for some i, where z is a zero of S. Prove that every subset of a nilpotent set is also nilpotent.

6.2.16 Let T be a nilpotent left ideal of S. Show that $T \cdot S$ is a nilpotent two-sided ideal.

SEVEN Finite-State Machines

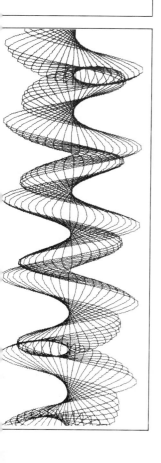

SEVEN | Finite-State Machines

The sequential machines treated in this chapter are abstract models of components and modules found within digital computers. We show that a sequential machine is associated with a set of transformations of a set of states. Because transformations are closely linked with semigroups, the appropriate mathematical theory for the study of sequential machines is the theory of semigroups. In the first section of this chapter we show how various properties of semigroups are analogous to properties of sequential machines.

For a historical point of view, we would like to begin this chapter by relating how various semigroup theorems have found application in the study of sequential machines. Unfortunately, there are notably few applications of this type. Algebraists have preferred to push semigroup theory in directions that generally are not applicable to sequential machines. In fact, the reverse direction of application has been more common. Several advances in the theory of sequential machines have been put to use by algebraists to advance semigroup theory. In this chapter we focus on the results of the greatest practical significance, and we omit some interesting algebraic properties of sequential machines.

One of the central concepts we discuss here is the idea of partitioning the state set of a sequential machine into classes that preserve some of the structure of the machine. From this discussion we derive an algorithm for state minimization, and a method for decomposing a machine into a cascade collection of smaller machines. We put to use several of the algebraic notions developed in Chapter One, particularly the notion of homomorphism. One subject of interest that arises from the study of machine homomorphisms is the study of machine simulation. One machine can simulate another if the second is a homomorphic image of the first.

The final section of this chapter deals with the use of sequential machines as sequence recognizers. Here a machine outputs a 1 if and only if the preceding input sequence is one of a designated set of sequences. The results of such studies have been widely applied to both computer hardware and programming systems. Portions of compilers and assemblers simulate sequential-machine recognizers. In this case, the task of a recognizer is to group the characters in the input stream into collections that have particular significance within the programming language. For example, some character groupings such as "1234" and

"5.43E+02" represents constants, whereas other groupings such as "DELTA" and "SUM" represent variables in the program. A recognizer in a compiler identifies the boundaries between these syntactic entities and passes the results of its analysis to the syntax-analyzer portion of the compiler. Consequently, sequential-machine theory is important to the systems programmer as well as to the computer designer.

7.1 Semigroups and Finite-State Machines

We begin with a very brief review of some basic facts that play an important role in our discussion of finite-state machines. Inspection of the internal structure of a typical digital computer shows that it is composed almost entirely of networks that compute binary functions of binary arguments. That is, each network accepts n binary signals at its inputs and produces m binary signals at its outputs, for some and n. Fig. 7.1.1 shows a typical network that computes a function from $\{0, 1\}^3$ into $\{0, 1\}^2$.

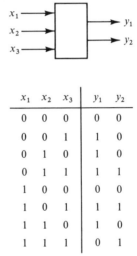

x_1	x_2	x_3	y_1	y_2
0	0	0	0	0
0	0	1	1	0
0	1	0	1	0
0	1	1	1	1
1	0	0	0	0
1	0	1	1	1
1	1	0	1	0
1	1	1	0	1

FIG. 7.1.1   A Typical Function from $\{0, 1\}^3$ into $\{0, 1\}^2$

If the output of a network depends only on the present states of its inputs, it is said to be a *combinational network*. The network of Fig. 7.1.1 is of this type. It also is possible to construct a network whose behavior at any instant of time depends on the past history of its inputs. A network whose output depends on prior inputs as well as its present inputs is said to be a *sequential network*. For example, the network in Fig. 7.1.2 is a binary counter. Whenever a 1 appears on the network input, the number represented on the output increases by 1.

We can construct the sequential network of Fig. 7.1.2 from a combinational network and a time-delay element by using a feedback

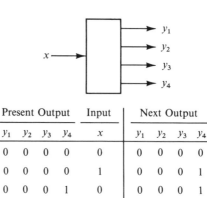

Present Output				Input	Next Output			
y_1	y_2	y_3	y_4	x	y_1	y_2	y_3	y_4
0	0	0	0	0	0	0	0	0
0	0	0	0	1	0	0	0	1
0	0	0	1	0	0	0	0	1
0	0	0	1	1	0	0	1	0
0	0	1	0	0	0	0	1	0
0	0	1	0	1	0	0	1	1
		⋮		⋮			⋮	

FIG. 7.1.2 A Sequential Network That Acts as a Binary Counter

circuit as shown in Fig. 7.1.3(*a*). In this figure, the adder is a combinational network that produces the binary representation of the sum of its two addends. In isolation, the adder produces an output that is independent of the previous inputs to the adder. When placed in the feedback circuit, the output is transmitted back to one of the inputs of the adder, so that it can be summed with the next input. Hence the sum circulates continuously through the feedback network, increasing by 1 whenever a 1 appears on the input. We assume that the feedback circuit has enough lines to carry the binary representation of the largest sum of interest.

In order to guarantee that the circuit of Fig. 7.1.3(*a*) is controlled correctly, we must add some other signals as shown in Fig. 7.1.3(*b*). The signal labeled "reset" disconnects the feedback loops to change the circulating sum to 0. This signal permits us to initialize the counter so that we can start it properly. The signal labeled "clock" forces the network to react to inputs at selected discrete points in time. The clock signal is merely a signal that alternates between 1 and 0 at a regular rate. Whenever the clock input is 0, no changes occur in the output of the network. Whenever the clock changes to 1, the network produces a new output, which depends on the current input and the previous output. The network is active only when the clock signal changes to 1— in effect, when the clock "ticks." Thus the input signal is free to change any time between ticks of the clock, and the network reacts only to the inputs present when the clock ticks. By constructing the feedback loop with time delays equivalent to the time between ticks of the clock, we can eliminate spurious behavior of the network due to inevitable variations in signal propagation time.

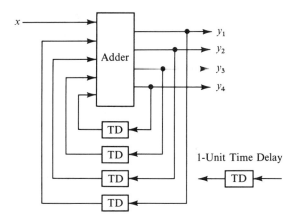

(a) The Basic Adder-and-Feedback Network

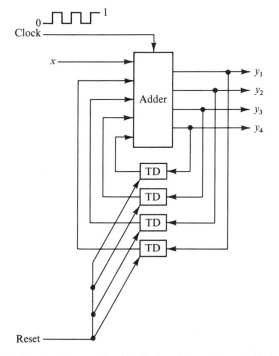

(b) The Basic Network with Clock and Reset Lines Added

FIG. 7.1.3   A Sequential Network Constructed from an Adder and Feedback Lines

A network of this type will function properly only if controlled by such reset and clock signals. Because the class of networks we treat here requires both a clock and a reset signal for proper operation, we presume that these signals are present in every such network, and we do not model them explicitly in following examples. In particular, whenever we deal with a network of the type shown in Fig. 7.1.3(*a*), the reader may safely assume that the actual network is similar to that shown in Fig. 7.1.3(*b*).

The essential difference between combinational and sequential circuits is that sequential circuits have memory whereas combinational circuits do not. The memory of a sequential circuit consists of the values stored on the unit delays in the feedback loop. We can model a combinational network mathematically by viewing it as a function from $\{0, 1\}^n$ into $\{0, 1\}^m$. The model of a sequential network inherently requires two functions, one that changes contents of the memory of the network and one that produces an output. In particular, we model sequential networks by algebraic structures known as finite-state machines.

Definition A *finite-state machine* is an algebraic structure $\langle S, I, Y, M, \delta \rangle$, where S, I, and Y are finite sets of states, inputs, and outputs, respectively, M is a mapping from $S \times I$ into S, and δ is a mapping from S into Y.

Fig. 7.1.4 illustrates this definition. The state of the machine is the symbol that is fed back to the input of M. At each clock tick, the *next-state function* $M(s, i)$ produces a new state that depends on the present state and the present input. Also at each clock tick, the *output function* δ maps the present state into an output symbol. In the figure, we have $s' = M(s, i)$ and $y = \delta(s)$. The output symbol $\delta(s')$ is produced one clock tick after the M accepts input i in state s. We usually name the finite-state machine by its next-state function, so the machine in Fig. 7.1.4 is called machine M. This is an abuse of notation, but it should lead to no confusion.

FIG. 7.1.4 An Abstract Model of a Finite-State Machine

We specify the behavior of a finite-state machine by specifying its next-state and output functions. For example, the tables in Fig. 7.1.5 describe a finite-state machine that produces a 1 output whenever the preceding four inputs form the sequence 0100. Before verifying that this machine does what we purport it does, we note that each input is associated with a transformation of the states. When we apply a sequence of inputs, the equivalent action is that of composing the individual input transformations. Hence, if $f_0 : S \rightarrow S$ and $f_1 : S \rightarrow S$ are the state transformations corresponding to input signals 0 and 1, respectively, then the input sequence 01 corresponds to the composition of f_0 with

Present State s	Next State $M(s, i)$		Output $\delta(s)$
	$i = 0$	$i = 1$	
s_1	s_2	s_1	0
s_2	s_2	s_3	0
s_3	s_4	s_1	0
s_4	s_5	s_3	0
s_5	s_2	s_3	1

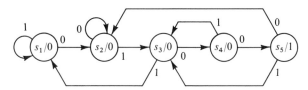

FIG. 7.1.5   State Tables and Graph for the Next-State and Output Functions of a Sequential Machine M

f_1. For convenience, we use *right* composition in the following discussion. Thus the input sequence 01 is associated with the state transformation $f_0 \cdot f_1$, which is defined to be f_0 followed by f_1. Fig. 7.1.6 shows $f_0 \cdot f_1$ and $f_1 \cdot f_0$ for the machine M of Fig. 7.1.5.

It now should be clear that, in this model of a finite-state machine, there is a semigroup associated with the next-state function. In particular, because the right composition of transformations is associative, the transformations f_0 and f_1 are generators of a set of transformations that forms a semigroup under function composition.

Semigroup graphs (Chapter Six) suggest a graphic way of describing the next-state function of a machine. Fig. 7.1.5 shows such a graph for machine M. In this graph, there is one node for each state in M and one edge leaving each node for each input symbol in I. The paths on this graph show state changes experienced by the machine for all input sequences. This type of graph is called a *state graph*.

s	$f_0 \cdot f_1(s)$	$f_1 \cdot f_0(s)$
s_1	s_3	s_2
s_2	s_3	s_4
s_3	s_3	s_2
s_4	s_3	s_4
s_5	s_3	s_4

FIG. 7.1.6   Right Composition of State Transformations for Machine M (Fig. 7.1.5)

The state graph in Fig. 7.1.5 shows quite clearly that M behaves in the manner that we have claimed for it. In effect, the states $s_1, s_2, \ldots, s_5$ remember how many symbols of the sequence 0100 have been received thus far. State s_1 corresponds to having received none of them, s_2 corresponds to having received the initial 0, s_3 occurs when the initial 0 and 1 are received, and so forth. We initialize M to start in state s_1. Then, if we apply the input sequence 0100, the machine advances through states s_2, s_3, s_4, and s_5, successively. As indicated on the graph by the notation $s/\delta(s)$ for each node, the output of the machine is 1 when it is in state s_5, and the output is 0 for all other states. If we input a sequence other than 0100, the machine does not advance sequentially through the states, but at each input moves to a state that indicates how much of the sequence 0100 exists among the most recent inputs. For example, when a 1 input is received in state s_4, the preceding four input symbols (including the 1 just received) must be the sequence 0101. The final two symbols, 01, are the first two symbols of 0100, and no longer initial subsequence of 0100 has been received at this point. Hence the machine returns to state s_3, the state corresponding to a received sequence of 01. If the next input is 1, the machine returns to state s_1, because the sequence 11 cannot be a part of the sequence 0100.

We mention earlier that the state transformations f_0 and f_1 are generators of a semigroup of transformations. The elements of this semigroup are state transformations induced by sequences of input symbols. Hence f_0, f_1, and $f_{01} = f_0 \cdot f_1$ are each distinct state transformations in the semigroup. Because of associativity, we have

$$f_i \cdot (f_j \cdot f_k) = (f_i \cdot f_j) \cdot f_k,$$

where i, j, and k are each input symbols in $\{0, 1\}$. We can restate this condition in the following form:

$$M(M(s, ij), k) = M(M(s, i), jk) \text{ for all } s \in S \text{ and for all } i, j, k \in I,$$

where $M(s, ij)$ refers to the state transformation of M that is induced by the input symbol i followed by the symbol j. We can extend M from the domain $S \times I$ to the domain $S \times I^+$ by defining $M(s, \alpha\beta) = M(M(s, \alpha), \beta)$ for all input sequences α and β in I^+. By defining $M(s, \Lambda) = s$ for the empty sequence Λ, we can extend M to the domain $S \times I^*$. Associativity guarantees that these extensions of the domain are well defined.

Because semigroups underlie the state behavior of finite-state machines, it is interesting to observe the ways that various properties of semigroups appear in finite-state machines. First, consider the machine M_1 whose state table is given in Fig. 7.1.7(a). The next-state mappings for the two inputs generate a semigroup of order 4; the maps for this semigroup are given in Fig. 7.1.7(b). We use the notation $[\alpha]$ to denote a class of input sequences, all of which induce the same transformation as does the sequence α. Every sequence in I^* transforms the states of M_1 in one of the four ways given in Fig. 7.1.7(b). For example, the sequence

0010 is in the class [10]. The reader can verify easily that the sequences 0, 00, 1, and 10 transform the states of M_1 as indicated in the table of transformations. Because the four functions listed are all of the possible functions from S into S, there are no other state transformations associated with sequences in I^*. Exercise 7.1.8 deals with a general algorithm for finding all of the state transformations in the semigroup generated by the transformations in the next-state table.

From the preceding example, we see that the state transformations of a sequential machine are the generators of the semigroup of state transformations, and we call that semigroup the *semigroup of the machine*. The mapping of the sequences of the semigroup I^* onto the state transformations in that semigroup is a homomorphism because $[\alpha] \cdot [\beta] = [\alpha\beta]$ for all input sequences α and β.

The example illustrates another interesting property of some sequential machines—a property that corresponds to a property of semigroups. The input symbol 1 forces M_1 into state s_2 regardless of its present state. Here we say that the input symbol 1 *resets* M_1 to state s_2. More generally, if an input symbol forces a machine to a specific state no matter what its present state, we call that symbol a *reset input*. For M_1, the state transformation f_1, which corresponds to the input symbol 1, has the property that $f_\alpha \cdot f_1 = f_1$ for all input sequences α, so f_1 is a right zero in the semigroup of M_1. For any machine, an input sequence resets the machine to a specific state if and only if the state transformation of that sequence is a right zero in the semigroup of the machine.

The semigroup of M_1 contains an ideal consisting of the input classes [1] and [10]. Inspection of the table reveals that, when an input sequence in either of these classes is preceded or followed by an arbitrary input sequence, the resulting state transformation must correspond to an input sequence in [1] or [10].

s	$M_1(s, i)$	
	$i = 0$	$i = 1$
s_1	s_2	s_2
s_2	s_1	s_2

(a) The State Table for M_1

s	Input Sequence			
	[0]	[00]	[1]	[10]
s_1	s_2	s_1	s_2	s_1
s_2	s_1	s_2	s_2	s_1

(b) The Next-State Mappings for M_1

FIG. 7.1.7   State Table and Next-State Mappings for a Machine M_1

Summary   A finite-state sequential machine is a mathematical model of a sequential network of the type commonly found in digital computers. We specify a sequential machine by giving its input, output, and state sets, and by specifying its next-state and output functions. In our model, the next state depends on both the present input and the present state, but the output depends only on the present state.

Because the next-state function of a machine has a state transformation for each input symbol, the study of sequential machines naturally involves the study of transformations of a set and compositions of these transformations. There is a natural way to extend the set of state transformations associated with individual input symbols to a set of transformations associated with input sequences. The set of distinct state transformations for a sequential machine is a semigroup because it is closed under function composition, and because function composition is associative.

The model we use here for finite-state machines is often called the *Moore model* because it is the outgrowth of a paper by E. F. Moore (1956). The *Mealy model* of finite-state machines (Mealy 1955) is an alternative model that is more widely used than the Moore model. The two models differ in their output functions. Recall that the Moore-machine output function is a map from S into Y. The Mealy-machine output function depends not only on the state of the machine, but on the present input as well. Hence it is a map from $S \times I$ into Y. It is not difficult to show that the two definitions lead to the same class of machines. The fundamental difference is that a Moore machine might need more states than a Mealy machine that performs the same computation. Because we are particularly interested in the next-state behavior of machines, it is most convenient to use the model with the simplest possible output function. Therefore we deal in this text with Moore machines exclusively.

The relation between sequential machines and semigroups did not play an important role in the development of the theory of sequential machines until 1962, when K. B. Krohn and J. L. Rhodes published the first of a sequence of papers on this topic. This first paper generated immediate interest in the algebraic aspects of machines and also stimulated work that advanced the theory of semigroups (see Krohn and Rhodes 1965).

Exercises

7.1.1 Construct the state table and state graph for a five-state sequential machine that gives a 1 output when the preceding four input symbols are 1110.

7.1.2 Consider the semigroup that is generated by the next-state transformations of the machine M_1 (Fig. 7.1.7). For any two sequences α and β in I^*, let R be the equivalence relation such that α R β if and only if the next-state mappings for α and β are identical. Show that this relation is a congruence with respect to composition in the semigroup.

7.1.3 Consider a machine with n states and the input alphabet $\{0, 1\}$. What is the maximal size of the semigroup of the machine? If the input alphabet is $\{0\}$, what is the maximal size of the semigroup? Assume that the inputs permute the states.

7.1.4 Prove that, if M is a machine with n states and if there is an input sequence that carries M from state s_1 to state s_2, then there exists such a sequence with fewer than n symbols.

7.1.5 Let M be a sequential machine with n states. Prove that, if each input map is a permutation of the n states, then the semigroup of M is a group.

7.1.6 Consider the machine M_2 whose state table is given in Fig. 7.1.8. Prove that its semigroup is S_n, the symmetric group of degree n.

7.1.7 Find the state transformations in the semigroup of machine M_3 whose state table is given in Fig. 7.1.9.

7.1.8 Construct an algorithm for finding the semigroup of a finite-state sequential machine.

s	$M_2(s, i)$	
	$i = 0$	$i = 1$
s_1	s_2	s_2
s_2	s_3	s_1
s_3	s_4	s_3
$\vdots$	$\vdots$	$\vdots$
s_{n-1}	s_n	s_{n-1}
s_n	s_1	s_n

FIG. 7.1.8   State Table for the Machine M_2 (Ex. 7.1.6)

s	$M_3(s, i)$	
	$i = 0$	$i = 1$
s_1	s_3	s_3
s_2	s_3	s_1
s_3	s_3	s_2

FIG. 7.1.9   State Table for the Machine M_3 (Ex. 7.1.7)

7.2 State Reduction and State Equivalence

A very brief investigation of finite-state machines reveals the interesting fact that different machines may be indistinguishable in the sense that, for each possible input sequence, they produce identical output sequences. Consequently, one aspect of the design of finite-state machines is the problem of designing the least expensive machine that will realize a specific behavior. The problem that we treat in this section is that of discovering and eliminating unnecessary states in a machine. Because the cost of constructing a real machine increases with the number of states in the machine, state reduction is a process of practical importance, even though recent advances in device technology have diminished costs substantially.

The machine M_1 shown in Fig. 7.2.1 is a trivial example of a machine with unnecessary states. The state graph for the machine makes it quite obvious why some states are unnecessary. If M_1 is initially in one of the states s_1, s_2, or s_3, then no input sequence can place it in state s_4 or s_5. Conversely, if M_1 is initially in state s_4 or s_5, then no input sequence can place it in s_1, s_2, or s_3. We say that a state t is *reachable* from a state s if there is some input sequence in I^* that carries M from s to t. We can divide the state set of M_1 into two subsets, $\{s_1, s_2, s_3\}$ and $\{s_4, s_5\}$. Any state is reachable from other states in its subset, but is not reachable from states in the other subset. Hence, if we know that M_1 can be started only in states belonging to one subset, then we need not include the states of the other subset when we construct M_1. This observation suggests that a first step in reducing the complexity of machines should involve determining the set of states reachable from the possible initial states. The following definitions help us make this task more precise. For these definitions, M is a machine with state set S.

	$M_1(s, i)$	
s	$i = 0$	$i = 1$
s_1	s_2	s_3
s_2	s_1	s_2
s_3	s_3	s_2
s_4	s_4	s_5
s_5	s_4	s_5

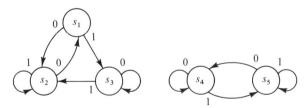

FIG. 7.2.1   State Table and Graph for a Sequential Machine M_1

Definition M is said to be *strongly connected* if, for every state s_i in S, every state s_j in S is reachable from s_i. A *strongly connected submachine* of M is a subset T of S for which every state in T is reachable from every other state in T by state transitions that never leave T.

Definition   A state s in a machine M is said to be *transient* if there is no nonempty input sequence that carries M from s back to s. A state s is said to be *conditionally transient* if there exists an input sequence that carries M from state s to some state t from which there is no nonempty input sequence that carries M back to s.

Machine M_2 in Fig. 7.2.2 illustrates these definitions. In this machine, states s_1, s_2, and s_3 form a strongly connected submachine of M_2, as do the states s_4 and s_5. State s_6 is a transient state, whereas states s_4 and s_5 are conditionally transient.

We can attempt to reduce the number of states in a machine whenever some subset of the state set is specified as the set of possible starting states. In fact, we can remove all of the states that are not reachable from at least one starting state.

In machine M_1 (Fig. 7.2.1), if all of the starting states lie within one of the disconnected submachines, then the other submachine may be eliminated. In machine M_2 (Fig. 7.2.2), because s_1, s_2, and s_3 are reachable from every state, they cannot be eliminated from the machine. If M_2 can be started in s_4 or s_5, then we can eliminate neither s_4 nor s_5 because they lie in a strongly connected submachine, but we can eliminate s_6. If s_6 is a starting state, then we can make no state reductions of this type because every state is reachable from s_6.

s	$M_2(s, i)$	
	$i = 0$	$i = 1$
s_1	s_2	s_3
s_2	s_1	s_2
s_3	s_3	s_2
s_4	s_3	s_5
s_5	s_4	s_5
s_6	s_5	s_4

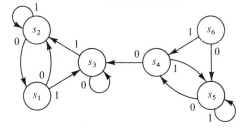

FIG. 7.2.2   State Table and Graph for a Sequential Machine M_2

Algorithms for detecting unreachable states are easy to construct (see Ex. 7.2.1).

Another important state-reduction process is possible when there are states that are indistinguishable from other states. Machine M_3 in Fig. 7.2.3 illustrates this situation. You can verify easily that an observer who sees only the input and output of the machine will be unable to distinguish state s_1 from state s_4. Because these states cannot be distinguished, we can replace every occurrence of s_4 in the state table by an s_1, thus forming the equivalent machine M_3' in which s_4 has been eliminated.

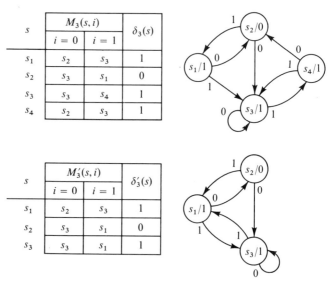

s	$M_3(s, i)$		$\delta_3(s)$
	$i = 0$	$i = 1$	
s_1	s_2	s_3	1
s_2	s_3	s_1	0
s_3	s_3	s_4	1
s_4	s_2	s_3	1

s	$M_3'(s, i)$		$\delta_3'(s)$
	$i = 0$	$i = 1$	
s_1	s_2	s_3	1
s_2	s_3	s_1	0
s_3	s_3	s_1	1

FIG. 7.2.3   A Sequential Machine M_3 and an Equivalent Reduced Machine M_3'

The following definitions make this notion of equivalence more precise.

Definition   Two states s_i and s_j of a machine M are said to be *equivalent* if, for every input sequence, the output sequence of M when started in state s_i is identical to the output sequence of M when started in state s_j. That is, for all α in I^*,

$$\delta\big(M(s_i, \alpha)\big) = \delta\big(M(s_j, \alpha)\big).$$

The relation of equivalence established by this definition is an equivalence relation. To distinguish this relation from other equivalence relations on the states, we call this relation the *state-equivalence relation*.

Definition   Two states s_i and s_j are said to be *k-equivalent* if, for every input sequence with k or fewer symbols, M produces the identical output sequence whether started in s_i or s_j.

Like state equivalence, k-equivalence is an equivalence relation on the states of a machine. From the definition of k-equivalence, we see immediately that, if two states are k-equivalent, they are r-equivalent for all $r < k$. Hence, if two states are in the same equivalence class of a k-equivalence partition, they are in the same class of the r-equivalence partitions for $r < k$. This observation leads to the following definition.

Definition An equivalence relation R on a set X is said to *refine* an equivalence relation R' on X if x R y implies x R' y for all x and y in X.

Observe that R refines R' if the equivalence classes of R are formed by breaking up the classes of R' into smaller classes. For example, k-equivalence refines r-equivalence for $r \leq k$.

The machine M_4, whose next-state and output functions are given in Fig. 7.2.4(a), illustrates the various types of equivalence. The 0-equivalence classes identify the states that give identical outputs for input sequences of length 0. This equivalence relation has two classes—one for the states with a 0 output, and the other for states with a 1 output. Hence the 0-equivalence classes of M_4 are $[s_1, s_2, s_5, s_8]_0$ and $[s_3, s_4, s_6, s_7, s_9, s_{10}]_0$, where we use the subscript 0 outside the brackets to denote 0-equivalence.

Two states are in the same 1-equivalence class if they give identical outputs for input sequences of length 1. Consider states s_1 and s_2, which are in the same 0-equivalence class. With a 1 input, the successor of state s_1 is s_6, and the successor of s_2 is s_3. States s_6 and s_3 are in the same 0-equivalence class, both giving an output of 0. Therefore, if M_4 starts in state s_1 or s_2, the input sequence 1 yields an output sequence of 10. However, with a 0 input, the successor of state s_1 is s_5, and the successor of s_2 is s_4. States s_5 and s_4 are in different 0-equivalence classes.

s	$M_4(s,i)$		$\delta_4(s)$	s	$M_4'(s,i)$		$\delta_4'(s)$
	$i = 0$	$i = 1$			$i = 0$	$i = 1$	
s_1	s_5	s_6	1	s_1	s_5	s_4	1
s_2	s_4	s_3	1	s_2	s_4	s_3	1
s_3	s_1	s_3	0	s_3	s_1	s_3	0
s_4	s_3	s_9	0	s_4	s_3	s_9	0
s_5	s_9	s_7	1	s_5	s_9	s_7	1
s_6	s_{10}	s_9	0	s_7	s_2	s_3	0
s_7	s_2	s_{10}	0	s_9	s_1	s_4	0
s_8	s_5	s_4	1				
s_9	s_8	s_6	0				
s_{10}	s_8	s_3	0				

FIG. 7.2.4 A 10-State Machine M_4 and an Equivalent Reduced Machine M_4'

Thus the input sequence 0 yields an output sequence of 11 if M_4 starts in state s_1 or an output sequence of 10 if M_4 starts in state s_2. Hence states s_1 and s_2 are not 1-equivalent. In general, two states are 1-equivalent if and only if (i) they are 0-equivalent, and (ii) their successors under any possible input are 0-equivalent.

Following this line of reasoning, we find that the 1-equivalence classes of M_4 are $[s_1, s_8]_1$, $[s_2, s_5]_1$, $[s_3, s_7, s_9, s_{10}]_1$, and $[s_4, s_6]_1$. Note how the class $[s_1, s_2, s_5, s_8]_0$ breaks up into $[s_1, s_8]_1$ and $[s_2, s_5]_1$ in this refinement of the 0-equivalence classes.

Extending this process to the search for 2-equivalence classes, observe, for example, that the successors of states s_3 and s_7 with a 0 input are not 1-equivalent. Hence s_3 and s_7 are not 2-equivalent, as we can verify by comparing outputs resulting from the input sequence 10. The 2-equivalence classes of M are $[s_1, s_8]_2$, $[s_3, s_{10}]_2$, $[s_4, s_6]_2$, $[s_2]_2$, $[s_5]_2$, $[s_7]_2$, and $[s_9]_2$. Note that these classes are a refinement of the 1-equivalence classes. In general, we find that two states are k-equivalent if and only if (i) they are $(k - 1)$-equivalent, and (ii) their successors under any single input symbol are $(k - 1)$-equivalent.

Carrying out this process for $k = 3$, we discover that the 3-equivalence partition is identical to the 2-equivalence partition. Moreover, when we repeat the process on the 3-equivalence partition, no further refinement can occur. We show later that the 2-equivalence partition is the state-equivalence partition for this machine. In fact, whenever a k-equivalence partition is unrefinable in the sense used here, it is the state-equivalence partition. Fig. 7.2.4(b) shows the reduced version of M_4, with equivalent states combined.

The preceding example illustrates several properties of state equivalence that are useful in identifying equivalent states. Next we state and prove each of these properties.

Lemma Two states are $(k + 1)$-equivalent if and only if they are k-equivalent and their successors under each input are k-equivalent.

Proof: We mentioned earlier that $(k + 1)$-equivalence implies k-equivalence. Hence, to prove this lemma it is sufficient to show that, if each input symbol carries a pair of k-equivalent states to k-equivalent states, then the states are $(k + 1)$-equivalent.

If two states s_1 and s_2 are k-equivalent, then they are 0-equivalent, so they have the same output for an empty input sequence. If s_1 and s_2 have k-equivalent successors for every input, then they yield the same output sequence for every input sequence of length 1, because the successor states must have identical outputs if they are k-equivalent. Every input sequence of length $k + 1$ or less therefore will yield identical output sequences for a start in either state, because the outputs will be identical after one input, and the k-equivalence of the successor states implies that the outputs to the next k or fewer input symbols will be identical. $\square$

Lemma   If the k-equivalence partition of the states of a machine M is identical to the $(k + 1)$-equivalence partition, then the k-equivalence partition is the same as the r-equivalence partition for all $r > k$.

Proof: We use induction to show that the k-equivalence partition is an r-equivalence partition for all $r > k$ when k-equivalence and $(k + 1)$-equivalence are identical relations.

We begin with the basis step, in which we show that the partition is a $(k + 2)$-equivalence relation. By definition of $(k + 1)$-equivalence, every pair of $(k + 1)$-equivalent states has k-equivalent successors under each input. Because k-equivalence is the same as $(k + 1)$-equivalence, every pair of $(k + 1)$-equivalent states has $(k + 1)$-equivalent successors under each input. Hence, from the preceding lemma, every pair of $(k + 1)$-equivalent states is $(k + 2)$-equivalent.

For the induction step, we must show that, if the partition is an r-equivalence partition for $r > k$, then it is an $(r + 1)$-equivalence partition. Proceeding as before, we assert that every pair of r-equivalent states has r-equivalent successors under each input, and therefore that every pair of r-equivalent states is $(r + 1)$-equivalent.   □

The preceding lemma suggests that we can compute the equivalence partition of states by computing 0-equivalence, 1-equivalence, etc., until we discover that the k-equivalence partition is the same as the $(k + 1)$-equivalence partition for some k. At this point we must terminate because no further refinement is possible. Fortunately, the final partition is indeed the state-equivalence partition that we seek. Moreover, we can bound the number of partitions that we have to construct.

Lemma   If k is the least integer for which the k-equivalence partition is the same as the $(k + 1)$-equivalence partition, then the k-equivalence partition is the state-equivalence partition.

Proof: In this case, any two k-equivalent states are indistinguishable, not only for all input sequences of length k or less, but also for all input sequences of length $k + 1$ or more, because k-equivalence implies r-equivalence for all $r > k$.   □

Lemma   Let k be the least k for which k-equivalence coincides with $(k + 1)$-equivalence. Then $k \leq n - 2$, where n is the number of states in the machine M, and where $n \geq 2$.

Proof: The proof depends on the idea that we cannot refine the 0-equivalence partition more than $n - 2$ times. Note that, if there is only one 0-equivalence class, then 0-equivalence coincides with state equivalence, because all states produce the same output and $k = 0$. When there is more than one 0-equivalence class, each refinement involves splitting at least one class into two disjoint classes, so the number of classes increases by at least one for each refinement. Therefore, if there are two 0-equivalence classes, refinement produces at least three 1-equivalence classes. If we can refine these classes, we obtain at least four 2-equivalence classes, and so forth. Because M has n states, there can be no more than n equivalence classes, so we cannot refine more than $n - 2$ times, and the refinement must terminate with the m-equivalence partition, where $m \leq n - 2$. Hence $m = k \leq n - 2$. $\square$

Let us look at an example in which this bound is attained. Suppose that the 0-equivalence partition is $[s_1]_0$, $[s_2, s_3, \ldots, s_n]_0$, and the 1-equivalence partition splits the second class into two classes by distinguishing s_2 from its 0-equivalent relatives. Hence we have $[s_1]_1$, $[s_2]_1$, $[s_3, s_4, \ldots, s_n]_1$. Similarly, each successive refinement splits off one state, until all states are distinguished by the last refinement for the $(n - 2)$-equivalence partition. The machine shown in Fig. 7.2.5 has this property and cannot be reduced.

We can summarize the state-reduction techniques in the form of an algorithm for reducing the number of states in a machine.

Step 1.   Determine which subset of S is the set of possible starting states, and call this subset T.

Step 2.   If $T \neq S$, then delete from S any state that is not reachable from some t in T.

Step 3.   Partition the remaining states into 0-equivalence classes, and let $k = 0$.

Step 4.   Refine the k-equivalence classes into classes of states that have k-equivalent successor states under each input. These are the $(k + 1)$-equivalence classes. (Step 4 refines the k-equivalence classes, where k increases by 1 each time the step is performed.)

Step 5.   If no refinement occurs in step 4, then go on to step 6. Otherwise, set k to $k + 1$ and repeat step 4.

s	$M(s, i)$		$\delta(s)$
	$i = 0$	$i = 1$	
s_1	s_1	s_1	1
s_2	s_2	s_1	0
s_3	s_3	s_2	0
s_4	s_4	s_3	0
$\vdots$	$\vdots$	$\vdots$	
s_i	s_i	s_{i-1}	0
$\vdots$	$\vdots$	$\vdots$	
s_n	s_n	s_{n-1}	0

FIG. 7.2.5   A Machine with n States and with $k = n - 2$

Step 6. Select one state from each class of equivalent states, and construct a new table for M with this reduced set of states.

Although this algorithm is rather efficient, it is not the most efficient algorithm known for the computation. In the algorithm we have given, the amount of computation time required to refine a partition is bounded from above by a time proportional to the product $|S| \cdot |I|$, where $|S|$ is the number of states and $|I|$ is the number of inputs. Hence the time required to compute the state-equivalence classes is bounded above by a time proportional to $|S|^2 \cdot |I|$. To determine the states reachable from a given state, we compute for a time no greater than a time proportional to $|S| \cdot |I|$, because this is the number of different state transitions in a machine. We must make this computation for at most $|S|$ states, and we find that the time for checking reachability does not grow faster than the time to check equivalence. The complete algorithm therefore requires a time no greater than $c \cdot |S|^2 \cdot |I|$ for some constant c.

Summary We can reduce the number of states in a sequential machine under at least two different conditions. We can eliminate a state if it is unreachable from all starting states of the machine, or if it is equivalent to another state in the machine. In the latter case, we mean by "equivalent" that two states are equivalent if it is impossible to distinguish between them by observing the inputs and outputs of the machine. Equivalence defined in this way is an equivalence relation, and the partition associated with this relation is called the state-equivalence partition.

We can compute the state-equivalence partition by computing a sequence of partitions starting with 0-equivalence, 1-equivalence, 2-equivalence, etc. Here, two states are k-equivalent if it is impossible to distinguish between them by observing the outputs generated from input sequences of length k or less. The $(k + 1)$-equivalence partition is easily computed from the k-equivalence partition for all k, and if the k-equivalence partition is the same as the $(k + 1)$-equivalence partition, then it is the state-equivalence partition.

The notions of state equivalence and state reduction were treated in each of the papers by D. A. Huffman (1954), G. H. Mealy (1955), and E. F. Moore (1956), which together established the foundations of the theory of sequential machines. Paull and Unger (1959) treated the problem of reducing the number of states in a sequential machine whose behavior is only partly specified. This problem still is not well solved in the sense that the known algorithms require a time that grows exponentially with the number of states. Hopcroft (1970) reported an efficient algorithm for state reduction of a completely specified machine. The time needed for the Hopcroft algorithm is proportional to $|S| \log_2 |S|$ (where S is the state set of the machine), rather than to $|S|^2$ as is the algorithm given in this section.

Exercises

7.2.1 Construct an algorithm suitable for implementation on a digital computer that finds all the states reachable from a specified state s of a given machine M.

7.2.2 Let M be the machine whose next-state table is given in Fig. 7.2.6. Identify the strongly connected submachines, the transient states, and the conditionally transient states of M.

7.2.3 Every strongly connected submachine has some input sequence that will carry it through each of its states at least once. Derive an upper bound on the shortest such sequence for a strongly connected n-state machine.

7.2.4 Show that k-equivalence and state equivalence are equivalence relations.

7.2.5 Find the equivalent-state partitions for machines M_1, M_2, and M_3 (Fig. 7.2.7) and construct the corresponding reduced machines.

7.2.6 Let M be an n-state machine in which no two states are equivalent. Prove that every pair of states give different output sequences for some input sequence of length $n - 2$ or less.

7.2.7 Let T be any subset of states of a machine M, where M has n states, no two of which are equivalent. Prove that, if T has m states, then there is a pair of states in T that give different output sequences for some input sequence of length $n - m$ or less.

	$M(s, i)$	
s	$i = 0$	$i = 1$
A	I	E
B	J	D
C	B	D
D	C	D
E	I	E
F	G	G
G	F	G
H	I	H
I	H	I
J	G	F

FIG. 7.2.6   Next-State Table for a Sequential Machine M (Ex. 7.2.2)

s	$M_1(s, i)$		$\delta_1(s)$
	$i = 0$	$i = 1$	
A	F	B	1
B	A	C	1
C	A	E	0
D	A	F	1
E	D	C	1
F	D	B	1

s	$M_2(s, i)$		$\delta_2(s)$
	$i = 0$	$i = 1$	
A	K	E	0
B	F	H	0
C	K	D	0
D	D	E	1
E	C	G	0
F	B	F	0
G	B	A	0
H	H	A	0
K	C	D	0

s	$M_3(s, i)$		$\delta_3(s)$
	$i = 0$	$i = 1$	
A	C	E	1
B	H	D	1
C	E	F	1
D	E	G	1
E	A	B	0
F	G	D	0
G	H	C	0
H	F	C	0

FIG. 7.2.7 Next-State and Output Tables for Three Sequential Machines (Ex. 7.2.5)

7.3 Machine Homomorphisms and Machine Simulation

The state-reduction algorithm of Section 7.2 illustrates that many different finite-state machines may be indistinguishable to an observer who can see only the inputs and outputs of the machines. In that section we concerned ourselves with the construction of the machine with the fewest states possible to realize a specified input/output behavior. In this section we adopt a different viewpoint. We investigate the conditions under which a machine M_2 can simulate a different machine M_1. We may wish to stock an inventory of "standard" machines, chosen so that we can use some standard machine or network of standard machines to simulate any particular machine M. Later in this section we investigate ways to simulate a machine by interconnecting two smaller machines.

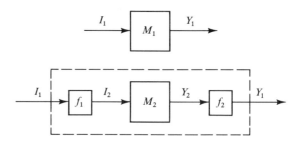

FIG. 7.3.1 Machine M_2 with Maps f_1 and f_2 Can Simulate Machine M_1

To begin our study, consider the machines shown in Fig. 7.3.1. M_1 is the machine we wish to simulate, and M_2 is the machine we have in our inventory. It so happens that M_1 accepts inputs from I_1 and produces outputs in Y_1, whereas M_2 accepts inputs from I_2 and produces outputs in Y_2. Hence our simulation must use maps on the inputs and outputs of M_2, so that M_2 can operate in its accustomed environment and yet be part of a composite machine whose external world is identical to that of M_1. Hence we introduce the functions $f_1 : I_1 \to I_2$ and $f_2 : Y_2 \to Y_1$ to change the environment for M_2. We say that M_2 *simulates* M_1 if an observer who can see only the inputs and outputs of the machines cannot distinguish between the two configurations. That is, if M_1 is started in any state s, then there is some state s' in M_2 such that

$$\delta_1\big(M_1(s, \alpha)\big) = f_2\big(\delta_2(M_2(s', f_1(\alpha)))\big)$$

for all input sequences α in I_1^*. In this equation, f_1 is a function from I_1^* to I_2^* that is the natural extension of its definition on I_1 to I_2. To be precise, for any x and y in I_1^*, the sequence xy in I_1^* is mapped into the sequence of the same length in I_2^*, $f_1(x) \cdot f_1(y)$, where the operation $\cdot$ is concatenation of sequences. Thus $f_1(xy) = f_1(x) \cdot f_1(y)$, and the function f_1 is a homomorphism from I_1^* into I_2^*.

As an example of a machine simulation, consider the machine M_1 in Fig. 7.3.2. M_1 is a modulo-3 adder. Each of its states is associated with an element of $\mathbf{Z}_3$, and each time that it receives an input symbol 1 it forms $1 +_3 s$, where s is its present state. We can simulate M_1 with machine M_2, a modulo-6 adder. M_2 forms the sum $1 +_6 s'$, where its present state s' is some element of $\mathbf{Z}_6$.

By using the functions f_1 and f_2 given in Fig. 7.3.2, we discover that the state pairs $\{0, 3\}$, $\{1, 4\}$, and $\{2, 5\}$ in M_2 simulate the states 0, 1, and 2, respectively, in M_1. For example, if M_2 is in either state 0 or state 3, it is simulating M_1 in state 0.

Although we have introduced the notion of machine simulation in terms of the maps f_1 on the inputs of M_1 and f_2 on the outputs of M_2, the example shows that the states of M_2 must in some way correspond to the states of M_1. In particular, if we reduce M_1 so that no two of its states are equivalent, it is obvious that for each state in M_1, there must be one or more states in M_2 that simulate it.

The central issue in the machine-simulation problem is to determine the correspondence between the next-state functions of M_1 and M_2. After this correspondence is established, it is relatively simple to determine the appropriate output map for M_2. Consequently, we focus our attention here on simulation of the state behavior of machines. This is equivalent to the assumption that both δ_1 and δ_2 are identity maps, so that the states of both M_1 and M_2 are observable at the outputs of the machines. The example of the modulo-3 and modulo-6 adders illustrates this assumption.

s	$M_2(s,i)$		$\delta_2(s)$
	$i=0$	$i=1$	
0	0	1	0
1	1	2	1
2	2	3	2
3	3	4	3
4	4	5	4
5	5	0	5

s	$M_1(s,i)$		$\delta_1(s)$
	$i=0$	$i=1$	
0	0	1	0
1	1	2	1
2	2	0	2

$i_1 \in I_1$	$f_1(i_1) \in I_2$
0	0
1	1

$y_2 \in Y_2$	$f_2(y_2) \in Y_1$
0	0
1	1
2	2
3	0
4	1
5	2

FIG. 7.3.2 Machine M_2 Can Simulate Machine M_1 with the Aid of Maps f_1 and f_2

Essential to the notion of machine simulation is the idea that the familiar concept of homomorphism applies to the algebraic structure of finite-state machines.

Recall that our concept of homomorphism is that of a structure-preserving map. We view a sequential machine homomorphism as a map that preserves the structure of the next-state function of the machine. Specifically, we do not require that a machine homomorphism preserve the structure of the output function in this discussion, although some authors do place such a restriction on machine homomorphisms.

Earlier in this text we deal with homomorphisms of sets with binary operations, such as homomorphisms of the structure $\langle G, \cdot \rangle$. Here a homomorphism is a single map with G as its domain, and we require that $f(g_1) \circ f(g_2) = f(g_1 \cdot g_2)$ for all g_1 and g_2 in G. For a machine homomorphism, the next-state function is a function from $S \times I$ into S, and therefore the function is not a binary operation on a single set. In this context, we require two maps to make a homomorphism—one map with S as its domain and the other with I as its domain. Hence a machine homomorphism is a pair of maps—one from a state set into a state set, and the other from an input set into an input set.

If a machine M_1 is a homomorphic image of a machine M_2, then the next-state function of M_1 has some (and possibly all) of the structure of the next-state function of M_2. If the structures are not identical, then machine M_2 always has the richer structure. We show later that machine M_2 can simulate machine M_1, its homomorphic image. In dealing with machine simulations, we occasionally find that a sub-machine of M_2 can simulate M_1, so that the corresponding machine homomorphism maps a submachine of M_2 into M_1. For this reason, we let our machine-homomorphism functions h_1 and h_2 be defined on subsets of the state set and input set of M_2. That is, the functions need not necessarily be defined on the entire machine M_2 when we look for homomorphisms from M_2 to M_1.

Next we state a formal definition of a machine homomorphism, and then show formally that a machine can simulate a homomorphic image of itself.

Definition   A machine M_1 is said to be a *homomorphic image* of a machine M_2 if there exist two surjections h_1 and h_2 such that

$$h_1 : J \rightarrow I_1, \quad \text{where } J \subseteq I_2,$$

and

$$h_2 : T \rightarrow S_1, \quad \text{where } T \subseteq S_2,$$

and such that

$$h_2\big(M_2(t,j)\big) = M_1\big(h_2(t), h_1(j)\big)$$

for all j in J and t in T. We say that the functions h_1 and h_2 form a *machine homomorphism*.

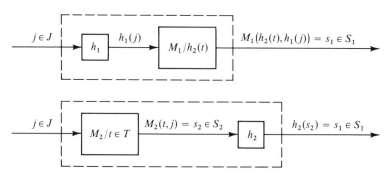

FIG. 7.3.3   Machine M_1 Is a Homomorphic Image of Machine M_2

Fig. 7.3.3 illustrates the definition of machine homomorphism. The dashed lines enclose machines that are indistinguishable to an observer who can see only the inputs and outputs of the dashed-line boxes.

A modulo-3 adder is a homomorphic image of a modulo-6 adder, because the functions h_1 and h_2 given in Fig. 7.3.4 satisfy the definition of machine homomorphism.

$i_2 \in I_2$	$h_1(i_2)$	$s_2 \in S_2$	$h_2(s_2)$
0	0	0	0
1	1	1	1
		2	2
		3	0
		4	1
		5	2

FIG. 7.3.4   Functions h_1 and h_2 Are the Machine Homomorphism Mappings Under Which a Modulo-3 Adder Is a Homomorphic Image of a Modulo-6 Adder

Fig. 7.3.5 clarifies the relation of machine homomorphism to machine simulation. If we operate on the inputs of the two machines in Fig. 7.3.3 with h_1^{-1}, then we have the machine structures shown in Fig. 7.3.5. Because $h_1^{-1} \circ h_1$ is an identity map, it is not shown in the figure. We recognize Fig. 7.3.5 as the machine structure that we encountered when we first introduced the notion of machine simulation (Fig. 7.3.1). Hence we can say that M_2 can simulate M_1 as shown in Fig. 7.3.5 if and only if the maps h_1 and h_2 constitute a machine homomorphism.

We must not overlook a minor problem in such a statement. Note that h_1^{-1} is not necessarily a function, because h_1 may be a many-to-one mapping. However, when h_1 is many-to-one, there are several inputs in I_2 that can simulate a particular input in I_1. Because it is sufficient for a simulation to have just one input in I_2 for each element of I_1, we can define h_1^{-1} to map each input in I_1 onto any input in I_2 that can simulate

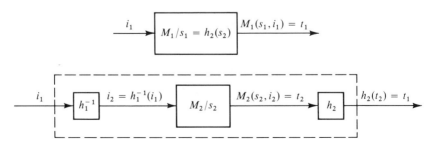

FIG. 7.3.5   Machine M_2 Is in an Environment in Which It Can Simulate Its Homomorphic Image M_1

it. Under these conditions h_1^{-1} is well defined because it is single valued for every element in its domain.

Machines M_1 and M_2 in Fig. 7.3.6 illustrate the use of nontrivial functions h_1 and h_2 for machine simulation. Careful inspection of the tables reveals that the functions h_1 and h_2 form a machine homomorphism from M_2 to M_1. The machine simulation is done as in Fig. 7.3.5, using the function h_1^{-1} that is given in Fig. 7.3.6. The reader should verify that h_1 and h_2 do indeed satisfy the conditions of machine homomorphism. A much more difficult problem is that of finding a pair of functions that map M_2 homomorphically into M_1 when none are given. In this instance, we could discover h_1 and h_2 through a series of judicious guesses and some trial and error. The homomorphism from M_2 to M_1 is not unique in this example (Ex. 7.3.1 will prove illuminating at this point).

s	$M_1(s,i)$	
	$i=0$	$i=1$
1	3	3
2	2	1
3	1	3

s	$M_2(s,i)$		
	$i=a$	$i=b$	$i=c$
A	D	B	A
B	A	D	D
C	B	C	B
D	A	B	A

$j \in J \subset I_2$	$h_1(j)$
a	1
b	0

$t \in T = S_2$	$h_2(t)$
A	3
B	1
C	2
D	3

$i_1 \in I_1$	$h_1^{-1}(i_1)$
0	b
1	a

FIG. 7.3.6   Machine M_2 Can Simulate Machine M_1 Using the Maps h_1^{-1} and h_2

In the remainder of this section we explore the conditions under which we can find a cascade-machine structure as shown in Fig. 7.3.7 that simulates a given machine. To simplify our discussion, we confine our attention to the fundamental problem of simulating state behavior. In effect, we assume that the state of the machine is observable at the output of the machine. Thus we deal only with output functions that are one-to-one from the state set to the output set.

Cascade decomposition of machines is of practical interest for at least two reasons. First, the cascade interconnection is one of the ways in which we might wish to interconnect standard machines to create a machine that is not otherwise in our inventory. Second, for any machine that can be simulated by a cascade structure, such a structure tends to be an economical way of constructing the machine.

Consider the cascade-machine structure shown in dashed lines in

Fig. 7.3.7. The *head machine* M_1 changes state according to its present state and the present input. The *tail machine* M_2 uses as its input both the present state of M_1 and the present input. The output of the network is a combinational (memoryless) function of the present states of both M_1 and M_2.

Because the network enclosed by the dashed lines has inputs, outputs, and internal memory, it is a finite-state machine. The state set of the cascade machine is $S_{M_1} \times S_{M_2}$, where S_{M_1} and S_{M_2} are the state sets of M_1 and M_2, respectively. If the present state of the cascade machine is the ordered pair $\langle s_1, s_2 \rangle$, then its next state is the ordered pair whose components are the next states of M_1 and M_2. Thus, for input i, the next state of the cascade machine is the ordered pair

$$\langle M_1(s_1, i), M_2(s_2, \langle s_1, i \rangle) \rangle.$$

Here we observe that M_1 acts on its input independently of the action of M_2. However, M_2 depends on M_1, because the input for M_2 is an ordered pair whose elements are the present state of M_1 and the present input. We assume that the two machines act simultaneously at the tick of a clock. Thus, when M_1 is in state s_1, its transition to the state $M_1(s_1, i)$ occurs at the same instant that M_2 changes to the state $M_2(s_2, \langle s_1, i \rangle)$.

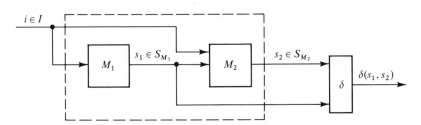

FIG. 7.3.7   A Cascade Machine as a Network of Two Simple Machines

Next we look at an example of a cascade connection of machines, before turning to a discussion of particular situations for which a next-state function of a machine M can be realized from the cascade connection of two smaller machines M_1 and M_2.

Fig. 7.3.8 illustrates a cascade decomposition of the machine M. Because M_1 has two states and M_2 has four states, the composite machine has eight states—one for each state in $S_{M_1} \times S_{M_2}$. From the state tables for M_1 and M_2 and the composite machine M, the reader can verify that M does indeed behave identically to the cascade connection of M_1 and M_2.

Notice that each state of M is associated with exactly one state pair in $S_{M_1} \times S_{M_2}$. In the remainder of this section we assume that every cascade decomposition has this property, although such an assumption

s_1	$M_1(s_1,i)$	
	$i=0$	$i=1$
s	s	t
t	t	s

s_2	$M_2(s_2,\langle s_1,i\rangle)$ when $\langle s_1,i\rangle$ is			
	$\langle s,0\rangle$	$\langle s,1\rangle$	$\langle t,0\rangle$	$\langle t,1\rangle$
a	b	a	b	a
b	c	b	d	c
c	d	c	c	b
d	a	d	a	d

$\langle s_1,s_2\rangle$	$M(\langle s_1,s_2\rangle,i)$	
	$i=0$	$i=1$
$\langle s,a\rangle$	$\langle s,b\rangle$	$\langle t,a\rangle$
$\langle s,b\rangle$	$\langle s,c\rangle$	$\langle t,b\rangle$
$\langle s,c\rangle$	$\langle s,d\rangle$	$\langle t,c\rangle$
$\langle s,d\rangle$	$\langle s,a\rangle$	$\langle t,d\rangle$
$\langle t,a\rangle$	$\langle t,b\rangle$	$\langle s,a\rangle$
$\langle t,b\rangle$	$\langle t,d\rangle$	$\langle s,c\rangle$
$\langle t,c\rangle$	$\langle t,c\rangle$	$\langle s,b\rangle$
$\langle t,d\rangle$	$\langle t,a\rangle$	$\langle s,d\rangle$

FIG. 7.3.8   Machines M_1 and M_2 Form a Cascade Decomposition of the Machine M

is not necessary in general. Because machine-decomposition theory is so rich, the machine decompositions grow quite numerous when this assumption is relaxed.

Now we turn to the problem of finding the conditions under which a machine M can be realized as the cascade connection of smaller machines M_1 and M_2. This topic is closely related to the concept of machine homomorphism. In particular, the subject depends heavily on state partitions with a property known as the substitution property. These partitions are related to homomorphic machine images.

Definition   Let M be a machine with state set S_M and input set I_M. Consider a partition P of the state set, where the P-equivalence class containing state s is denoted $[s]_P$. Then the partition P is said to have the *substitution property* if, for every pair of P-equivalent states s_1 and s_2, the successor of s_1 is P-equivalent to the successor of s_2 under each input.

The machine M shown in Fig. 7.3.9 provides an example of a partition with the substitution property. Observe that the partition

$$P = \{[1,3]_P, [2,4]_P, [5]_P\}$$

has the substitution property. For example, the input 1 carries states 2 and 4 into states 3 and 1, respectively, which are P-equivalent states.

s	$M(s,i)$	
	$i = 0$	$i = 1$
1	2	5
2	4	3
3	2	5
4	4	1
5	4	1

s'	$M'(s',i)$	
	$i = 0$	$i = 1$
$[1,3]_P$	$[2,4]_P$	$[5]_P$
$[2,4]_P$	$[2,4]_P$	$[1,3]_P$
$[5]_P$	$[2,4]_P$	$[1,3]_P$

FIG. 7.3.9   Machine M' Has P-Equivalence Classes of M As Its States, Where P Is a Partition with the Substitution Property

When a partition of the state set has the substitution property, we can treat the classes of the partition as states in a new machine M', which uses the same input set as M. The substitution property guarantees that the next-state mapping of M' is well defined. That is,

$$M'([s]_P, i) = [M(s,i)]_P.$$

Hence, for the partition P on machine M, we can construct the 3-state machine M' shown in Fig. 7.3.9. The state set $S_{M'}$ of M' is the set of equivalence classes in the partition P of the state set S_M of M. Using the notation introduced in Chapter One for quotient structures, we say $S_{M'} = S_M/P$, read as "S_M modulo P."

We shall show that, whenever there exists a partition P with the substitution property for a machine M, there exists a cascade decomposition in which the machine M' on the P-equivalence classes is the front machine. Before we prove this result, we note that the substitution property is related to machine homomorphisms.

Theorem   A partition P of the states of a machine M has the substitution property if and only if there exists a machine homomorphism that maps the states S_M into the P-equivalence classes.

Proof: To prove the forward implication, we assume that partition P has the substitution property, and we construct the machine homomorphism. Let the machine M' be the machine with state set $S_{M'} = S_M/P$ and with input set I_M, the same as that of M. For each s' in $S_{M'}$ and each i in I_M, we define $M'(s',i) = [M(s,i)]_P$, where $s' = [s]_P$ for some s in S_M. As we mention earlier, this function is well defined when P has the substitution property. The maps h_1 and h_2 for the machine homomorphism are the maps such that $h_1(i) = i$ for all i, and $h_2(s) = [s]_P$ for all s in S_M. Then

$$h_2(M(s,i)) = M'(h_2(s), h_1(i)),$$

so M' is a homomorphic image of M.

To prove the reverse implication, we assume that M' is a homomorphic image of M, and we find a partition with the substitution property. Let the homomorphism functions be h_1 and h_2. For the equivalence relation P, we say that two states s_1 and s_2 in S_M are P-equivalent if and only if $h_2(s_1) = h_2(s_2)$ in $S_{M'}$. Clearly this is an equivalence relation. Then, for all i in I_M and for all s in S_M,

$$M'([s]_P, i) = M'(h_2(s), i) = h_2(M(s, i)) = [M(s, i)]_P.$$

The successors of P-equivalent states in M under every input are P-equivalent, so P has the substitution property. □

Every machine has two trivial partitions with the substitution property: the partition in which every state is P-equivalent to every other state, and the partition in which every state is P-equivalent only to itself. Because neither of these partitions leads to an interesting cascade decomposition, we limit our discussion here to the nontrivial partitions.

The central theorem in our study of cascade decompositions states that there is a one-to-one correspondence between cascade decompositions and partitions with the substitution property.

Theorem A machine M has a cascade decomposition into a head machine M_1 and a tail machine M_2 if and only if there exists a partition P of the state set of M such that P has the substitution property.

Proof: To prove the forward implication, we show that, if M has a cascade decomposition into M_1 and M_2, then M_1 is a homomorphic image of M. If the cascade connection of M_1 and M_2 is a decomposition of a machine M, then each state of M is associated with one state pair in $S_{M_1} \times S_{M_2}$. Let the state s in S_M correspond to the state $\langle s_1, s_2 \rangle$ in $S_{M_1} \times S_{M_2}$. Because M_1 and M_2 form a decomposition of M, we must have

$$M(s, i) = \langle M_1(s_1, i), M_2(s_2, \langle s_1, i \rangle) \rangle$$

for each i in I and each s in S_M. Then, if $M(s, i) = t$, we must have $M_1(s_1, i) = t_1$, where t is represented by $\langle t_1, t_2 \rangle$. Now consider the map $h: S_M \to S_{M_1}$ that maps s into s_1 if s is represented by $\langle s_1, s_2 \rangle$. We have

$$h(M(s, i)) = t_1 = M_1(s_1, i) = M_1(h(s), i).$$

Then M_1 is a homomorphic image of M. From the preceding theorem, there must exist a corresponding partition P with the substitution property.

To prove the reverse implication, we show how to construct the machines M_1 and M_2 that simulate M. Let M_1 be the machine with state set S_M/P and with the next-state function given by

$$M_1([s]_P, i) = [M(s, i)]_P.$$

Then M_1 will correctly compute the first coordinate of the next-state function. The machine M_2 is constructed so that, if $M(s, i) = t$, then $M_2(s_2, \langle s_1, i \rangle) = t_2$, where s and t are represented by $\langle s_1, s_2 \rangle$ and $\langle t_1, t_2 \rangle$, respectively. To complete the proof, it is sufficient to show how each state of M is represented by a pair in $S_{M_1} \times S_{M_2}$, and how with this representation we can construct the next-state function for M_2.

We let the number of states in M_2 be equal to the size of the largest P-equivalence class. Then we can assign each s in S_M to some state pair $\langle s_1, s_2 \rangle$ so that, for two states s and t, we have $s_1 = t_1$ if and only if s and t are P-equivalent. Because the next-state function of M_2 is a map from $S_{M_2} \times (S_{M_1} \times I)$ into S_{M_2}, we define $M_2(s_2, \langle s_1, i \rangle)$ to be t_2 if $M(s, i) = t$. With this definition, if $s = \langle s_1, s_2 \rangle$ and $t = M(s, i)$, then

$$M(\langle s_1, s_2 \rangle, i) = \langle M_1(s_1, i), M_2(s_2, \langle s_1, i \rangle) \rangle = \langle t_1, t_2 \rangle = t,$$

as required for the machine simulation by a cascade decomposition. $\square$

s_1	s_2	
	A	B
$a = [1, 3]$	1	3
$b = [2, 4]$	2	4
$c = [5]$	5	—

S_{M_1}
$a = [1, 3]_P$
$b = [2, 4]_P$
$c = [5]_P$

$$s \in S_M = \langle s_1, s_2 \rangle$$

FIG. 7.3.10   The Map $S_M \to S_{M_1} \times S_{M_2}$ That Represents Each State of M as a 2-tuple

To clarify the construction process given in the theorem, let us re-examine the machine M (Fig. 7.3.9), which has a nontrivial partition P with the substitution property. Because there are three P-equivalence classes and no class has more than two elements, M_1 and M_2 have three and two states, respectively. Fig. 7.3.10 shows the state of M that corresponds to each pair of states of M_1 and M_2. Each state s of M corresponds to the pair $\langle s_1, s_2 \rangle$, where $s_1 \in S_{M_1}$ identifies the P-equivalence class containing s, and $s_2 \in S_{M_2}$ identifies a particular component of a P-equivalence class. Note, for example, that state A in S_{M_2} identifies either state 1, 2, or 5 of M, depending on which P-equivalence class is identified by s_1.

With this correspondence established, we can construct the next-state functions for the cascade machine and for M_1 and M_2 as shown in Fig. 7.3.11.

The ten nonempty entries of M_2 are a rearrangement of the second coordinates of the entries of the table for M. For example, $M_2(A, \langle a, 1 \rangle)$ must be the second coordinate of $M(\langle a, A \rangle, 1)$, which is state A. The two empty entries in M_2 are unspecified because these conditions are unreachable if the cascade machine truly simulates M. In the actual construction of the cascade machine, a designer may choose to specify the two missing entries of M_2 so as to minimize the cost of M_2 or to be compatible with error-detection circuitry.

The trivial P-equivalence relations yield no useful decompositions, because for these relations either M_1 or M_2 must have as many states as M. Nontrivial decompositions, however, lead to potentially useful

$s = \langle s_1, s_2 \rangle$	$M(\langle s_1, s_2 \rangle, i)$	
	$i = 0$	$i = 1$
$1 = \langle a, A \rangle$	$\langle b, A \rangle$	$\langle c, A \rangle$
$2 = \langle b, A \rangle$	$\langle b, B \rangle$	$\langle a, B \rangle$
$3 = \langle a, B \rangle$	$\langle b, A \rangle$	$\langle c, A \rangle$
$4 = \langle b, B \rangle$	$\langle b, B \rangle$	$\langle a, A \rangle$
$5 = \langle c, A \rangle$	$\langle b, B \rangle$	$\langle a, A \rangle$

$s_1 = [s]_P$	$M_1(s_1, i)$	
	$i = 0$	$i = 1$
$a = [1, 3]_P$	b	c
$b = [2, 4]_P$	b	a
$c = [5]_P$	b	a

s_2	$M_2(s_2, \langle s_1, i \rangle)$ When $\langle s_1, i \rangle$ Is					
	$\langle a, 0 \rangle$	$\langle a, 1 \rangle$	$\langle b, 0 \rangle$	$\langle b, 1 \rangle$	$\langle c, 0 \rangle$	$\langle c, 1 \rangle$
A	A	A	B	B	B	A
B	A	A	B	A	—	—

FIG. 7.3.11   Next-State Functions for a Cascade Machine That Simulates Machine M of Fig. 7.3.9 and for Its Head and Tail Machines

decompositions, because both M_1 and M_2 must have fewer states than M in such cases. In some cases, the cascade machine is quite economical relative to other implementations of M.

Summary The central idea of this section is that of homomorphism. Machine homomorphisms, like homomorphisms of other algebraic structures, are closely related to simulation. A machine M_2 can simulate a machine M_1 if and only if M_1 is a homomorphic image of M_2.

We can simulate a machine M by a cascade connection of two other machines if we can find a partition of M that has the substitution property. Such a partition exists if and only if the next-state function of M is well defined on the classes of the partition. If the partition exists, the quotient structure of equivalence classes under the action of the next-state function of M is a homomorphic image of M itself. Moreover, whenever a partition of M under the action of the next-state function is a homomorphic image of M, then M has a partition with the substitution property.

To each partition of M with the substitution property, there corresponds a cascade decomposition in which the head machine is the quotient machine of the equivalence classes. Conversely, to each cascade decomposition that realizes M by a one-to-one state correspondence, there corresponds a homomorphism from M onto the head machine of the decomposition. The head machine is a machine on equivalence classes of states in a partition with the substitution property.

The theory of machine decompositions has been extended well beyond the material presented here. The most notable extension relaxes the assumption that each state of M corresponds to one state of the composite machine. Cascade decompositions under this relaxed assumption depend on the existence of "covers" of the states of M that have the substitution property. A cover is like a partition, except that the classes of states within a cover need not be disjoint. Because the restrictions on state covers are far less stringent than those on state partitions, decompositions based on covers are far more numerous than decompositions based on partitions. The book by Hartmanis and Stearns (1966) provides further details. Hennie (1968) has written an excellent text on the general theory of sequential machines and the study of machine decompositions.

Related to the work of Hartmanis and Stearns is the algebraic theory of machine decompositions. This theory is an outgrowth of a paper by Krohn and Rhodes (1962). Algebraic decompositions are cascade decompositions for which the semigroups of the head and tail machines are each subsemigroups of the machine that they simulate. The use of algebraic decompositions tends to be restrictive in the sense that some machines with nontrivial Hartmanis-Stearns decompositions have no Krohn-Rhodes decompositions, while the converse is never true. Nonetheless, the Krohn-Rhodes theory of machine decomposition is an elegant theory that attracted much interest.

To summarize the Krohn-Rhodes theory briefly, we say that a machine is a *group machine* if each input permutes the states of the machine. The semigroup of a group machine is a group. The front machine of a cascade decomposition of a group machine is a homomorphic image of the group machine. Consequently, the semigroup of the front machine is a homomorphic image of the group of the group machine. From the fundamental theorem of group homomorphism, we know that, if G' is a homomorphic image of G, then G' is isomorphic to G/H, where H is a normal subgroup of G. Thus, if G is a simple group, then G has no nontrivial normal subgroup, and therefore G has no nontrivial Krohn-Rhodes decompositions. Otherwise, the possible cascade decompositions of G are in one-to-one correspondence with its normal subgroups.

We can iterate algebraic decompositions by replacing a head or tail machine with a cascade decomposition that can simulate it. We can iterate the algebraic decomposition of a group machine to the point where each component machine is a group machine whose semigroup is a simple group. Moreover, as a consequence of the famous Jordan-Holder theorem (see Van der Waerden 1931), we can prove that every complete cascade decomposition of a group machine has the same component machines, although possibly not in the same order.

Machines that are not group machines generally have richer algebraic decompositions. Such machines can be decomposed into a cascade of two-state machines and a group machine. Each of the two-state machines must be one of only three types, and the group machine has a decomposition structure as described earlier. A major drawback of the Krohn-Rhodes decomposition is that the number of states in the cascade decomposition may in some cases be substantially larger than the number of states in the original machine. For example, for each n, there exists a two-input, n-state machine whose semigroup is S_n. The Krohn-Rhodes decomposition of this machine consists of a two-state machine and a machine with $n!/2$ states that does computation in A_n, the alternating group of degree n. The latter machine cannot be further decomposed for $n \neq 4$, because A_n is a simple group. Thus, the theory replaces an n-state machine by a machine with $n!/2$ states, yet cannot tell us how to construct the latter machine economically.

Ginzburg (1968) provides one of the best expositions of the Krohn-Rhodes theory. Other works on this theory include those by Kalman et al. (1969) and Arbib (1969).

Exercises

7.3.1 For the machines shown in Fig. 7.3.6, find all of the homomorphisms from M_2 into M_1.

s	$M(s,i)$	
	$i = 0$	$i = 1$
1	6	2
2	5	8
3	8	4
4	3	6
5	8	7
6	1	5
7	1	8
8	3	1

FIG. 7.3.12 The Machine M for Ex. 7.3.2

7.3.2 Construct the cascade decomposition for the machine M shown in Fig. 7.3.12. Note that states 1, 3, and 5 are P-equivalent.

7.3.3 Construct the cascade decomposition for the machine M shown in Fig. 7.3.13. Note that states 1, 2, 3, and 4 are P-equivalent. Find a mapping from S_M to $S_{M_1} \times' S_{M_2}$ such that the next-state function of M_2 is independent of M_1.

7.3.4 The decomposition of M in Ex. 7.3.3 is called a *parallel-machine* decomposition, because M can be simulated by M_1 and M_2 as shown in Fig. 7.3.14. Show that a machine M has a parallel decomposition if and only if there exist two partitions P_1 and P_2 with the substitution property and such that no pair of P_1-equivalent states lie in the same P_2-equivalence class.

7.3.5 For the machine M shown in Fig. 7.3.15, find all of the nontrivial partitions with the substitution property.

7.3.6 Construct an algorithm for finding all of the state partitions with the substitution property for a given machine M.

s	$M(s, i)$	
	$i = 0$	$i = 1$
1	5	3
2	6	4
3	7	1
4	8	2
5	1	7
6	2	8
7	3	5
8	4	6

FIG. 7.3.13   The Machine M for Ex. 7.3.3

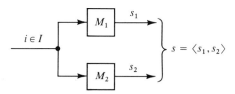

FIG. 7.3.14   A Parallel-Machine Decomposition for Machine M of Fig. 7.3.13

s	$M(s, i)$	
	$i = 0$	$i = 1$
1	5	2
2	8	3
3	7	4
4	6	1
5	3	6
6	2	7
7	1	8
8	4	5

FIG. 7.3.15   The Machine M for Ex. 7.3.5

7.3.7 The composition table for a group G may be viewed as the next-state table of a sequential machine M that has the elements of G as its states and as its inputs. In this case, $M(g, g') = g \cdot g'$. Prove that the state partitions of M with the substitution property are in one-to-one correspondence with the subgroups of G.

7.3.8 A *state cover C* with the substitution property for a machine M satisfies all of the properties of a state partition with the substitution property except that the subsets in the state cover need not be disjoint. Assume that C is a state cover of a machine M as described in Ex. 7.3.7, such that the identity state of M is in a unique subset of C. Prove that C is a collection of cosets relative to some subgroup of G.

7.3.9 In Ex. 7.3.8, prove that C is a collection of cosets relative to a subgroup of G if any state, not necessarily the identity state, is in a unique subset of C.

7.4 Sequential Machines as Sequence Recognizers

In this section we continue our study of sequential machines, but we take a viewpoint somewhat different from that in preceding sections. Here we investigate the application of sequential machines to the problem of sequence recognition. In abstract terms, we assume that the set of finite input sequences is partitioned into two classes, and we wish to construct a device that outputs a 1 if and only if an input sequence is in a particular equivalence class. In less abstract terms, this is precisely the task performed by a compiler in doing a syntax check of a program. The set of all possible programs is the set of input sequences; each program lies in one of two classes, depending solely on whether it is syntactically correct or incorrect.

Our major result in this section is a demonstration that a finite-state machine can recognize a set of input sequences if and only if the set is regular. A regular set is one that satisfies some simple defining properties. We show how to construct a finite-state-machine recognizer for each regular set and how to find the regular set recognized by each finite-state machine.

At the end of the section we show a practical application of the idea of regular sets. Compilers for programming languages normally simulate finite-state machines in order to accomplish part of the process of syntax checking. If the collection of syntactically correct programs for a given programming language were a regular set, then the syntax recognizer for that language could be a program that does nothing more than simulate a finite-state machine. However, the syntactically correct programs never form a regular set in practice. Yet most compilers do simulate finite-state machines in order to recognize certain subsequences in a program. These subsequences include such things as numbers, variable names, and reserved words. We conclude this section by showing a typical recognizer of this type.

We begin with a discussion that leads up to a formal definition of a regular set. Let I be a set of letters. Recall that I^* is the free monoid generated by I. That is, I^* contains all of the finite sequences of elements of I. We label this set I to signify that it is a set of input symbols for a finite-state machine.

Definition
A finite-state machine M is said to *recognize* a subset T of sequences in I^* if, when the machine is started in a specific initial state s_0, the machine produces a 1 output for each sequence in T, and otherwise produces a 0 output.

In this model, we recognize an input sequence by indicating whether or not it is in T. To do so, we reset M to the initial state s_0; then we excite M with the input sequence, one symbol at a time. When the last symbol in the sequence has forced a state transition of M, we observe the output of M. If the output is 1, then the sequence is in T; if the output is 0, then the sequence is not in T. More generally, a 1 output indicates that the input sequence up to that point is in T, and a 1 output may occur more than once during the input of a sequence when initial subsequences of the sequence are in T.

Fig. 7.4.1 shows two examples of sequence recognizers. For each machine, $I = \{0, 1\}$, and the initial states are labeled s_0. The double circles identify states that give a 1 output, which in these two examples happen to be the states labeled s_0. Machine M_1 recognizes the set of input sequences that contain no 0's and contain an even number of 1's. Machine M_2 recognizes the set of sequences that contain an even number of 1's and any arbitrary number of 0's, where the 0's and 1's can be interspersed in any manner. Both machines recognize the sequences 11, 1111, 111111, etc., and the null sequence Λ; neither machine recognizes 1, 111, 11111, etc. Machine M_2 recognizes the sequences 0110,

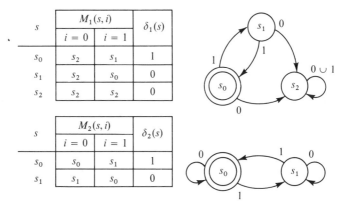

s	$M_1(s, i)$		$\delta_1(s)$
	$i = 0$	$i = 1$	
s_0	s_2	s_1	1
s_1	s_2	s_0	0
s_2	s_2	s_2	0

s	$M_2(s, i)$		$\delta_2(s)$
	$i = 0$	$i = 1$	
s_0	s_0	s_1	1
s_1	s_1	s_0	0

FIG. 7.4.1 Two Finite-State Machines That Act as Sequence Recognizers

1110100, and 1011001 but, because these sequences contain 0's, none of these sequences is recognized by M_1.

We assume that the output set for a finite-state-machine recognizer is $\{0, 1\}$. Thus an input sequence moves the machine from s_0 to a state with a 1 output if and only if the sequence is in the set of recognizable sequences. This observation suggests the next definition.

Definition A state in a finite-state-machine recognizer is called a *final state* if it has a 1 output. Then, according to our previous definition of sequence recognizer, an input sequence is said to be *recognized* by a finite-state machine if the sequence carries the machine from the initial state s_0 to any final state. We follow the convention of indicating final states by double circles, as in Fig. 7.4.1. Note that a state may be both initial and final.

Each of the machines shown in Fig. 7.4.1 recognizes an infinite set of sequences. To characterize these sets—or any sets of sequences recognizable by a sequential machine—we use a compact notation involving algebraic expressions known as regular expressions. We next define this term formally; then we show how each regular expression represents a set of input symbols.

Definition Let I be a finite set of input symbols. The following four rules describe all *regular expressions* that we can construct from the set I:

    (i)   each i in I is a regular expression;

    (ii) $\varnothing$ and Λ are regular expressions;

    (iii) if V and W are regular expressions, then $V \cup W$, VW, and V^* are regular expressions;

    (iv) the only regular expressions are those described by rules **(i)**, **(ii)**, and **(iii)**, where rule **(iii)** is applied at most a finite number of times.

We use regular expressions to represent sets of input sequences under the following conventions.

    (i)   For each i in I, the regular expression i represents the singleton set $\{i\}$.

    (ii) The regular expression $\varnothing$ represents the empty set of input sequences. The regular expression Λ represents $\{\Lambda\}$, the set containing the empty sequence.

    (iii) Let V and W be regular expressions, and let $||V||$ and $||W||$ be the sets of input sequences that they represent. Then $||V \cup W||$, $||VW||$, and $||V^*||$ are defined as

$$||V \cup W|| = ||V|| \cup ||W|| = \{v | (v \in V) \vee (v \in W)\},$$
$$||VW|| = \{vw | (v \in V) \wedge (w \in W)\}, \text{ and}$$
$$||V^*|| = ||\Lambda \cup V \cup VV \cup VVV \cup \ldots ||,$$

where v and w are individual input sequences.

Note that $V \cup W$ in regular-expression notation denotes the union of two sets of sequences. The notation VW indicates the concatenation of two sets of sequences, where the set concatenation is formed by concatenating each element of $||V||$ with each element of $||W||$. Finally, V^* represents the set of sequences composed of zero or more sequences in $||V||$ concatenated together. Simple examples of regular expressions for the set $I = \{0, 1\}$ include $0^*11^* \cup 010^*$ and $(00 \cup 1)^*110$. In the second example, the parentheses around $00 \cup 1$ indicate that $*$ applies to this whole expression. In the expression $00 \cup 1^*110$, the $*$ applies only to the regular expression 1. Table 7.4.1 lists some further examples of regular expressions, with word descriptions of the sets that they represent.

Not every set of input sequences can be described by a regular expression. For example, if 0^n denotes the sequence of n 0's, then no regular expression describes the set $\{0^n1^n | n = 0, 1, 2, \ldots\}$. Similarly, if α^R denotes the reversal of a sequence α, then the set $\{\alpha\alpha^R | \alpha \in I^*\}$ has no corresponding regular expression.

Technically we should distinguish between a regular expression V and the set $||V||$ of input sequences that it represents. However, again we slightly abuse notation and use V to denote both an expression and the set that it represents. Any ambiguities that arise from this abuse can always be resolved from the context of the expression.

TABLE 7.4.1 Examples of Regular Expressions Used to Describe Sets of Input Sequences

Regular Expression	Description of Set Represented
$(0 \cup 1)^*$	All sequences in $I^* = \{0, 1\}^*$.
$(11)^*$	All sequences containing an even number of 1's and containing no 0's. (These are the sequences recognized by machine M_1 of Fig. 7.4.1.)
$(0^*10^*1)^*0^*$	All sequences with an even number of 1's and an arbitrary number of 0's anywhere in the sequence. (These are the sequences recognized by machine M_2 of Fig. 7.4.1.)
$(101 \cup 000 \cup 10 \cup 1)^*$	All sequences formed by concatenating the subsequences 101, 000, 10, and 1 to each other in arbitrary order, with each subsequence appearing zero or more times. Included in this set are such sequences as Λ, 1, 11, 111, 1101, 1000, and 00010. Not included are such sequences as 100, 001, and 10100.
$((011 \cup 10)^*(111)^*)^*$	These sequences have zero or more repetitions of the subsequences in $(011 \cup 10)^*(111)^*$. The subsequences in turn have zero or more repetitions of 011 or 10, followed by zero or more repetitions of 111. Among the sequences in this set are Λ, 011, 10, 111, 011111, 10111, 01110111, 10011111, 1011110111, 01111110111, and 0111011110011111.

Definition   The set of sequences represented by a regular expression is called a *regular set*.

Note that $\emptyset$ and Λ both are regular expressions and that we distinguish between them. The regular expression $\emptyset$ represents the empty set of input sequences, whereas the regular expression Λ represents the set $\{\Lambda\}$, which is a set with one element (the empty sequence). We must distinguish between these expressions because different machines are required to recognize the corresponding regular sets. A machine that recognizes the regular set $\emptyset$ is a machine in which no final state is reachable from the initial state; any input causes the machine to give a 0 output. In contrast, a machine that recognizes the regular set Λ gives a 1 output in its initial state when no symbols have been received (the correct response to the null sequence) but never returns to the initial state and can reach no other final states. Fig. 7.4.2 shows machines that recognize $\emptyset$ and Λ.

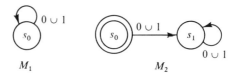

FIG. 7.4.2   Machine M_1 Recognizes $\emptyset$, and Machine M_2 Recognizes Λ

S. C. Kleene (1956) was the first to show that every finite-state machine recognizes a regular set, and that every regular set has a finite-state-machine recognizer. We are now ready to prove the first of the two theorems that describe the relation of regular sets to finite-state-machine recognizers. This is called the *Kleene analysis theorem*.

Theorem   The set of input sequences that carries a finite-state machine from any state s_i to any state s_j is a regular set. State s_i and state s_j may be distinct states or they may be the same state.

Proof: The proof is inductive. We consider the set of sequences that carry M from s_i to s_j such that M stays entirely within a k-state submachine that includes the states s_i and s_j. We prove that the set of such sequences is regular for $k = 1$. Then we show that, if the sets of sequences for $1 \leq k \leq n$ are regular for any $n \geq 1$, then so are the sets of sequences for $k = n + 1$.

Basis step, $k = 1$. Let T be the set of sequences that carry M from s_i to s_j such that M stays entirely within a 1-state submachine. Then $s_i = s_j$, and the sequences must leave M fixed in the state s_i. Thus T includes Λ and each input symbol that leaves M fixed in state s_i. If the input symbols $x_1, x_2, \ldots, x_r$ leave M in state s_i, then T is described by the regular expression $(x_1 \cup x_2 \cup \ldots \cup x_r)^*$. Fig. 7.4.3($a$) illustrates

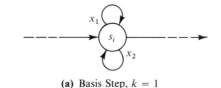

(a) Basis Step, $k = 1$

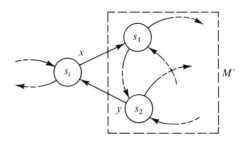

(b) Induction Case 1, $s_i = s_j$

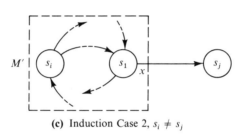

(c) Induction Case 2, $s_i \neq s_j$

FIG. 7.4.3 Examples of Situations Encountered in the Proof of Kleene's Analysis Theorem

this case. If no input symbol leaves M in s_i, then $T = \Lambda$, which also is a regular expression. Hence the induction hypothesis is true when $k = 1$.

Induction step. Now we assume that the set of sequences that carry M from s_i to s_j within a k-state submachine is regular for each k in the interval $1 \leq k \leq n$. We show that the set of sequences for $k = n + 1$ also must be regular. We consider two cases: first, $s_i = s_j$, and then $s_i \neq s_j$.

Case 1, $s_i = s_j$. We examine the set of sequences that carry M from s_i back to s_i for the first time, while M stays entirely within an $(n + 1)$-state submachine. Call this set of sequences T. Fig. 7.4.3(b) illustrates the situation for this case. The submachine M' has n states, and M' and s_i together comprise a submachine with $n + 1$ states.

The sequences that carry M from s_i back to s_i for the first time include single input letters that leave M fixed in state s_i. They also include sequences that move M from s_i to some state s_1 in M', then from s_1

(perhaps through other states) to some state s_2 in M', before a final transition moves from s_2 back to s_i.

Let T_{12} be the set of all sequences that move M' from s_1 to s_2, and note that T_{12} is a regular set by the induction hypothesis because M' has n states. Let x be the input symbol that carries M from s_i to s_1, and let y be the symbol that carries M from s_2 to s_i. Then $xT_{12}y$ is a set of sequences that carries M from s_i back to s_i for the first time, while staying within the $(n + 1)$-state submachine. Thus every sequence in T is described by an expression of the form $xT_{jk}y$ (a sequence through M') for s_j and s_k in M', or by a single input letter that leaves M fixed in s_i. Both of these forms are regular expressions, so the sets that they describe are regular.

Therefore T is a regular set, because T is the union of regular sets of single input letters and regular sets of the form $xT_{jk}y$. Then T^* is the set of all input sequences that carry M from s_i back to s_i, not necessarily for the first time, while M stays entirely within an $(n + 1)$-state submachine. Because T^* is a regular set, the induction step is proved for the case when $s_i = s_j$.

Case 2, $s_i \neq s_j$. We proceed as in the preceding case. Now we let T be the set of sequences that carry M from s_i to s_j for the first time, while M stays entirely within an $(n + 1)$-state submachine. Fig. 7.4.3(c) depicts this situation. Here M' is an n-state submachine, with s_i as one of its states. A typical sequence that carries M through M' to s_j for the first time carries M to some state s_1 in M' just prior to reaching s_j.

We can move M from s_i to s_j by first moving from s_i to s_1 and then using a single input symbol to move from s_1 to s_j. By the induction hypothesis, the set of sequences that move M' from s_i to s_1 is a regular set (because M' has n states). Let us call this set T_{i1}, and let x be the input symbol that carries M from s_1 to s_j. Then $T_{i1}x$ is the set of sequences that carry M from s_i to s_j for the first time through s_1 as the immediate predecessor of s_j, and $T_{i1}x$ is a regular set. Consequently, T (the set of all sequences that move M from s_i through M' to s_j for the first time) is a regular set, because it is the union of regular sets $T_{ik}x$, where s_k ranges over all of the states in M', and x ranges over all of the input symbols that move M from s_k to s_j.

Now let T' be the set of all sequences that carry M from s_j back to s_j for the first time while M stays entirely within the $(n + 1)$-state submachine. From Case 1, we know that T' is regular. Then $T(T')^*$ is the set of all sequences that carry M from s_i to s_j, not necessarily for the first time, while M stays within the $(n + 1)$-state submachine. This set is regular, and the proof of the theorem is complete. □

As an immediate consequence of the Kleene analysis theorem, we have the following important result.

Corollary The set of sequences recognized by a finite-state machine is regular.

Proof: Let s_0 be the starting state of a finite-state machine M.

The set of sequences that carry M to a particular final state s_f is a regular set. If M has several final states, then the set of sequences that carry M from s_0 to any final state in the set of final states also is regular, because this set is the union of regular sets. If M has no final state, then M recognizes $\varnothing$, which also is a regular set. $\qquad\qquad$ $\square$

We now turn our attention to the proof of the Kleene synthesis theorem, which states that every regular expression can be recognized by a finite-state machine. Our proof is a constructive one—that is, we show precisely how to build a recognizer for any given regular expression. When we carry out this construction, in some cases we obtain a recognizer that does not fit our definition of a sequential machine. However, such recognizers are equivalent to sequential machines, because for each such recognizer we can construct a finite-state machine that recognizes precisely the same sequences.

In our development of this topic, we first examine the class of recognizers that we obtain from the Kleene synthesis theorem, and we prove that each recognizer in this class is equivalent to a finite-state machine. Then we prove the Kleene synthesis theorem. The class of recognizers is described by the following definition.

Definition $\qquad$ A *transition system* $\langle S, I, t, \mathsf{R}_\Lambda \rangle$ is an algebraic structure with a set S of states, a set I of input symbols, a next-state function $t: S \times I \to \mathscr{P}(S)$, and a relation R_Λ on S.

Fig. 7.4.4 shows a typical transition system in both tabular and graphical form. A transition system is much like a sequential machine, but there are two distinguishing characteristics. In a sequential machine, each input symbol carries a state into a unique next state, whereas in a transition system each input symbol carries the system from a state into a subset of states. For example, in Fig. 7.4.4, note that a 0 input causes the system to move from s_1 to $\{s_2, s_3\}$.

The second essential difference concerns the relation R_Λ. We interpret the relation $s_i \, \mathsf{R}_\Lambda \, s_j$ to mean that the empty sequence Λ carries the

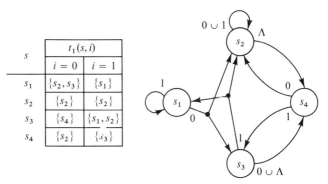

s	$t_1(s, i)$	
	$i = 0$	$i = 1$
s_1	$\{s_2, s_3\}$	$\{s_1\}$
s_2	$\{s_2\}$	$\{s_2\}$
s_3	$\{s_4\}$	$\{s_1, s_2\}$
s_4	$\{s_2\}$	$\{s_3\}$

FIG. 7.4.4 An Example of a Transition System with $\mathsf{R}_\Lambda = \{\langle s_2, s_4 \rangle, \langle s_3, s_4 \rangle\}$

transition system from state s_i to s_j. Using this interpretation, we show transitions labeled Λ in the state graph of Fig. 7.4.4 from s_2 to s_4 and from s_3 to s_4. Recall that the state transitions in sequential machines occur at unit intervals as each successive input symbol is received. A Λ-transition in a transition system is a state transition that does not coincide with the receipt of an input symbol, but rather occurs spontaneously. In essence, the machine responds to an input symbol at each clock tick, but it responds to a Λ-input without waiting for the next tick of the clock.

Because of the two differences between transition systems and finite-state machines, we have some difficulty in extending the next-state function t to the domain $S \times I^*$ as we previously extended the next-state function of a finite-state machine. For example, if we start the transition system of Fig. 7.4.4 in state s_3, then to what subset of states does the sequence 10 carry it? The input 1 carries the system to states s_1 and s_2, simultaneously. Now how are we to compute the effect of the 0 in the input sequence 10? Thus far we have defined the next-state function only for the system in a single state. We resolve this problem by defining

$$t_1(\{s_1, s_2\}, 0) = t_1(s_1, 0) \cup t_1(s_2, 0).$$

Thus the sequence 10 carries the system from state s_3 to the subset of states $\{s_2, s_3\}$. Note that the sequence 10 is the same as $\Lambda 10$, $1\Lambda 0$, $\Lambda 1\Lambda 0$, etc. Thus, when we take Λ-transitions into consideration, we find that $t_1(s_3, 10) = \{s_2, s_3, s_4\}$.

We now set out to prove that each transition system is equivalent to some sequential machine. We do this in two steps. First we show that every Λ-transition can be removed and replaced by ordinary transitions. Then we show that the next-state function $t: S \times I \rightarrow \mathcal{P}(S)$ extends naturally to a function $t': \mathcal{P}(S) \times I \rightarrow \mathcal{P}(S)$, and that t' is a next-state function of a finite-state machine. The state set of t' is $\mathcal{P}(S)$, the power set of the state set of S, so t' describes how subsets of states in S are mapped into subsets of states under the action of t.

We have not yet stated what we mean by equivalence of transition systems. Note that transitions in a transition system may carry the system from a single state into a subset of states. We cannot always speak of a transition system M in a state s_i, because M may be in a subset of states rather than in a single state. With this view, a transition system M is equivalent to a transition system M' if

(i)   each subset of states of M corresponds to a unique subset of states of M', and

(ii) for each input sequence α, if α carries M from state subset T_1 to state subset T_2, then α carries M' from state subset T_1' to state subset T_2', where T_1 and T_2 correspond to T_1' and T_2', respectively.

This definition of equivalence does not include the notion of recognizer because it does not involve initial and final states. We take up the question of initial and final states and sequence recognition after we establish the equivalence of transition systems and finite-state machines in the next two lemmas.

We begin by removing the Λ-transitions.

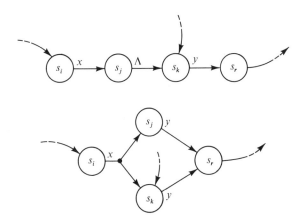

FIG. 7.4.5   Two Equivalent Transition Systems

Lemma   Every transition system is equivalent to a transition system without Λ-transitions.

Proof: Fig. 7.4.5 illustrates the method for eliminating Λ-transitions from a transition system. If an input symbol x carries s_i into s_j, and Λ carries s_j into s_k, then $x\Lambda = x$ carries s_i into s_k. Therefore we insert an x transition between s_i and s_k. Similarly, if the input symbol y carries s_k into s_r, then $\Lambda y = y$ carries s_j into s_r, and we insert a y transition between s_j and s_r.

In general, if there is a Λ-transition from s_j to s_k, then we can remove the Λ-transition by duplicating all of the input transitions into s_j as input transitions into s_k, and all of the output transitions from s_k as output transitions from s_j. It is not difficult to verify that this construction produces an equivalent transition system.   □

Following this procedure to remove the Λ-transitions from the system of Fig. 7.4.4, we obtain the system shown in Fig. 7.4.6. For example, note that the input sequences $0\Lambda = 0$ and $1\Lambda = 1$ in Fig. 7.4.4 will carry s_4 back to itself. Therefore, s_4 in Fig. 7.4.6 is carried back to itself by either a 0 or 1 input.

Now we are ready to show the equivalence of transition systems to finite-state machines.

Lemma   A transition system without Λ-transitions is equivalent to some finite-state machine.

s	$t_2(s, i)$	
	$i = 0$	$i = 1$
s_1	$\{s_2, s_3, s_4\}$	$\{s_1\}$
s_2	$\{s_2, s_4\}$	$\{s_2, s_3, s_4\}$
s_3	$\{s_2, s_4\}$	$\{s_1, s_2, s_3, s_4\}$
s_4	$\{s_2, s_4\}$	$\{s_3, s_4\}$

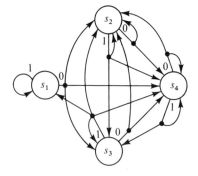

FIG. 7.4.6   A Transition System That Is Equivalent to the One in Fig. 7.4.4 But Has No Λ-Transitions

Proof: Consider a transition system with the next-state function $t: S \times I \to \mathscr{P}(S)$. We show how to construct a next-state function $t': \mathscr{P}(S) \times I \to \mathscr{P}(S)$ for a finite-state machine with state set $\mathscr{P}(S)$. Consider a typical subset $\{s_1, s_2, \ldots, s_r\}$ in $\mathscr{P}(S)$. For input symbol x, t' maps this subset into the union of subsets $\underset{i}{\cup} t(s_i, x)$. Observe that t' is well defined and is a function from $\mathscr{P}(S) \times I$ into $\mathscr{P}(S)$. Then t' is the next-state function of a finite-state machine with state set $\mathscr{P}(S)$ and input set I. Moreover, the state transitions in t' correspond in a one-to-one fashion with the transitions among subsets of states of t. $\square$

Fig. 7.4.7 shows a finite-state machine that is equivalent to the transition system of Fig. 7.4.6. The table was generated by the subset construction process described in the preceding theorem, but we list the next-state function only for the subsets of states reachable from $\{s_1\}$, $\{s_2\}$, $\{s_3\}$, and $\{s_4\}$. As an example of the construction of the table, note that t'_2 carries $\{s_2, s_4\}$ into $\{s_2, s_3, s_4\}$ under a 1 input because

$$t'_2(\{s_2, s_4\}, 1) = t_2(\{s_2\}, 1) \cup t_2(\{s_4\}, 1) = \{s_2, s_3, s_4\}.$$

Because we deal with transition systems as sequence recognizers, we generally specify an initial state and a set of final states in the transition system. When we construct an equivalent finite-state machine for a transition system, we must determine the initial and final states for the sequential machine. It is not difficult to prove the following lemmas (Ex. 7.4.7).

Lemma Let t_1 be a transition system, and let t_2 be an equivalent transition system with no Λ-transitions. Let a state of t_2 be an initial state if it corresponds to an initial state of t_1 or if it corresponds to a state that is reachable from an initial state of t_1 by a sequence of Λ-transitions.

s	$t_2'(s, i)$	
	$i = 0$	$i = 1$
$A = \{s_1\}$	$\{s_2, s_3, s_4\}$	$\{s_1\}$
$B = \{s_2\}$	$\{s_2, s_4\}$	$\{s_2, s_3, s_4\}$
$C = \{s_3\}$	$\{s_2, s_4\}$	$\{s_1, s_2, s_3, s_4\}$
$D = \{s_4\}$	$\{s_2, s_4\}$	$\{s_3, s_4\}$
$E = \{s_2, s_4\}$	$\{s_2, s_4\}$	$\{s_2, s_3, s_4\}$
$F = \{s_3, s_4\}$	$\{s_2, s_4\}$	$\{s_1, s_2, s_3, s_4\}$
$G = \{s_2, s_3, s_4\}$	$\{s_2, s_4\}$	$\{s_1, s_2, s_3, s_4\}$
$H = \{s_1, s_2, s_3, s_4\}$	$\{s_2, s_3, s_4\}$	$\{s_1, s_2, s_3, s_4\}$

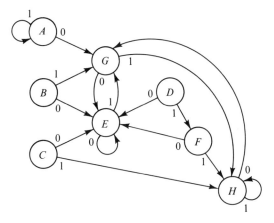

FIG. 7.4.7   A Finite-State Machine That Is Equivalent to the Transition System of Fig. 7.4.6

Let a state of t_2 be a final state if it corresponds to a final state in t_1. Then the set of input sequences that move t_1 from any initial state to any final state is the same as the set of input sequences that move t_2 from any initial state to a final state.

Lemma   Let t_2 be a transition system without Λ-transitions, and let t_2' be the equivalent finite-state machine. Let the initial state of t_2' be the unique subset of S that contains just the initial states of t_2, and let the set of final states of t_2' be the set of subsets of S that contain one or more final states of t_2. Then t_2' recognizes the set of sequences that move t_2 from some initial state to some final state.

To illustrate these lemmas, consider the machines shown in Figs. 7.4.4, 7.4.6, and 7.4.7. Table 7.4.2 shows the initial and final states of t_2 and t_2' that correspond to particular choices of initial and final states for t_1.

TABLE 7.4.2   Initial and Final States for Machines t_2 (Fig. 7.4.6) and t_2' (Fig. 7.4.7) That Correspond to Particular Initial and Final States for t_1 (Fig. 7.4.4)

Initial States			Final States		
t_1	t_2	t_2'	t_1	t_2	t_2'
s_1	s_1	A	s_1	s_1	A, H
s_2	s_2, s_4	E	s_2	s_2	B, E, G, H
s_2, s_3	s_2, s_3, s_4	G	s_1, s_2	s_1, s_2	A, B, E, G, H
s_1, s_2, s_3	s_1, s_2, s_3, s_4	H			

Now we are ready to state and prove the *Kleene synthesis theorem.*

Theorem   Every regular set can be recognized by some finite-state sequential machine.

Proof: We show how to construct a transition-system recognizer for any regular set. From the previous lemmas, we know that this transition system is equivalent to some finite-state machine.

The following construction process parallels the definition of a regular expression.

(i)   Fig. 7.4.8(*a*) shows the state graph for a machine that recognizes the input letter x.

(ii)   Fig. 7.4.2 shows the state graphs for machines that recognize $\varnothing$ and Λ.

(iii)   Let V and W be regular sets, and let M_V and M_W be their finite-state-machine recognizers. Then M_V and M_W each have one initial state. Fig. 7.4.8(*b*) shows a transition-system recognizer for $V \cup W$, Fig. 7.4.8(*c*) for VW, and Fig. 7.4.8(*d*) for V^*. The Λ-transitions to the inputs of M_V and M_W are transitions to the initial state of each machine. The output transitions from M_V and M_W come from each of the final states of the respective machines. In Fig. 7.4.8, a double circle indicates the unique final state of each transition system.

Each of the transition systems described can be replaced by an equivalent finite-state machine. If necessary, the construction then can be iterated using these machines. In this way, we can construct a finite-state machine that recognizes any given regular set.   □

The construction described in the proof of the synthesis theorem is of theoretical interest, but it has only limited appeal in practice. The primary disadvantage of the construction is that it tends to produce finite-state machines with unnecessarily large numbers of states, because the change from a transition system to a finite-state machine results in an exponential increase in the number of states. Of course, we may be

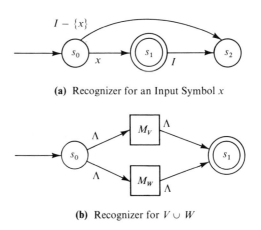

(a) Recognizer for an Input Symbol x

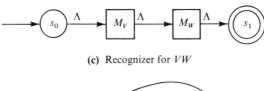

(b) Recognizer for $V \cup W$

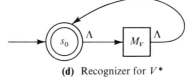

(c) Recognizer for VW

(d) Recognizer for V^*

FIG. 7.4.8   Transition-System Recognizers for Regular Sets

able to use the state-minimization procedure to remove some of these states, but the number of removable states may not be sufficiently great to warrant the use of the algorithm. Another fundamental problem in the practical use of the construction algorithm arises from the fact that each regular set may be described by many different regular expressions. Although they describe the same regular set, different regular expressions may lead to finite-state-machine realizations of vastly different costs. As a trivial example, note that the regular expressions (11)* and (11 $\cup$ 1111)* describe the same regular set, but that the algorithm leads to a much more expensive recognizer for the second expression.

Early in the development of the theory of regular expressions, researchers held high hopes for practical application of the theory to machine design, but other techniques for design of sequential machines have won out in the final analysis. It is rather interesting to discover that the theory of regular expressions has influenced the design of compilers, however. We turn now to the use of regular-expression recognizers in compilers, and we give realistic examples of the design of software recognizers.

In essence, a programming language such as FORTRAN, ALGOL, or PL/I consists of a collection of grammatical rules that specify how to construct a syntactically correct program. Also associated with the language are rules that specify how to assign meaning to grammatical constructions. A program in such a language is a long sequence of characters, each of which is drawn from an alphabet that includes letters, digits, and various punctuation characters (including the blank character). A compiler for a programming language scans the text of a program one character at a time from the beginning to the end. During the scan, the compiler determines which grammatical rules have been used, discovers the meaning for the program, and creates an output program with the identical meaning but in a form that can be executed directly by a computer.

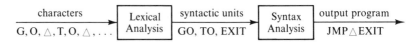

FIG. 7.4.9   The Phases of a Typical Compiler

The compiler normally carries out the compilation process in phases. Fig. 7.4.9 shows a conceptual view of the various phases in operation on a fragment of a program. To set the context for this discussion, suppose that the compiler is scanning the character sequence

$$GO \triangle TO \triangle EXIT$$

where we have used $\triangle$ to indicate the blank character. The first phase of a compiler receives one character at a time, just as a finite-state machine observes one input symbol at a time. Hence the first phase receives the characters "G," "O," "$\triangle$," "T," . . . in sequence. This phase performs sequence recognition to the extent that it identifies the various constants, identifiers, and reserved words in the program. The output of this phase is the sequence of words "GO," "TO," and "EXIT." In Fig. 7.4.9, we call this first phase *lexical analysis.*

The second phase operates on the syntactic entities produced by the first phase, and determines the way in which the grammatical rules have been applied. In this example, the second phase recognizes that the words "GO," "TO," and "EXIT" in this order mean to branch to the label EXIT. The compiler then produces the output "JMP $\triangle$ EXIT," which is an instruction that has the identical meaning in the language of a specific computer. In most cases, the compiler output is not text but a binary n-tuple.

We call the second phase *syntax analysis* because it has a slight functional similarity to syntax analysis of natural language. Because the grammatical constructions of programming languages usually are more complex than the construction rules for regular sets, the set of sequences recognized by a compiler is not regular. Hence syntax analysis

TABLE 7.4.3   Regular Expressions That Represent Syntactic Units in a Programming Language

Syntactic Unit	Regular Expression
Identifier	$\triangle^*L(L \cup D)^*\triangle$
Integer	$\triangle^*DD^*\triangle$
Decimal Number	$\triangle^*(D^*D.D^* \cup D^*.DD^*)\triangle$
Floating-Point Number	$\triangle^*(D^*D.D^* \cup D^*.DD^*)(+ \cup -)DD\triangle$

must depend on tools that are more powerful than the regular-expression recognizers we have constructed previously. However, in most cases, the items recognized during lexical analysis do form regular sets, and most lexical analyzers simulate finite-state machines.

Consider the lexical analysis for a typical ALGOL or PL/I program. The recognizer must determine the boundaries in the input sequence between identifiers, constants, and other syntactic units. Typically, we can describe each of these sets of units by a regular expression. Table 7.4.3 gives an example, in which we use L to represent any alphabetic character (letter) A, B, ..., Z; we use D to represent any numeric character (digit) 0, 1, ..., 9; we use $\triangle$ to represent a blank; and we use the period, plus sign, and minus sign to represent the characters themselves. The table gives regular expressions that describe the sets of identifiers, integers, decimal numbers, and floating-point numbers. We can construct a finite-state-machine recognizer for these four quantities by using the techniques described earlier in this section. However, by using some good intuition, we can obtain a reasonably

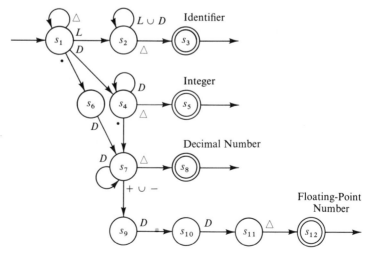

FIG. 7.4.10   A Finite-State-Machine Recognizer for a Typical Lexical Analyzer

compact machine for this simple example, whereas the machine obtained through use of our earlier algorithm is quite complex. Fig. 7.4.10 shows the state diagram for our simplified machine. The figure omits state transitions for incorrect sequences. For example, the sequence . . . DL . . . is invalid for the recognizer wherever it appears, and detection of this sequence should cause a transition to an error state.

A simulation of the machine in Fig. 7.4.10 in ALGOL or PL/I might appear in part as follows:

```
STATEONE:  GET NEW CHAR;
           IF CHAR = BLANK THEN GO TO
              STATEONE

           ELSE IF CHAR = LETTER THEN GO TO
              STATETWO

           ELSE IF CHAR = DIGIT THEN GO TO
              STATEFOUR

           ELSE IF CHAR = POINT THEN GO TO
              STATESIX

           ELSE GO TO ERROR;
STATETWO:  GET NEW CHAR;
           IF CHAR = LETTER OR CHAR =
              DIGIT THEN GO TO STATETWO

           ELSE IF CHAR = BLANK THEN GO TO
              STATETHREE

           ELSE GO TO ERROR;
```

Summary There is a strong correspondence between regular sets and finite-state machines. Every regular expression is recognized by some finite-state machine, and all sets of sequences recognizable by finite-state machines are regular sets. The Kleene analysis and synthesis theorems prove these notions constructively. Three types of operations on regular sets preserve regularity: the operations of set union, set concatenation, and the * operation of set iteration. The synthesis procedure uses one particular construction for each of these operations. The construction procedure described here yields a transition system rather than a finite-state machine, but each transition system is equivalent to a finite-state machine.

Kleene's work appeared in 1956, but it is fair to say that the importance of these ideas does not stand out in the context of his presentation. Copi et al. (1958) restated Kleene's results (with many fresh ideas of their own) in a paper of unusual clarity. This paper deserves much credit for bringing Kleene's work to the attention of the computing community. Rabin and Scott (1959) and Myhill (1957) have produced other work of interest in the theory of finite-state-machine recognizers. The construction procedure in the proof of Kleene's synthesis theorem is due to Ott and Feinstein (1961). Brzozowski (1962) surveyed the early contributions to the theory of regular expressions and commented on applications to computer design. Gries (1971) discusses the recognition of regular expressions during the lexical phase of compilation. The material of this section is treated in considerable detail in textbooks by Harrison (1965), Booth (1967), Nelson (1968), and Hennie (1968).

Exercises

7.4.1 Let T, V, and W be regular expressions. Verify the following identities:
(a) $(V^*)^* = V^*$;
(b) $T(V \cup W) = TV \cup TW$;
(c) $(V \cup W)^* = (V^* W^*)^*$.

7.4.2 Let α be a sequence of symbols in I^*, and let α' be the sequence derived from α by reversing the order of the symbols in α. Prove that the set $R = \{\alpha\alpha' \,|\, \alpha \in I^*\}$ is not a regular set. HINT: Prove that a machine that recognizes this set must have an infinite number of states because, if a recognizer has only n states, then it must fail to recognize at least one sequence of length $2n + 1$ in R.

7.4.3 Let R be a regular set. Prove that $I^* - R$ is a regular set.

7.4.4 Give word descriptions of the regular sets represented by the following regular expressions:
(a) 1^*;
(b) 10^*;
(c) $(10)^*$;
(d) $0^*(10)^*$;
(e) $0^*(10)^*0$;
(f) $0^*(10)^*0^*$;
(g) $0^*(10 \cup 0)^*0^*(10 \cup 0)^*$;
(h) $0^*(10^* \cup 0^*)(10^* \cup 0^*)$.

7.4.5 Use the procedure in the proof of the Kleene synthesis theorem to construct transition-system recognizers for the following regular sets:
(a) 1;
(b) 1101;
(c) 10^*110^*1;
(d) $1(0^* \cup 11)^*(0^* \cup 1)$.

7.4.6 Let t_1 be a transition system, and let t_2 be an equivalent transition system without Λ-transitions. Let a state of t_2 be an initial state if it is initial in t_1 or is reachable from an initial state in t_1 by a sequence of Λ-transitions. Let a state of t_2 be a final state if and only if it is final in t_1. Prove that the set recognized by t_2 is the same set as that recognized by t_1.

7.4.7 Let t_2 be a transition system without Λ-transitions, and let t_2' be an equivalent finite-state machine. Let the initial state of t_2' be the state corresponding to the subset of initial states of t_2, and let a state of t_2' be final if and only if it corresponds to a subset of states of t_2 that contains a final state. Prove that t_2 and t_2' recognize the same set.

7.4.8 Prove that the transition-system constructions shown in Fig. 7.4.8 for $V \cup W$, VW, and V^* do indeed recognize these regular sets.

s	$M_1(s, i)$		$\delta_1(s)$
	$i = 0$	$i = 1$	
s_0	s_1	s_2	0
s_1	s_2	s_3	0
s_2	s_0	s_3	1
s_3	s_3	s_3	1

FIG. 7.4.11   Machine M_1 for Exs. 7.4.9 and 7.4.10

7.4.9 For the machine M_1 shown in Fig. 7.4.11, find a regular expression that represents the set of sequences that carry M_1 from s_0 back to s_0.

7.4.10 Find a regular expression for the set recognized by machine M_1 of Fig. 7.4.11.

s	$M_2(s, i)$		$\delta_2(s)$
	$i = 0$	$i = 1$	
s_0	s_1	s_3	0
s_1	s_2	s_4	0
s_2	s_2	s_0	1
s_3	s_4	s_1	0
s_4	s_3	s_0	1

FIG. 7.4.12   Machine M_2 for Ex. 7.4.11

7.4.11 Find a regular expression for the set recognized by machine M_2 of Fig. 7.4.12.

7.4.12 Construct a finite-state-machine recognizer for the regular set

$$\big((0 \cup 11)^*0\big)^*.$$

Follow the procedure given in the proof of the Kleene synthesis theorem to obtain a transition-system recognizer; then remove the Λ-transitions, and finally construct an equivalent finite-state machine.

7.4.13 Construct a recognizer for the regular set of Ex. 7.4.12, this time using state-minimization procedures and intuition to obtain a machine with the smallest possible number of states.

7.4.14 Let M be a finite-state machine with a unique initial state and a unique final state. Give an algorithm for finding the regular set recognized by M. HINT: The algorithm is similar to the proof of the Kleene analysis theorem.

EIGHT Rings and Fields

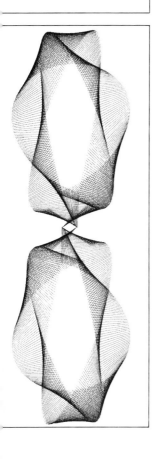

EIGHT | Rings and Fields

In preceding chapters we examine algebraic structures with one operation, and we investigate several applications of such structures to computers. In this chapter we initiate the study of algebraic structures with two operations, and we discuss their practical applications in Chapter Nine.

We are motivated to study structures with two operations because the real number system has two operations: addition and multiplication. The operations of subtraction and division are defined in terms of addition and multiplication, because they are operations involving additive and multiplicative inverses, respectively. Because the real number system has two fundamental arithmetic operations (not just one), the theories of groups and semigroups are inadequate to describe the rich structure of arithmetic in this system. Thus our study of the real number system (or of any structure with properties similar to those of the real number system) requires a study of algebraic structures with two operations.

From previous discussions, we know that computer arithmetic systems have only a finite number of distinct elements and therefore are inherently different from real arithmetic systems. Nevertheless, the general results we obtain in this chapter for algebraic structures with two operations pertain both to computer arithmetic and to real arithmetic. Our study reveals the properties shared by computer arithmetic systems and the real number system, and it partially explains how a finite algebraic structure can be a satisfactory imitation of an infinite algebraic structure.

The first section of this chapter is devoted to rings (which are among the least complex of the algebraic structures with two operations) and to the several structures of varying complexity that are derived from rings. The section defines several different structures and indicates their relations to one another. This discussion provides an overview of systems that arise in a variety of practical circumstances. Because our major applications in Chapter Nine require the use of the structure known as an algebraic field, we spend the major portion of Chapter Eight in an investigation of the theory of finite fields, preparing a background for the following chapters.

The applications discussed in Chapter Nine are to the study of a special class of finite-state machines. In practical circumstances, we find that the theory aids us in the design of sequence generators, coders and decoders for error-correcting codes, in measuring interplanetary distances, and in the application of Pólya theory to the enumeration of classes of switching functions.

8.1 Algebraic Structures with Two Operations

In preparation for our study of linear sequential machines in Chapter Nine, we introduce here several related algebraic structures. Of these structures, the most general is the algebraic ring. The other structures all are derived from rings, in the sense that they are rings that satisfy additional axioms. Of primary interest for our later applications are fields, vector spaces, and algebras. We postpone discussion of applications until the next chapter. Here we give the defining properties of these structures, and in later sections of this chapter we undertake a deeper investigation of their properties.

We begin with the definition of rings, and we proceed through the section to progressively richer structures.

Definition A *ring* $R = \langle R, +, \cdot \rangle$ is an algebraic structure such that R is a set, $+$ (addition) and $\cdot$ (multiplication) are compositions on R, and the following axioms are satisfied:

(i) $\langle R, + \rangle$ is an abelian group;

(ii) $\langle R, \cdot \rangle$ is a semigroup; and

(iii) multiplication distributes over addition—that is,

$$x \cdot (y + z) = (x \cdot y) + (x \cdot z) \text{ and } (y + z) \cdot x = (y \cdot x) + (z \cdot x)$$

for all x, y, z in R.

Note that we continue to abuse notation by calling the structure by the name of the set on which it is defined.

There are many familiar examples of rings. In fact, each of the sets $\mathbf{Z}$, $\mathbf{Q}$, $\mathbf{R}$, and $\mathbf{C}$ is a ring under the operations of addition and multiplication. $\mathbf{Z}_n$ also is a ring under addition and multiplication modulo n.

Rings possess sufficient structure that we immediately can say something about the structure of $\langle R, \cdot \rangle$ in a ring $\langle R, +, \cdot \rangle$.

Definition In a ring $\langle R, +, \cdot \rangle$, the unique identity in $\langle R, + \rangle$ is called *zero* and denoted 0.

Theorem In any ring, 0 is a two-sided zero in the semigroup $\langle R, \cdot \rangle$.

Proof: We have $0 \cdot x = (0 + 0) \cdot x = (0 \cdot x) + (0 \cdot x)$ for all x in R. Adding the inverse of $0 \cdot x$ to both sides of this equation, we obtain

$$0 = (0 \cdot x) - (0 \cdot x) = (0 \cdot x) + [(0 \cdot x) - (0 \cdot x)] = 0 \cdot x,$$

so that 0 is a left zero in $\langle R, \cdot \rangle$. By symmetry, it also is a right zero. Hence it is a two-sided zero. $\qquad \square$

We see that the ring axioms guarantee the presence of a multiplicative zero in a ring. What other properties can we deduce from the ring axioms? Because $\langle R, + \rangle$ is an abelian group, the axioms of closure, associativity, identity, inverses, and commutativity are satisfied. There is virtually nothing more that we can say about $\langle R, + \rangle$ within the mathematical framework that we have developed. The interaction of addition and multiplication in a ring is partially specified by the axiom of distributivity, and we have not specified whether or not addition distributes over multiplication.

The substructure whose properties are not fully specified is $\langle R, \cdot \rangle$. For this structure, we know that closure and associativity are satisfied and that 0 is a zero. Consequently, there remains only the specification of the properties of $\langle R - \{0\}, \cdot \rangle$. If we say nothing about this structure, then we can say only that R is a ring. If this structure satisfies any or all of the axioms of closure, identity, inverses, or commutativity, then we obtain an algebraic structure with special properties. These structures are of varying degrees of interest.

Fig. 8.1.1 illustrates the algebraic structures that we obtain as we add axioms to the system of ring axioms. Specifically, we obtain the following structures:

    (i)   *commutative ring*—R is a ring and $\langle R - \{0\}, \cdot \rangle$ is commutative;

    (ii) *ring with identity*—R is a ring and there is an identity (1) in $\langle R - \{0\}, \cdot \rangle$;

   (iii) *ring without divisors of* 0—R is a ring and $\langle R - \{0\}, \cdot \rangle$ is closed;

   (iv) *integral domain*—R is a ring and $\langle R - \{0\}, \cdot \rangle$ satisfies the axioms of identity, commutativity, and closure;

    (v) *skew field*—R is a ring and $\langle R - \{0\}, \cdot \rangle$ is a group;

   (vi) *field*—R is a ring and $\langle R - \{0\}, \cdot \rangle$ is an abelian group.

Note that closure of $\langle R - \{0\}, \cdot \rangle$ is *not* satisfied if and only if there exist x, y in R such that $x \cdot y = 0$, because $\langle R, \cdot \rangle$ is closed. Hence closure in $\langle R - \{0\}, \cdot \rangle$ is satisfied if and only if there are no divisors of 0.

When R has more than one element, $\langle R, \cdot \rangle$ cannot be a group because a nontrivial group cannot have a zero element. However, $\langle R - \{0\}, \cdot \rangle$ can be a group, as indicated by the definitions of integral domain, skew field, and field.

We can illustrate the preceding definitions with some familiar algebraic systems. $\langle \mathbf{Z}, +, \cdot \rangle$ is an integral domain but is not a field. $\langle \mathbf{Q}, +, \cdot \rangle$ and $\langle \mathbf{R}, +, \cdot \rangle$ are fields. The structure $\langle \mathbf{Z}_n, +_n, \cdot_n \rangle$ is always a commutative ring with identity, but it is a field if and only if n is prime.

Among algebraists there exists some disagreement about the definition of integral domain. We require here that the axiom of identity hold for $\langle R - \{0\}, \cdot \rangle$ in an integral domain, but some algebraists prefer to omit this axiom from the definition of integral domain. This difference in definitions is not crucial because the familiar examples of rings that satisfy commutativity and closure in $\langle R - \{0\}, \cdot \rangle$ also have a multi-

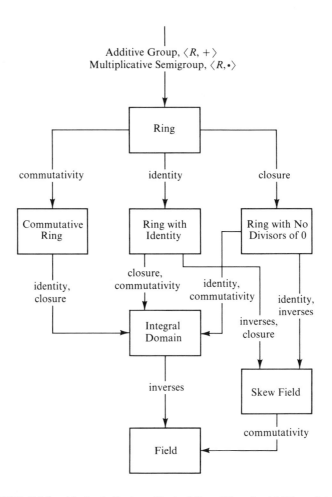

FIG. 8.1.1   Algebraic Systems Derived from Rings by Addition of Axioms for the Structure $\langle R - \{0\}, \cdot \rangle$

plicative identity. Hence they are integral domains under either definition. The next theorem shows that finite integral domains are fields. Hence, because all of our applications deal with finite sets, we need not consider integral domains separately from fields.

Theorem   If $\langle R, +, \cdot \rangle$ is an integral domain and R is finite, then R is a field.

Proof: It is sufficient to show that every nonzero element r in R has a multiplicative inverse. First we show that each nonzero r in R has a right inverse with respect to multiplication. For each nonzero r, the map $f_r : R - \{0\} \to R - \{0\}$, where $f_r(t) = r \cdot t$, is the map corresponding to left multiplication by r. If this map is not an injection, then there exist nonzero t_1 and t_2 such that $r \cdot t_1 = r \cdot t_2$ or

$$r \cdot t_1 - r \cdot t_2 = r \cdot (t_1 - t_2) = 0,$$

and $R - \{0\}$ has nonzero divisors of 0. Because R is an integral domain, this is a contradiction, so that left multiplication by r must be an injection. Because, by hypothesis, R is finite, we can use the pigeonhole principle to show that left multiplication by r is a surjection of $R - \{0\}$. Hence, for every nonzero r, there is a unique solution in $R - \{0\}$ of the equation $r \cdot x = 1$. Because multiplication is commutative, $r \cdot x = x \cdot r$, so that the right multiplicative inverse of r is a two-sided multiplicative inverse. □

The structure $\langle \mathbf{C}, +, \cdot \rangle$ is a field, but it also has the interesting property that it is a two-dimensional structure over $\mathbf{R}$. That is, each complex number $a + ib$, where $i = \sqrt{-1}$, may be treated as a 2-tuple $\langle a, b \rangle$ with a and b in $\mathbf{R}$. The notion of multidimensional algebraic structures is common throughout the science and engineering disciplines. We encounter such structures not only in the complex number system, but also in the use of vectors and scalars for modeling physical systems. In the remainder of this section, we use the concepts of rings and fields to build up axiom systems for vector spaces and other multidimensional algebraic structures.

Definition A set V is a *module over a ring K* (sometimes called a *K*-module) if $\langle V, K, +, \cdot \rangle$ is an algebraic structure where $\cdot$ maps $K \times K$ into K and $K \times V$ into V, where $+$ maps $K \times K$ into K and $V \times V$ into V, and where

    (i) $\langle K, +, \cdot \rangle$ is a ring (the ring of *scalars*),
    (ii) $\langle V, + \rangle$ is an abelian group (the group of *vectors*), and
    (iii) for all a, b in K and all u, v in V we have

$$a \cdot (u + v) = a \cdot u + a \cdot v,$$
$$(a + b) \cdot u = a \cdot u + b \cdot u, \text{ and}$$
$$a \cdot (b \cdot u) = (a \cdot b) \cdot u.$$

The definition of module does not define addition and multiplication for all types of operands. Scalar addition and scalar multiplication have the structure of the ring K. Vector addition is defined through the abelian group V. We do not define addition of a scalar and a vector, nor do we necessarily define multiplication of two vectors. In fact, practical applications of modules concentrate heavily on the group properties of vectors and on the notion of multiplication of vectors by scalars.

The set of complex numbers $\mathbf{C}$ forms a module over the real numbers $\mathbf{R}$. Another example of a module is the set V of all vectors $v = \langle v_1, v_2, \ldots, v_n \rangle$ that are n-tuples with components in a ring K. We define addition in V to be componentwise addition. That is, if $u = \langle u_1, u_2, \ldots, u_n \rangle$ and $v = \langle v_1, v_2, \ldots, v_n \rangle$, then

$$u + v = \langle u_1 + v_1, u_2 + v_2, \ldots, u_n + v_n \rangle.$$

If k is some scalar in K, then multiplication of a vector by a scalar is defined by $k \cdot u = \langle k \cdot u_1, k \cdot u_2, \ldots, k \cdot u_n \rangle$, where the multiplication of the scalar components of $k \cdot u$ is multiplication in K.

Because K may not only be a ring, but also may satisfy axioms other than the ring axioms, we must consider the additional structure that is imposed on V if it is a module over such a K.

First we let K be a ring with identity. By convention we label the (unique) multiplicative identity as 1, and we call K a *ring with unity*. The most reasonable axiom to require in this case is that 1 be a multiplicative identity with respect to vectors.

Definition V is a *unitary* module (or a unitary K-module) if $\langle V, K, +, \cdot \rangle$ is a module, and if

(i) K is a ring with unity, and
(ii) $1 \cdot u = u$ for all u in V.

Because **R** is a ring with unity, **C** is a unitary module over **R**. Next we let K be a field.

Definition V is a *vector space* over K if $\langle V, K, +, \cdot \rangle$ is a unitary module and K is a field.

Because **R** is a field, **C** is a vector space over **R**. A typical element of **C** has the form $a + ib$, where a and b are in **R** and $i = \sqrt{-1}$.

If we define multiplication of two vectors in a vector space, we obtain the following structure.

Definition V is a *linear associative algebra* over K if V is a vector space over K and if

(i) $\langle V, +, \cdot \rangle$ is a ring, and
(ii)   (*bilinearity*) for all a, b in K and all u, v, w in V, we have

$$u \cdot (a \cdot v + b \cdot w) = a \cdot u \cdot v + b \cdot u \cdot w, \text{ and}$$
$$(a \cdot v + b \cdot w) \cdot u = a \cdot v \cdot u + b \cdot w \cdot u.$$

From the preceding definition, we see that **C** is a linear associative algebra over **R**. We define the multiplication in **C** to be

$$(a + ib) \cdot (c + id) = (ac - bd) + i(ad + bc).$$

Fig. 8.1.2 summarizes the axiom systems for the algebraic structures that are derived from modules.

Exercises

8.1.1   Prove that $\langle \mathbf{Z}_n, +_n, \cdot_n \rangle$ is a field if and only if n is prime.

8.1.2   Let R be a commutative ring with no nonzero divisors of 0. Show that, if R is finite, then R is a field.

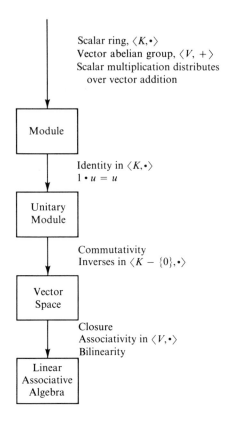

FIG. 8.1.2 Algebraic Structures Derived from a Module V over a Ring K

8.1.3 Let R be the set of all $m \times m$ matrices with coefficients in **R**. Prove that R is a ring with respect to matrix addition and matrix multiplication. Determine which axioms hold for $R - \{0\}$ in this ring.

8.1.4 Let R be the set of $m \times m$ matrices with entries in $\mathbf{Z}_n$. Show that this set is a ring with respect to matrix addition and matrix multiplication if all arithmetic operations are done modulo n.

8.1.5 Let R be a ring, and consider the set of polynomials of the form $r_0 + r_1 \cdot x + r_2 \cdot x^2 + \cdots$, where x is an indeterminate quantity. Prove that the set of polynomials in x with coefficients in R is a module over R with respect to addition of polynomials and multiplication of polynomials by a ring element. Determine whether this structure is a unitary module, a vector space, or a linear associative algebra.

8.1.6 Let r, s, t, and u be elements of a ring R. Prove that the following identities hold:
(a) $(r + s) \cdot (t + u) = r \cdot t + s \cdot t + r \cdot u + s \cdot u;$
(b) $(r - s) \cdot (t - u) = (r \cdot t + s \cdot u) - (r \cdot u + s \cdot t).$

8.1.7 Consider $\mathscr{P}(X)$, the power set of X. Define the operation $\oplus$ on $\mathscr{P}(X)$ to be $Y \oplus Z = (Y - Z) \cup (Z - Y)$, where Y and Z are subsets of X. Prove that $\langle \mathscr{P}(X), \oplus, \cap \rangle$ is a ring.

8.1.8 Let $\langle R, +, \cdot \rangle$ be a ring. For each integer $n \geq 1$, define $n \cdot r$ to be $(r + r + \cdots + r)$ where the sum has n terms and r is in R. Prove that, if $n = mp$ where m and p are positive integers, then $n \cdot r = (m \cdot r) \cdot (p \cdot r)$ for all r in R.

8.1.9 The *characteristic* of an integral domain R is the order of 1 in the additive group of the integral domain. Thus, if we define $n \cdot 1$ in R to be $(1 + 1 + \cdots + 1)$ where the sum has n terms, the characteristic of R is the least integer n for which $n \cdot 1 = 0$ in R. Prove that an integral domain has a characteristic that is either a prime integer or infinite. HINT: If the integer n is equal to mp where m and p are integers, then in the integral domain $n \cdot 1 = (m \cdot 1) \cdot (p \cdot 1)$. There are no nonzero divisors of 0 in an integral domain.

8.1.10 Let $\langle R, + \rangle$ be a group, and let $\cdot$ be an associative composition on R that distributes over $+$ and for which the element 1 of R is an identity. Prove that $\langle R, + \rangle$ is a commutative group and that $\langle R, +, \cdot \rangle$ is a ring. HINT: Use the equality

$$(r + s) \cdot (1 + 1) = (r + s) \cdot 1 + (r + s) \cdot 1.$$

8.1.11 Prove that, if R is a commutative ring, then $(r + s)^2 = r^2 + 2 \cdot r \cdot s + s^2$.

8.1.12 Let R be a noncommutative ring. Find the expansion of the product $(r + s)^3$ for r and s in R. What is the general form of the expansion of the product $(r + s)^n$?

8.2 Finite Fields In the previous section we set forth definitions for a potpourri of algebraic structures. In this section we concentrate on the study of finite fields. In Chapter Nine, we investigate applications of this study to the theory of linear sequential machines. We also touch upon a major application of this study to error-correcting codes for reliable data communication and computation.

Our treatment of fields differs substantially from the usual mathematical treatments. Most mathematical texts discuss such classic results as the proof that equations of the fifth degree or higher cannot be solved in general by ordinary arithmetic and root extraction, or the fact that it is impossible to trisect an arbitrary angle with a ruler-and-compass construction. These results shed little light on the applications of finite fields to computer science.

The principal result of this chapter is that there exists a field of order q if and only if $q = p^n$ for some integer n and for some prime integer p. Perhaps the most significant aspect of fields for our later applications is that every finite field is isomorphic to an algebra of polynomials. This result is analogous to our previous observations that every group is isomorphic to a permutation group and that every semigroup is isomorphic to a transformation semigroup. Thus we manipulate polynomials to imitate field operations, just as we manipulate permutations and transformations to imitate group and semigroup operations. To determine the product $\alpha\beta$ of two field elements α and β, we multiply their polynomial representations, and the product polynomial is the representation of $\alpha\beta$. For reasons discussed later in this section, we do polynomial multiplication modulo $p(x)$, where $p(x)$ is a polynomial. For polynomial algebras, this operation is analogous to multiplication modulo n for integers.

Our approach in this section is to show how to construct fields of order p^n for all $n \geq 1$ and all prime p. First we show that $\langle \mathbf{Z}_p, +_p, \cdot_p \rangle$ is a field for prime p, and then we show how to construct a field of order p^n from a field of order p. The latter construction involves polynomial algebras. After showing how to construct fields of order p^n, we then prove that no other finite fields exist. This is done by showing that every finite field is isomorphic to a linear algebra over a scalar coefficient field of prime order.

We begin by finding fields of order p for all prime p. We know that $\mathbf{Z}_n$, the set of integers modulo n, is a ring with respect to addition and multiplication modulo n for all $n \geq 2$. When n is prime, $\langle \mathbf{Z}_n, +_n, \cdot_n \rangle$ is a field.

Theorem $\langle \mathbf{Z}_n, +_n, \cdot_n \rangle$ is a field if and only if n is prime.

Proof: Because every finite integral domain is a field and $\langle \mathbf{Z}_n, +_n, \cdot_n \rangle$ is a finite ring for all n, it is sufficient to show that $\langle \mathbf{Z}_n, +_n, \cdot_n \rangle$ is an integral domain if and only if n is prime. From the definition of integral domain, we know that we must show that the axioms of commutativity and identity hold for $\langle \mathbf{Z}_n - \{0\}, \cdot_n \rangle$ and that there are no nonzero divisors of 0 in $\langle \mathbf{Z}_n, \cdot_n \rangle$. It is trivial to show that the axioms of commutativity and identity hold for all n, so the proof reduces to the proof that there are no nonzero divisors of 0 in $\langle \mathbf{Z}_n, \cdot_n \rangle$ if and only if n is prime. If n is not prime, then $n = xy$ for x and y in $\mathbf{Z}_n - \{0\}$, so that $x \cdot_n y = 0$, and thus $\langle \mathbf{Z}_n, +_n, \cdot_n \rangle$ is not an integral domain. Conversely, if $\langle \mathbf{Z}_n, +_n, \cdot_n \rangle$ is not an integral domain, then there exist x and y in $\mathbf{Z}_n - \{0\}$ such that $x \cdot_n y = 0$, or xy is a multiple of n. But if $xy = kn$ for some k, then n is divisible by $gcd(x, y)$ if $gcd(x, y) \neq 1$, and by either x or y if $gcd(x, y) = 1$. In either case, n is not prime. $\square$

Because $\langle \mathbf{Z}_p, +_p, \cdot_p \rangle$ is a field for prime p, we denote this field as $GF(p)$, the *Galois field of order* p.

We have proved the existence of a finite field of prime order for every prime. Hence we can use these fields (or any other finite fields) to build linear associative algebras of any dimension. In fact, we now show

how to construct fields of order p^n for all $n \geq 1$. In particular, consider the set of polynomials in x of degree less than n, whose coefficients lie in a field K of order q. There are q^n polynomials in this set. We define addition of polynomials in the usual way: we obtain the sum $f(x) + g(x)$ by adding the coefficients of like powers of x in $f(x)$ and $g(x)$. Multiplication of a polynomial by a field element also is defined in the usual way: each coefficient of the polynomial is multiplied by the field element. You can readily verify that the structure obtained under these definitions is a vector space over the field K.

Now we turn the vector space into a linear associative algebra by introducing polynomial multiplication. If we start with two polynomials $f(x)$ and $g(x)$ of degree less than n, then their product $f(x) \cdot g(x)$ when formed in the usual way is not necessarily a polynomial of degree less than n. In order to satisfy closure, we force $f(x) \cdot g(x)$ to have degree less than n by using modular polynomial multiplication.

Before we discuss modular polynomial multiplication, we first review the Euclidian division algorithm for polynomials whose coefficients lie in a field. In dealing with $\mathbf{Z}_n$ under addition and arithmetic modulo n, we have made implicit use of the Euclidian division algorithm for the integers. Euclid was the first to prove that, if n and x are integers in $\mathbf{P}$, then there are unique integers q and r that satisfy the equality $x = qn + r$, where $0 \leq r < n - 1$. Normally we compute these quantities using ordinary division, where q is the *quotient* of the division and r is the *remainder*. It is this result that we use when doing arithmetic modulo n. For example, to compute $y +_n z$ or $y \cdot_n z$, we perform the computation using ordinary arithmetic and obtain a result x. If x is in $\mathbf{Z}_n$, then x is the true result. If not, then $x = qn + r$, and r is in $\mathbf{Z}_n$. Moreover, $x \equiv r \bmod n$ because $x - r$ is a multiple of n. Hence r is the result of the operation.

Now consider any two polynomials $f(x)$ and $g(x)$ with coefficients that lie in a field K. We use the notation $deg[f(x)]$ to denote the degree of the polynomial $f(x)$. Because the coefficients of the polynomials lie in a field, we can divide $f(x)$ by $g(x)$ using ordinary polynomial division. Euclid's division algorithm has a counterpart for polynomial division, as indicated by the *Euclidian division theorem for polynomials*, which follows.

Theorem   If $f(x)$ and $g(x)$ are polynomials with coefficients in a field K, then there are unique polynomials $q(x)$ and $r(x)$, with coefficients in K, for which $f(x) = q(x) \cdot g(x) + r(x)$, where $0 \leq deg[r(x)] < deg[g(x)]$.

Proof: We can compute $q(x)$ and $r(x)$ as the quotient and remainder, respectively, of the polynomial division of $f(x)$ by $g(x)$. To perform the division, it is sufficient to require that the coefficients of $f(x)$ and $g(x)$ lie in the field K, and that the coefficients of $q(x)$ and $r(x)$, in general, lie in K as well. We assume that the details of the polynomial division algorithm are well known to the reader and that no further discussion of the algorithm is required in this proof. Note that the algorithm yields $r(x)$ such that $deg[r(x)] < deg[q(x)]$.

To prove that $q(x)$ and $r(x)$ are unique, assume that

$$f(x) = q_1(x) \cdot g(x) + r_1(x) = q_2(x) \cdot g(x) + r_2(x),$$

where $0 \le deg[r_1(x)], deg[r_2(x)] < deg[g(x)]$. Then

$$q_1(x) \cdot g(x) - q_2(x) \cdot g(x) = r_2(x) - r_1(x),$$

or

$$(q_1(x) - q_2(x)) \cdot g(x) = r_2(x) - r_1(x).$$

If $q_1(x) \ne q_2(x)$, then the polynomial on the left side of the last equation has degree at least as large as the degree of $g(x)$, whereas the polynomial on the right side has degree less than that of $g(x)$. This is a contradiction, and therefore we have $q_1(x) = q_2(x)$. In this case, $r_1(x) = r_2(x)$, so that $q(x)$ and $r(x)$ are unique as claimed. □

The reader should verify that application of the division algorithm to the polynomials $f(x) = x^4 + x^3 + 2x + 2$ and $g(x) = x^2 - 1$ (with coefficients in $\mathbf{Z}$) yields $q(x) = x^2 + x + 1$ and $r(x) = 3x + 3$. Next we define polynomial multiplication to be analogous to multiplication modulo n.

Definition   Let $f(x)$ and $g(x)$ be two polynomials in the vector space of polynomials of degree less than n over a field K. Then $f(x) \cdot g(x)$ is defined to be $[f(x)g(x)] \bmod (x^n - 1)$, where $f(x)g(x)$ denotes the product obtained by using ordinary polynomial multiplication with coefficients computed in the field K.

To compute $[f(x)g(x)] \bmod (x^n - 1)$, we divide $f(x)g(x)$ by $x^n - 1$, using the Euclidian division algorithm, and the remainder polynomial $r(x)$ is the result we seek. Thus the product of $x^3 + 2$ and $x + 1$ in the vector space of polynomials modulo $(x^4 - 1)$ is

$$[(x^3 + 2)(x + 1)] \bmod (x^4 - 1) = [x^4 + x^3 + 2x + 2] \bmod (x^4 - 1)$$
$$= x^3 + 2x + 3.$$

It is easy to verify that the multiplication operation defined above turns the vector space of polynomials into a linear associative algebra of polynomials. We call this algebra the *algebra of polynomials modulo* $(x^n - 1)$.

There is nothing particularly exceptional about $x^n - 1$ in this construction. In fact, any polynomial of degree n will do.

If we define $f(x) \cdot g(x)$ to be $[f(x)g(x)] \bmod p(x)$, where $p(x)$ is any polynomial over K of degree n, this definition of multiplication also yields a linear associative algebra, which we call the *algebra of polynomials modulo* $p(x)$.

In the algebra of polynomials modulo $(x^n - 1)$ with coefficients in $\mathbf{Z}$, the polynomials form a ring with respect to addition and multiplication. Moreover, this ring is a commutative ring with an identity, but it is not an integral domain because

$$(x - 1) \cdot (x^{n-1} + x^{n-2} + \cdots + 1) = x^n - 1 = 0$$

in the ring, and thus the ring has divisors of zero. If we replace $x^n - 1$ by another polynomial $p(x)$ of degree n, then the resulting associative algebra may be an integral domain if $p(x)$ has specific properties. It so happens that, whenever the algebra is an integral domain, it also is a field. We already have encountered an analogous theorem when we discovered that $\mathbf{Z}_n$ is a field with respect to arithmetic modulo n if and only if n is prime. For polynomials, the notion that corresponds to prime number is the notion of irreducibility.

Definition A polynomial $p(x)$ of degree $n > 0$ over K is *irreducible* over K if and only if there do not exist polynomials $f(x)$ and $g(x)$ of degree greater than 0 over K such that $f(x)g(x) = p(x)$, where multiplication is ordinary polynomial multiplication with coefficient operations in K.

Theorem In the algebra of polynomials modulo $p(x)$, where $p(x)$ is a polynomial of degree n over a field K, the polynomials form a field with respect to polynomial addition and multiplication if and only if $p(x)$ is irreducible.

Proof: The proof of this theorem is similar to the proof of the theorem on prime fields. In order to have closure in the group of non-zero polynomials, we cannot have divisors of 0. Therefore, $p(x)$ must be irreducible. If $p(x)$ is irreducible, then we can show that inverses exist in the polynomial ring. The full proof is left as an exercise for the reader (Ex. 8.2.2). □

At this point, it is instructive to illustrate the preceding theorem with the field of complex numbers. Consider the polynomial $x^2 + 1$, where we assume that the coefficients lie in $\mathbf{R}$. Because we know that the roots of the polynomial are $\pm i$, we know that it cannot be factored into the product of two polynomials of degree 1 with coefficients in $\mathbf{R}$. Hence $x^2 + 1$ is irreducible over $\mathbf{R}$. Now consider the algebra of polynomials over $\mathbf{R}$ modulo $(x^2 + 1)$. How do we do multiplication in this algebra? The product of the polynomials $ax + b$ and $cx + d$ is the polynomial

$$\left[acx^2 + (ad + bc)x + bd\right] \bmod (x^2 + 1) = (ad + bc)x + (bd - ac).$$

But this is exactly analogous to the multiplication of the complex number $ai + b$ by the complex number $ci + d$. Hence, the algebra of polynomials modulo the irreducible polynomial $x^2 + 1$ is isomorphic to the complex numbers. This observation suggests the following definition.

Definition A field K' is said to be an *extension field* of degree n of a field K if K' is isomorphic to an algebra of polynomials over K modulo an irreducible polynomial of degree n. K is called the *ground field* of K'.

The complex-number field is an extension of degree 2 of the real-number field.

The previous theorem enables us to construct fields from polynomial algebras, but we must first be able to determine if a polynomial $q(x)$ is irreducible. We next show how to do this. In so doing, we discover that there are irreducible polynomials over $GF(p)$ of degree n for all $n \geq 1$. Hence we find a way to construct fields of order p^n for all prime p and for all $n \geq 1$.

How can we determine if a polynomial $q(x)$ is irreducible? Because irreducible polynomials are analogous to prime numbers, we can determine if $q(x)$ is irreducible in much the same way that we determine if a number is prime. Suppose, for example, that $q(x)$ has degree n. If it is reducible, then it factors into $q_1(x)q_2(x)$, where each of the factors has degree less than n. Because at least one of the two factors has a degree no greater than $n/2$, we can determine if $q(x)$ is irreducible if we know all of the irreducible polynomials of degree no greater than $n/2$. If n is odd, we need only inspect polynomials with degrees up to $(n - 1)/2$. Note that this observation is analogous to the fact that, if an integer r factors into $r = q_1 q_2$, then at least one of the two factors does not exceed $\sqrt{r}$.

Before we construct a table of irreducible polynomials, it is worthwhile to observe that $q(x)$ is irreducible if and only if $kq(x)$ is irreducible, where k is any field element in the coefficient field. Consequently, if $q(x)$ has a leading coefficient k, then $k^{-1}q(x)$ has a leading coefficient 1, and $q(x)$ is irreducible if and only if $k^{-1}q(x)$ is irreducible. Thus, in our test for irreducibility, we can limit our attention to polynomials with a leading coefficient 1.

Definition   A polynomial $p(x)$ of degree n is a *monic* polynomial if the coefficient of x^n is unity.

Now we can show how to construct a table of monic irreducible polynomials. The method is a sieve method, similar to methods used for the construction of tables of prime numbers.

To construct a small table of monic irreducible polynomials, consider polynomials over $GF(2)$. The irreducible polynomials of degree 1 are x and $x + 1$. If a polynomial of degree 2 factors, then its factors must be x or $x + 1$. Hence, the reducible polynomials of degree 2 must be $x^2 = x \cdot x$, $x^2 + x = x \cdot (x + 1)$, and $x^2 + 1 = (x + 1) \cdot (x + 1)$. The only polynomial of degree 2 that is not in this list is $x^2 + x + 1$, so it must be irreducible. Similarly, we consider how we can combine the irreducible factors of degree 2 or less to form products of degree 3. We have, for example, $x^3 = x \cdot x \cdot x$ and $x^3 + x = x \cdot (x + 1)^2$. There are four polynomials of degree 3 that have just the polynomials x and $x + 1$ as factors, and two polynomials of degree 3 that have $x^2 + x + 1$ as a factor. Hence six of the eight polynomials of degree 3 are reducible, and the remaining two must be irreducible. Table 8.2.1 shows these irreducible polynomials, and the table also lists irreducible polynomials over $GF(3)$.

In the next section, we prove that there are irreducible polynomials

TABLE 8.2.1   Irreducible Polynomials over $GF(2)$ and $GF(3)$

Degree	$GF(2)$	$GF(3)$
1	x	x
	$x + 1$	$x + 1$
		$x + 2$
2	$x^2 + x + 1$	$x^2 + 1$
		$x^2 + 2x + 2$
		$x^2 + x + 2$
3	$x^3 + x + 1$	$x^3 + 2x + 1$
	$x^3 + x^2 + 1$	$x^3 + 2x^2 + 1$
		$x^3 + x^2 + 2x + 1$
		$x^3 + 2x^2 + x + 1$
		$x^3 + 2x + 2$
		$x^3 + x^2 + 2$
		$x^3 + x^2 + x + 2$
		$x^3 + 2x^2 + 2x + 2$
4	$x^4 + x^3 + x^2 + 1$	$x^4 + x + 2$
	$x^4 + x + 1$	$x^4 + 2x^3 + 2$
	$x^4 + x^3 + 1$	$x^4 + 2x + 2$
		$\cdot$
		$\cdot$
		$\cdot$

of degree n over $GF(p)$ for all $n \geq 1$. For an intuitive justification of this statement, note that there are p^n monic polynomials of degree n over $GF(p)$. The number of irreducible polynomials over $GF(p)$ grows sufficiently slowly with n that we cannot find p^n different ways of combining irreducible polynomials of degree less than n to obtain p^n distinct product polynomials of degree n. Hence some monic polynomials of degree n must lack irreducible factors of degree less than n.

Now we can construct a field of order 4 by constructing an extension field from the algebra of polynomials over $GF(2)$ modulo the irreducible polynomial $x^2 + x + 1$. Let us name the field elements 0, 1, α, and β, and let us identify the elements with the polynomials 0, 1, x, and $x + 1$, respectively. Fig. 8.2.1 shows the addition and multiplication tables for this field.

+	0	1	α	β
0	0	1	α	β
1	1	0	β	α
α	α	β	0	1
β	β	α	1	0

$\cdot$	0	1	α	β
0	0	0	0	0
1	0	1	α	β
α	0	α	β	1
β	0	β	1	α

FIG. 8.2.1   Addition and Multiplication Tables for a Field of Order 4

Note that $\alpha + \beta = 1$ because $(x) + (x + 1) = 1$ in this algebra. Similarly, $\alpha \cdot \beta = 1$ because $(x) \cdot (x + 1) = x^2 + x$, but

$$x^2 + x \equiv 1 \bmod (x^2 + x + 1).$$

Because the polynomial representation of a finite field lends itself naturally to arithmetic operations, there is no need to construct the composition tables for field arithmetic when we use this representation. Later we show that every finite field has a representation as an algebra of polynomials.

Up to this point we have investigated ways of constructing finite fields, and we now can construct fields of order p^n. In the remainder of this section, we prove that there are no other finite fields. The next definition is needed for our proof.

Definition The *characteristic* of any field (finite or infinite) is the order of 1 in the additive group of the field.

Theorem The characteristic of any field is either prime or infinite.

Proof: For each integer n, we define $n \cdot 1$ to be $(1 + 1 + \cdots + 1)$, where the sum has n terms. If the additive order of 1 is n, then $n \cdot 1 = 0$. Suppose that n is not prime and that it can be factored into $m \cdot q$. Then $n \cdot 1 = (m \cdot 1)(q \cdot 1) = 0$. That is, the product of the sum of p ones and the sum of q ones must be 0. But in a field there are no nonzero divisors of 0, so either $m \cdot 1$ or $q \cdot 1$ is 0. However, this contradicts the assumption that 1 has order n. The contradiction forces us to conclude that the characteristic of the field must be either infinite or a prime integer. $\square$

Now we know that every finite field has a prime characteristic. In other words, the element 1 of a finite field generates an additive group isomorphic to $\langle \mathbf{Z}_p, +_p \rangle$. The proof of the preceding theorem contains an even stronger result. If $m \cdot 1$ and $q \cdot 1$ are arbitrary elements of the additive group generated by 1, then $(m \cdot 1) \cdot (q \cdot 1) = n \cdot 1$, where $n = mq$ and where $n \cdot 1$ is an element generated by 1. Hence the group of prime order generated by 1 is closed under multiplication. This observation is sufficient to prove that the elements in the group generated by 1 form a subfield isomorphic to the field $\langle \mathbf{Z}_p, +_p, \cdot_p \rangle$. Therefore, every finite field has a subfield of prime order, whose order is the characteristic of the field. The next theorem proves that every finite field has order p^n, where p is prime. The strategy of the proof is to show that every finite field is a linear associative algebra over its prime subfield, although this assertion is not stated explicitly in the proof.

Theorem The order of every finite field is a power of a prime.

Proof: By convention, the elements in a finite field of the additive subgroup generated by 1 are denoted by the integers $0, 1, \ldots, p - 1$, and the subgroup is denoted by P. Let x_1 be any nonzero field element. Then the set $P \cdot x_1$ has p members, because the invertibility of x_1 with respect to multiplication precludes the possibility that there can be two integers m and n in P such that $m \cdot x_1 = n \cdot x_1$. If the set $P \cdot x_1$ exhausts the field, then the field has order p. If not, there is some element x_2 not in $P \cdot x_1$.

Consider the set $P \cdot x_1 + P \cdot x_2$. $P \cdot x_1$ has p elements and $P \cdot x_2$ has p elements, so $P \cdot x_1 + P \cdot x_2$ has p^2 or fewer distinct elements. If there are fewer than p^2 distinct elements in $P \cdot x_1 + P \cdot x_2$, then there must be integers m_1, m_2, n_1, and n_2 in P that satisfy

$$m_1 \cdot x_1 + n_1 \cdot x_2 = m_2 \cdot x_1 + n_2 \cdot x_2,$$

or $\qquad (m_1 - m_2) \cdot x_1 + (n_1 - n_2) \cdot x_2 = 0.$

Thus there must exist m and n in P such that $m \cdot x_1 = n \cdot x_2$, where $m = m_1 - m_2$ and $n = n_2 - n_1$, and where at least one of m and n is nonzero. We claim that both m and n must be nonzero. For example, if $n = 0$, then $m \cdot x_1 = 0$, and both m and x_1 are nonzero field elements. Because a field has no nonzero divisors of zero, this is impossible.

Therefore $m \cdot x_1 = n \cdot x_2$, and m and n are nonzero. Then

$$n^{-1} \cdot m \cdot x_1 = x_2,$$

and $n^{-1} \cdot m$ is in P because m and n are nonzero elements of P. Hence x_2 is in the set $P \cdot x_1$, which contradicts the assumption that x_2 is not an element of $P \cdot x_1$. Thus we conclude that $P \cdot x_1 + P \cdot x_2$ has precisely p^2 elements.

If we have exhausted the elements in the field, then the field has p^2 elements. If not, then there is some x_3 that is not in $P \cdot x_1 + P \cdot x_2$. Using the method of reasoning of the preceding paragraphs, we show that $P \cdot x_1 + P \cdot x_2 + P \cdot x_3$ has p^3 elements. Proceeding inductively, we show that $P \cdot x_1 + P \cdot x_2 + \cdots + P \cdot x_n$ has p^n elements. Hence the order of any finite field is a power of its characteristic, and the theorem is proved. $\qquad\square$

Summary

There exist finite fields of order q if and only if q is a power of a prime. $\mathbf{Z}_p$ under addition and multiplication modulo p is a field if and only if p is prime. Finite fields can be constructed as polynomial algebras of dimension n with polynomial coefficients in a prime field.

The proof that finite fields must have order p^n for prime p depends on the fact that a finite field must be an algebra over a prime field, and the number of field elements in this algebra is a power of the order of the prime field. The prime field in this case is the subset of field elements generated from the element 1 under addition in the field.

Exercises

8.2.1 Prove that, if K' is a subfield of order p^m of a field K of order p^n, then m divides n. HINT: Show that, if x_1 and x_2 are nonzero elements of K, then the set $K' \cdot x_1 + K' \cdot x_2$ has p^{2m} elements, and use induction to show that K must have p^{rm} elements for some integer r.

8.2.2 Prove that, if $p(x)$ is irreducible over a finite field K, then the algebra of polynomials modulo $p(x)$ is a field. HINT: Use the pigeonhole principle.

8.2.3 The following statement is a well-known result from the theory of numbers: if p and q are relatively prime integers, then there exist integers a and b in $\mathbf{Z}$ such that $ap + bq = 1$. The analogous result for polynomials over a field K is the following: if $p(x)$ and $q(x)$ are relatively prime polynomials, then there exist polynomials $a(x)$ and $b(x)$ such that $a(x)p(x) + b(x)q(x) = 1$. Use this result to prove that the algebra of polynomials over any field K modulo an irreducible polynomial $p(x)$ is a field.

8.2.4 Construct a table of irreducible polynomials of degree 2 over $GF(4)$ and $GF(5)$.

8.2.5 Construct a multiplication table for the nonzero elements of $GF(8)$. Represent $GF(8)$ as an algebra of polynomials modulo $(x^3 + x + 1)$ over $GF(2)$ to construct this table. What group is isomorphic to the multiplicative group of $GF(8)$?

8.2.6 Let $q(x)$ be an irreducible polynomial of degree n over $GF(p)$. What is the order of the field of polynomials over $GF(p)$ modulo $q(x)$?

8.2.7 Prove that $x^2 + 1$ is irreducible over $GF(7)$.

8.2.8 Prove that, if $p(x)$ and $q(x)$ are relatively prime polynomials over $GF(p)$, then they are relatively prime over every field that contains $GF(p)$ as a subfield.

8.3 The Structure of Finite Fields

In the previous section, our discussion centers primarily on techniques for constructing finite fields. In this section we examine the detailed structure of finite fields. Among the important results of this section is the fact that the multiplicative group of each finite field is cyclic. We also characterize the elements of finite fields as roots of certain polynomials. From this characterization comes the central result that there exist finite fields of order p^n for every prime p and every $n \geq 1$.

Although this chapter is devoted primarily to the mathematical aspects of finite fields, we should mention that many of the theorems as well as the proofs find their way into applications in the next chapter. In particular, the fact that the multiplicative group of a field is cyclic is used in the synthesis of sequence generators that can produce extremely long sequences with interesting properties of randomness. The polynomial factorization properties are used in the analysis of linear sequential machines. In addition to these applications that are covered in the next chapter, the material in this section also has been applied extensively to algebraic coding theory and to mathematical systems theory.

We begin our study by showing in the next three theorems that the multiplicative group of a finite field is cyclic.

Theorem Let $f(x)$ be a polynomial of degree n over any field K. Then $f(x)$ has at most n distinct roots in K.

Proof: Suppose that $r_1, r_2, \ldots, r_n$, and $r_{n+1} \in K$ are $n + 1$ distinct roots of $f(x)$. That is, $f(r_i) = 0$ for each i. We show that each of the polynomials $x - r_i$ is a divisor of $f(x)$. This result follows from the Euclidian division theorem because $f(x) = (x - r_i)q_i(x) + d_i(x)$, where $d_i(x)$ is a polynomial of degree 0—that is, a field element. Then, because r_i is a root of $f(x)$,

$$0 = f(r_i) = (r_i - r_i)q_i(r_i) + d_i(r_i) = 0 + d_i(r_i) = d_i(r_i).$$

Then each d_i is 0, and each polynomial $x - r_i$ is a divisor of $f(x)$.

Now we use the Euclidian division theorem to show that

$$f(x) = c(x - r_1)(x - r_2) \ldots (x - r_n)$$

for some constant c. This follows because

$$f(x) = q(x)(x - r_1)(x - r_2) \ldots (x - r_n) + d(x),$$

but, because both the polynomials $f(x)$ and $(x - r_1)(x - r_2) \ldots (x - r_n)$ have degree n, the polynomial $q(x)$ must have degree 0, and therefore $q(x)$ is a field element which we can call c. Because $(x - r_1)$ is a divisor of $f(x)$, we have

$$f(x)/(x - r_1) = c(x - r_2)(x - r_3) \ldots (x - r_n),$$

and $d(x)$ must be 0. Thus we have

$$f(x) = c(x - r_1)(x - r_2) \ldots (x - r_n).$$

Because $f(r_{n+1}) = 0$, we must have

$$c(r_{n+1} - r_1)(r_{n+1} - r_2) \ldots (r_{n+1} - r_n) = 0.$$

But this is impossible unless $r_{n+1} = r_i$ for some i, because in the field K there are no divisors of 0 except 0. $\square$

Theorem In a finite field of order p^n, every nonzero element of the field satisfies the equation $x^{p^n - 1} - 1 = 0$.

Proof: The nonzero elements of a field of order p^n form a multiplicative group of order $p^n - 1$. But the order of every element of a group divides the order of the group, so that $x^{p^n - 1} = 1$ and $x^{p^n - 1} - 1 = 0$ for all nonzero x. $\square$

Theorem The nonzero elements of a field of order p^n form a cyclic group of order $p^n - 1$ under multiplication in the field.

Proof: The multiplicative group of a field is an abelian group of order $p^n - 1$. Let m be the exponent of that group, and note that

$m \leq p^n - 1$. Every element of the multiplicative group satisfies the equation $x^m - 1 = 0$, and so this equation has $p^n - 1$ distinct roots in the field. From a previous theorem we know that $p^n - 1$ cannot exceed m, so $m = p^n - 1$. Hence, some element has an order equal to the order of the multiplicative group, and therefore this element must generate the group. Hence the group must be cyclic. $\square$

As an illustration of the cyclic multiplicative-group structure of a finite field, consider the algebra of polynomials over $GF(2)$ modulo the irreducible polynomial $x^3 + x + 1$. Recall that the algebra is a field because it is modulo an irreducible polynomial. To show that the multiplicative group is cyclic, we show that there is a nonzero field element that generates the nonzero elements of the field as it is raised to successively higher powers. Because the field contains all of the nonzero polynomials of degree less than 3, the field has $2^3 = 8$ elements, and the multiplicative group must be the cyclic group with $2^3 - 1 = 7$ elements, or C_7. Because seven is prime, every element of C_7 generates C_7, so that every nonzero field element generates the multiplicative group. Let us pick the polynomial x to be the field element α. Then we have

$$\alpha = x,$$
$$\alpha^2 = x^2,$$
$$\alpha^3 = x^3 = x + 1,$$
$$\alpha^4 = x^2 + x,$$
$$\alpha^5 = x^3 + x^2 = x^2 + x + 1,$$
$$\alpha^6 = x^3 + x^2 + x = x^2 + 1, \text{ and}$$
$$\alpha^7 = x^3 + x = 1.$$

We see that the first seven powers of α are distinct and that they are all of the nonzero elements of the field.

We now make a significant change in our conceptualization of field elements. Thus far, our only models of finite fields are polynomial algebras modulo irreducible polynomials. However, this model is only a representation. We can also view the field elements as abstract entities that we manipulate much like numbers. This approach is perfectly natural in the context of complex numbers, for example, because every complex number $z = a + ib$ also is a polynomial of degree less than 2 in the variable i. Yet we freely manipulate complex numbers, knowing that the polynomial representation provides a convenient way of computing the sum or product of two complex numbers. Similarly, we now treat elements of finite fields as abstract entities, and we use their polynomial representations as convenient tools for the manipulation of the elements.

In the remainder of this section, we examine polynomials over finite fields and learn as much as possible about their roots. Several important facts emerge, eventually leading to the proof that all fields of order p^n

are isomorphic. The intermediate results of this section and the techniques used in the proofs are applied to several of the topics investigated in Chapter Nine.

The next two theorems are basic tools for the discussion that follows.

Theorem Let K be a field with characteristic p. Then $(\alpha + \beta)^p = \alpha^p + \beta^p$ for all α and β in K.

Proof: From the binomial theorem we have

$$(\alpha + \beta)^p = \alpha^p + p\alpha^{p-1}\beta + \cdots + p\alpha\beta^{p-1} + \beta^p.$$

Every coefficient except those of α^p and β^p is a multiple of p and thus is 0 in the field K. Hence $(\alpha + \beta)^p = \alpha^p + \beta^p$. $\qquad\square$

For any field of order p, we know that the multiplicative group is a cyclic group of order $p - 1$. Hence $\alpha^{p-1} = 1$ and $\alpha^p = \alpha$ in such a field. Then $(\alpha + \beta)^p = \alpha^p + \beta^p = \alpha + \beta$, which is somewhat stronger than the result stated in the theorem. However, it is *not true* that $\alpha^p = \alpha$ for all α in fields of order p^n for $n \geq 2$.

Theorem Let $f(x)$ be a polynomial over a field K of characteristic p. Then $f(x)^p = f(x^p)$.

Proof: The proof can be done by induction on the degree of the polynomial $f(x)$. Much of the proof is similar to the proof of the previous theorem, and this proof therefore is left as an exercise for the reader (Ex. 8.3.1). $\qquad\square$

Because we have frequent occasion to discuss extension fields over $GF(p)$, we denote such extension fields as $GF(p^n)$, where p^n is the order of the extension field. We have not yet shown that all fields of the same order are isomorphic, but this is true. Therefore the notation $GF(p^n)$ refers to a unique field for particular values of p and n.

Next we investigate the relation between the elements of a finite field and the roots of polynomials.

One of the problems that we encounter in Chapter Nine requires that we find a polynomial with coefficients in $GF(p)$ that has a given element of $GF(p^n)$ as one of its roots. The solution to this problem enables us to generate sequences of specified lengths, and to find simple coders and decoders for error-correcting codes. The problem is easy to solve when we try to find polynomials that have real coefficients and that have specified complex numbers as their roots. For example, to find a polynomial with real coefficients and with a root $a + ib$, we note that complex roots of polynomials occur in conjugate pairs. Hence $a - ib$ is another root of the polynomial, and a satisfactory polynomial is

$$(x - a - ib)(x - a + ib) = x^2 - 2ax + a^2 - b^2.$$

In dealing with finite fields, we shall discover that roots occur in conjugate sets, which are not necessarily sets with two elements. The polynomial associated with each conjugate set is irreducible and is called the minimum polynomial of each element in the set. In the next few theorems we derive the properties of the conjugate sets. We begin with a formal definition of minimum polynomial.

Definition Let α be an element of a field K of characteristic p. The *minimum polynomial* of α, denoted by $M_\alpha(x)$, is the monic polynomial of least degree over $GF(p)$ that has α as a root.

This definition is somewhat troublesome. We must be sure that every α in K is a root of some polynomial over $GF(p)$. Only then can we select a monic polynomial of least degree. We also must be sure that in no case will there be two distinct monic polynomials of least degree with α as a root.

We dispose of these two problems with the following arguments. Because every nonzero element of K is a root of the polynomial $x^{p^n-1} - 1$ if K has order p^n, then we are guaranteed that every element of K satisfies some polynomial equation over $GF(p)$. Now we show by contradiction that the minimum polynomial is unique. Suppose that α is a root of $f(x)$ and $g(x)$, which are both monic polynomials over $GF(p)$ of equal degree, and that this is the minimal degree of any polynomial over $GF(p)$ with α as a root. Then the polynomial $f(x) - g(x)$ has α as a root and is of smaller degree. If $f(x) - g(x)$ is not monic, then we divide it by the leading coefficient to make it monic and still have α as a root. Thus we conclude that the minimum polynomial of α is well defined.

Now let us reexamine the field of order 8 that is the algebra of polynomials over $GF(2)$ modulo $(x^3 + x + 1)$. Table 8.3.1 shows each nonzero field element and its minimum polynomial. It is not surprising to find that the two minimum polynomials of degree 3 each have three roots in the field, and that the polynomial of degree 1 has one root in the field. You can easily verify that the field elements are indeed roots of the polynomials in the table. Later theorems establish that the polynomials given in the table are minimal and tell us how to find such polynomials in general. To eliminate a possible source of confusion, we have used

TABLE 8.3.1 A Polynomial Representation of $GF(8)$ as a Polynomial Algebra Modulo $(x^3 + x + 1)$

Field Element	Minimum Polynomial
$\alpha = x$	$\xi^3 + \xi + 1$
$\alpha^2 = x^2$	$\xi^3 + \xi + 1$
$\alpha^3 = x + 1$	$\xi^3 + \xi^2 + 1$
$\alpha^4 = x^2 + x$	$\xi^3 + \xi + 1$
$\alpha^5 = x^2 + x + 1$	$\xi^3 + \xi^2 + 1$
$\alpha^6 = x^2 + 1$	$\xi^3 + \xi^2 + 1$
$\alpha^7 = 1$	$\xi + 1$

the indeterminate x in the polynomial representations of the field elements and the indeterminate ξ in the minimum polynomials.

To see that α^3 is a root of $\xi^3 + \xi^2 + 1$, let $\beta = \alpha^3$, and compute

$$\beta^3 = \alpha^9 = \alpha^2 = x^2,$$
$$\beta^2 = \alpha^6 = x^2 + 1,$$

and therefore

$$\beta^3 + \beta^2 + 1 = (x^2) + (x^2 + 1) + (1) = 0,$$

where we add coefficients of like powers of x in $GF(2)$. Hence α^3 is a root as claimed.

The next theorem shows that the minimum polynomial of a field element is irreducible.

Theorem Let K be an extension field of $GF(p)$, and let α in K have the minimum polynomial $M_\alpha(x)$ over $GF(p)$. Then $M_\alpha(x)$ is irreducible over $GF(p)$.

Proof: Suppose that $M_\alpha(x)$ factors over $GF(p)$ into $f(x) \cdot g(x)$, where the degree of both factors is greater than 0. Then $f(\alpha) \cdot g(\alpha) = 0$ in the field K but, because there are no nonzero divisors of 0 in K, either $f(\alpha) = 0$ or $g(\alpha) = 0$. Hence α is a root of either $g(x)$ or $f(x)$, both of which have degree less than the degree of $M_\alpha(x)$. But this contradicts the definition of $M_\alpha(x)$. Hence $M_\alpha(x)$ is irreducible over $GF(p)$ $\square$

Corollary Every polynomial that is irreducible over its coefficient field is the minimum polynomial of each of its roots.

Proof: See Ex. 8.3.11. $\square$

In Table 8.3.1, note that each of the minimum polynomials is given as irreducible in Table 8.2.1, in agreement with the preceding theorem.

Theorem Let $f(x)$ be any polynomial over $GF(p)$ that has α as a root. Then $f(x)$ is divisible by $M_\alpha(x)$.

Proof: Use the Euclidian algorithm to compute $q(x)$ and $r(x)$ such that $f(x) = M_\alpha(x)q(x) + r(x)$, where $r(x)$ has degree less than $M_\alpha(x)$. Because $f(\alpha) = M_\alpha(\alpha) = 0$, we have $r(\alpha) = 0$. But the only polynomial over $GF(p)$ with degree less than $M_\alpha(x)$ and with α as a root is the 0 polynomial. Hence $r(x) = 0$ and $f(x) = M_\alpha(x) \cdot q(x)$. $\square$

Theorem Let $M_\alpha(x)$ be the minimum polynomial over $GF(p)$ of α, an element of $GF(p^n)$. Then $\alpha^p, \alpha^{p^2}, \ldots, \alpha^{p^i}, \ldots, \alpha^{p^{n-1}}$ also are roots of $M_\alpha(x)$.

Proof: Because $f(x)^p = f(x^p)$, we have $M_\alpha(\alpha^p) = M_\alpha(\alpha)^p = 0^p = 0$, and α^p is a root of $M_\alpha(x)$. Because α^p is a root, $(\alpha^p)^p = \alpha^{p^2}$ must be a root. Similarly, $\alpha^{p^3}, \alpha^{p^4}, \ldots$ are roots. Because $\alpha^{p^n} = \alpha$, we obtain up to n roots from this relation. $\square$

In the examples of minimum polynomials for $GF(8)$ in Table 8.3.1, note that α, α^2, and α^4 are the three roots of one polynomial, and the other polynomial has roots $\beta = \alpha^3$, $\beta^2 = \alpha^6$, and $\beta^4 = \alpha^{12} = \alpha^5$.

Because every irreducible polynomial is the minimum polynomial of each of its roots, the preceding theorem tells how to find several roots of an irreducible polynomial if we know one of its roots. The counterpart of this theorem for polynomials over the real numbers is the observation that, if $a + ib$ is a root of an irreducible polynomial, then so is $a - ib$, the complex conjugate of $a + ib$.

The preceding theorem states that $\alpha^p, \alpha^{p^2}, \ldots$ are among the roots of $M_\alpha(x)$, but it so happens that these are *all* of the roots of $M_\alpha(x)$. Thus, if we know one root of an irreducible polynomial, we can find all of the roots by taking pth powers. We proceed to prove this fact in the following few theorems. Our method is first to show that the roots of an irreducible polynomial $f(x)$ of degree d over $GF(p)$ must lie in a field of order p^d, but not in any smaller field. Thus we know that each root satisfies $x^{p^d} = x$, because every element of $GF(p^d)$ satisfies this equation. This shows that there are no more than d distinct pth powers of a root of $f(x)$. Then we show that there are exactly d distinct pth powers, so that these must be the d roots of $f(x)$.

Theorem Let $f(x)$ of degree d be an irreducible polynomial over $GF(p)$, and let α be a root of $f(x)$. Then α lies in a finite field of order p^d, and is not an element of a field of smaller order.

Proof: First we show that α lies in $GF(p^d)$. Consider the polynomials modulo $f(x)$ over $GF(p)$. This set of polynomials forms a ring under polynomial addition and multiplication and, because $f(x)$ is irreducible, the ring is a field. This field contains p^d elements. There is an element of this field whose minimum polynomial is $f(x)$, namely the polynomial x. Then we can represent α by the polynomial x, and therefore α is in $GF(p^d)$.

To show that no smaller field contains α, we observe that, because α is in $GF(p^d)$, it must satisfy the equation $\alpha + \alpha + \cdots + \alpha = 0$, where the left side of the equation has p terms. Consequently, any field that contains α must have characteristic p. Hence let us assume that α is in $GF(p^e)$, where $e < d$. This field contains $GF(p)$ as a subfield.

Because the minimum polynomial of α has degree d, there is no polynomial over $GF(p)$ of degree less than d that has α as a root. Now consider the linear combinations of powers of α of the form

$$b_0 + b_1\alpha^1 + b_2\alpha^2 + \cdots + b_{d-1}\alpha^{d-1} + \alpha^d,$$

where each b_i is in $GF(p)$, and the coefficient of α^d is unity.

As mentioned earlier, $GF(p)$ is a subfield of $GF(p^e)$ for all $e \geq 1$, because $GF(p)$ is the subfield generated by 1 under addition. Therefore the linear combination just given involves coefficients in $GF(p^e)$ and powers of α, which also are in $GF(p^e)$. Because $GF(p^e)$ is closed under addition and multiplication, the linear combinations of powers of α must be elements of $GF(p^e)$. There are p^d linear combinations of powers of α altogether, because there are p^d ways of picking the coefficients. Because $p^e < p^d$, there must be two linear combinations that are equal to the same field element in $GF(p^e)$. Let these combinations be $g(\alpha)$ and $h(\alpha)$, respectively. Then $g(\alpha) - h(\alpha) = 0$, and therefore α is a root of the polynomial $g(x) - h(x)$ of degree less than d with coefficients over $GF(p)$. This is a contradiction, and hence the smallest field containing α is $GF(p^d)$. □

This theorem also has a counterpart for polynomials over the real numbers: if the roots of an irreducible polynomial are not real, then they lie in a larger field, the field of complex numbers. It so happens that all irreducible polynomials of first degree have real roots, and those of second degree have complex roots.

We still are interested in showing that, if $f(x)$ is an irreducible polynomial of degree d and has a root α, then all of its roots are α^{p^i}, where $0 \le i \le d - 1$. Thus far, we know that the pth powers of α are among the roots and that all of the roots lie in $GF(p^d)$. The following theorem proves that the pth powers of α are all of the roots.

Theorem Let $f(x)$ be an irreducible polynomial of degree d over $GF(p)$, and let α be a root of $f(x)$. Then the d distinct roots of $f(x)$ are α^{p^i}, where $0 \le i \le d - 1$.

Proof: A preceding theorem shows that the pth powers of α are among the roots of $f(x)$. If we show that there are d distinct roots among the pth powers of α, then we have found all of the roots. Suppose that $\alpha^{p^i} = \alpha^{p^j}$, where $0 \le i, j \le d - 1$. Without loss of generality, assume $i < j$. Then, because

$$x^{p^i} + y^{p^i} = (x + y)^{p^i},$$

we have

$$\alpha^{p^j} - \alpha^{p^i} = (\alpha^{p^{j-i}} - \alpha)^{p^i} = 0.$$

Thus, if α is not 0, then it is a root of $x^{b-1} - 1$, where $b = p^{j-i}$. Then α belongs to $GF(p^{j-i})$, which is smaller than $GF(p^d)$. However, this contradicts the preceding theorem. Hence there are d distinct pth powers of α, and these are the d roots of $f(x)$. □

Now let us take another look at $GF(2^3)$ in the light of results obtained thus far. Because every nonzero field element satisfies the equation $x^7 = 1$, these field elements must be the seven distinct roots of the polynomial $x^7 - 1$. Suppose that α is a field element that generates the multiplicative group. Then α, α^2, and α^4 are the roots of an irreducible factor of $x^7 - 1$, and α^3, $(\alpha^3)^2 = \alpha^6$, and $(\alpha^3)^4 = \alpha^{12} = \alpha^5$ are the roots of another irreducible factor of $x^7 - 1$. Hence, two of the irreducible factors of $x^7 - 1$ have degree 3. The last factor has 1 as a root, so this is a polynomial of degree 1. It is not difficult to check that $x^7 - 1$ factors into

$$(x^3 + x + 1)(x^3 + x^2 + 1)(x + 1),$$

where the irreducible factors are the minimum polynomials we have derived earlier.

This exercise shows that the factorization of $x^{p^{n}-1} - 1$ into irreducible factors over $GF(p)$ gives us the minimum polynomials of all the nonzero field elements in $GF(p^n)$. Conversely, the structure of $GF(p^n)$ tells us something about the factorization of $x^{p^{n}-1} - 1$. By studying this factorization, we can show that fields of order p^n exist for all prime p and $n \geq 1$, and that all fields of the same order are isomorphic. To obtain these important results, we make use of the next two lemmas.

Lemma The polynomial $x^m - 1$ divides $x^n - 1$ over any field if and only if m divides n.

Proof: If m divides n, then we can easily show by polynomial division that $x^m - 1$ divides $x^n - 1$. If m does not divide n, say $n = qm + r$ where $r < m$, then the polynomial division algorithm yields

$$x^n - 1 = x^r(x^{qm} - 1) + x^r - 1.$$

Because $x^{qm} - 1$ is divisible by $x^m - 1$, we can rewrite this expression as

$$x^n - 1 = a(x)(x^m - 1) + b(x).$$

In the latter equation, the degree of $b(x)$ is less than m, and we obtain the polynomials $a(x)$ and $b(x)$ by polynomial division. Comparison of the right sides of the two equations shows that

$$a(x) = x^r(x^{qm} - 1)/(x^m - 1) \quad \text{and} \quad b(x) = x^r - 1$$

because $x^m - 1$ divides $x^{qm} - 1$. Then $x^m - 1$ divides $x^n - 1$ if and only if $b(x) = 0$; that is, if and only if $r = 0$, which occurs if and only if m divides n. ☐

Lemma Let q be any integer greater than 1. Then $q^m - 1$ divides $q^n - 1$ if and only if m divides n.

Proof: This result follows directly from the preceding lemma. ☐

Now we are ready to state the major result concerning the factorization of the polynomial $x^{p^{n}-1} - 1$ over $GF(p)$. We do this by examining the factorization of $x^{p^n} - x = x(x^{p^{n}-1} - 1)$. Note that this polynomial has as its roots all of the elements of $GF(p^n)$, including 0.

Theorem The irreducible factors of $x^{p^n} - x$ over $GF(p)$ are all of the irreducible polynomials whose degrees are divisors of n.

Proof: The irreducible polynomial x clearly is one factor, and its degree is a divisor of n. The proof of this theorem therefore rests upon two assertions: **(i)** if $f(x) \neq x$ is irreducible over $GF(p)$ and if

$$deg[f(x)] = d$$

and if d divides n, then $f(x)$ is a divisor of $x^{p^{n}-1} - 1$; and **(ii)** every irreducible factor of $x^{p^{n}-1} - 1$ has a degree that divides n.

(i) Assume that $f(x)$ is irreducible over $GF(p)$ and has degree d. Then every root of $f(x)$ is a root of $x^{p^{d}-1} - 1$, so $f(x)$ divides $x^{p^{d}-1} - 1$. From the preceding two lemmas, we conclude that $x^{p^{d}-1} - 1$ divides

$x^{p^{n}-1} - 1$ if and only if d divides n. Therefore $f(x)$ is a factor of $x^{p^{n}-1} - 1$.

(ii) Let $f(x)$ be any irreducible factor of $x^{p^{n}-1} - 1$ of degree d. Suppose that d is not a divisor of n. Then every root of $f(x)$ is a root of both of the polynomials $x^{p^{n}-1} - 1$ and $x^{p^{d}-1} - 1$. Assume that $n = qd + r$, where $0 \le r < d$. Then

$$\alpha = \alpha^{p^{n}} = \alpha^{p^{qd+r}} = (\alpha^{p^{qd}})^{p^{r}} = (\alpha)^{p^{r}},$$

because $\alpha^{p^{qd}} = \alpha$. Then

$$\alpha^{p^{r}} - \alpha = \alpha(\alpha^{p^{r}-1} - 1) = 0,$$

and therefore α is a root of $x^{p^{r}-1} - 1$. Then α lies in a field of order p^{r}, which is smaller than p^{d}. This is a contradiction, and hence d divides n. ☐

From the preceding theorem we now can show that there exist irreducible polynomials of all degrees, and that we can find representations of finite fields of size p^{n} for all primes p by using polynomial algebras modulo irreducible polynomials.

Definition Let $I_{p}(n)$ be the number of irreducible polynomials of degree n over $GF(p)$.

Theorem $p^{n} = \sum_{d \mid n} d \cdot I_{p}(d)$.

Proof: There are p^{n} roots of $x^{p^{n}} - x$. We factor $x^{p^{n}} - x$ into irreducible polynomials and count the roots. For each d that divides n, there are $I_{p}(d)$ irreducible factors, each of which has d roots. This observation proves the theorem. ☐

The preceding theorem permits us to compute by a recursive calculation the number of irreducible polynomials over $GF(p)$ for every degree. For example, consider the calculation of $I_{2}(n)$ for $n = 1, 2, \ldots, 6$. We begin this calculation with $I_{2}(1) = 2$, because the polynomials x and $x + 1$ are both irreducible. Then we have

$$I_{2}(2) = (2^{2} - I_{2}(1))/2 = (4 - 2)/2 = 1,$$
$$I_{2}(3) = (2^{3} - I_{2}(1))/3 = (8 - 2)/3 = 2,$$
$$I_{2}(4) = (2^{4} - I_{2}(1) - 2I_{2}(2))/4 = (16 - 2 - 2)/4 = 3,$$
$$I_{2}(5) = (2^{5} - I_{2}(1))/5 = (32 - 2)/5 = 6,$$
$$I_{2}(6) = (2^{6} - I_{2}(1) - 2I_{2}(2) - 3I_{2}(3))/6 = 54/6 = 9.$$

Note that these results agree with Table 8.2.1 as far as that table goes. In Chapter Nine we discover that there is an interesting relation between the number of irreducible polynomials and the cyclic structure of feedback registers.

We assert previously that there is an irreducible polynomial of degree n over $GF(p)$ for each $n \ge 1$, and we used this assertion to claim the existence of $GF(p^{n})$ for each prime p and each integer $n \ge 1$. From the recurrence relation above for $I_p(n)$, we can show relatively easily that $I_p(n) \geq 1$ for all $n \geq 1$. First, if n is prime, then

$$I_p(n) = (p^n - I_p(1))/n = (p^n - (p - 1))/n,$$

which is not less than 1 for all $n \geq 1$. If n is not prime, then its largest divisor is no larger than $n/2$, and n has no more than $n/2$ distinct divisors. Note that we are grossly overestimating the number of distinct divisors, and in many cases we are overestimating the size of the largest divisor of n. Observe also that, for each divisor d of n, $I_p(d)$ cannot exceed p^d. Now to obtain a lower bound for $I_p(n)$, we subtract all of the over-estimates for $I_p(d)$ for each divisor d from p^n. Then $I_p(n)$ is bounded below by

$$I_p(n) \geq \left(p^n - \sum_{d=1}^{n/2} p^d \right) n, \tag{8.3.1}$$

or

$$I_p(n) \geq \left(p^n - ((p^{(p/2)+1} - 1)/(p - 1)) \right)/n. \tag{8.3.2}$$

In Eq. 8.3.2. we make use of the following equality for the sum of a geometric series:

$$\sum_{i=0}^{m} p^i = (p^{i+1} - 1)/(p - 1). \tag{8.3.3}$$

Because 1, 2, and 3 are prime, $n = 4$ is the least integer of interest for which we can apply Eq. 8.3.2. For $n = 4$, Eq. 8.3.2 becomes

$$I_p(4) \geq (p^4 - p^3 + 1)/4 \geq 2$$

for all prime p. For $n > 4$, Eq. 8.3.2 becomes larger because p^n/n grows much faster than does $p^{n/2}/n$, so that Eq. 8.3.2 is at least 1 for all $n \geq 4$. Thus $I_p(n) \geq 1$ for all $n \geq 1$, and we have established the existence of fields of order p^n for all prime p and $n \geq 1$.

From this point, it is straightforward but tedious to show that all finite fields of the same order are isomorphic. We state the theorem and outline a proof.

Theorem All finite fields of the same order are isomorphic.

Proof: Let K and K' be two fields of order p^n, and let α and α' be generators of their respective multiplicative groups. The field elements of the two fields are powers of α and α', respectively, and these powers are all of the roots of the polynomial $x^{p^n-1} - 1$. It so happens that arithmetic in K is determined uniquely by the minimum polynomial of α, because the representation of K is the algebra of polynomials over $GF(p)$ modulo $M_\alpha(x)$, and this representation is determined completely by $M_\alpha(x)$. From this we discover that the minimum polynomials of all of the field elements of K are determined once we have associated α with a particular minimum polynomial. Similarly, $M_{\alpha'}(x)$ determines arithmetic in K'. Now let r be an integer such that the minimum polynomial of α^r is $M_{\alpha'}(x)$. Then we claim that the mapping $f:K' \to K$, such that $f[(\alpha')^i] = \alpha^{ir}$, is an isomorphism of the fields.

To prove this claim, we must prove three things. We must show that f preserves addition, preserves multiplication, and is a bijection. To show that f preserves addition, we must show that, for all i, j, and k, if $(\alpha')^i + (\alpha')^j = (\alpha')^k$, then $\alpha^{ir} + \alpha^{jr} = \alpha^{kr}$. This follows because both α^r and α' have the same minimum polynomial, and the minimum polynomial uniquely determines the addition table for all powers of its roots. Hence, powers of α^r and of α' must behave identically under addition. We can prove similarly that f preserves multiplication, but this can be proved more directly because

$$ f[(\alpha')^i] \cdot f[(\alpha')^j] = \alpha^{ir}\alpha^{jr} = \alpha^{(i+j)r} = f[(\alpha')^{i+j}]. $$

The final step of the proof is to show that f is a bijection. For each i, $(\alpha')^i$ has the same minimal polynomial as α^{ir} because all arithmetic operations on powers of α' and α^r are identical. Because there are $p^n - 1$ distinct powers of α', and these are the $p^n - 1$ distinct roots of the polynomial $x^{p^n-1} - 1$, then the powers of α^r include all of the $p^n - 1$ distinct roots of $x^{p^n-1} - 1$. Hence the mapping is one-to-one. Because the field is finite, the mapping is a surjection by the pigeonhole principle, so it is a bijection. Thus the two fields are isomorphic. □

Many of the details that have been omitted from the proof just given are made clear by examples that appear in the following section. The crux of the proof rests on the properties that the arithmetic operations in a field are determined uniquely by the minimum polynomial of one of the generators of its multiplicative group, and that the order of the field determines the set of minimum polynomials for the field elements.

Summary We find that the multiplicative group of the nonzero field elements of a finite field is a cyclic group. This result is extremely useful for constructing representations of fields, because we can identify some field element as a generator of the cyclic group, and all of the nonzero field elements are powers of the generator.

A second important result is that the roots of an irreducible polynomial of degree d over $GF(p)$ are pth powers of one of the roots. Thus the d roots are $\alpha, \alpha^p, \alpha^{p^2}, \ldots, \alpha^{p^{d-1}}$ for some field element α. From this last property, we are able to prove that the factors of $x^{p^n} - x$ over $GF(p)$ are all of the irreducible polynomials over $GF(p)$ whose degrees divide n. Because we are able to determine the factors of $x^{p^n} - x$, we are also able to find the number of irreducible polynomials of degree n over $GF(p)$, and we discover that this number is at least 1 for all $n \geq 1$. Thus, for each n and each prime p there exists an irreducible polynomial over $GF(p)$ of degree n. Thus we can construct a field as the algebra of polynomials modulo that irreducible polynomial. Hence we prove that fields of order p^n exist for every $n \geq 1$ and prime p.

Finally, because each of the p^n elements of a field of order p^n is a root of $x^{p^n} - x$, we are able to find an isomorphism between two different fields of order p^n. The isomorphism is a map that maps α in K onto α' in K', where α and α' are both roots of the same irreducible polynomial. The isomorphism is completely determined if α and α' are each generators of the multiplicative group of the field.

The historical development of the material in this chapter is of interest. Évariste Galois, of course, initiated the study of fields, and he proved several of the specific results set forth in this chapter. For example, he proved that $f(x^p) = [f(x)]^p$ in a field of characteristic p, and that, if α is a root of $f(x)$ in this field, then so are $\alpha^p, \alpha^{p^2}, \ldots, \alpha^{p^i}$ (Galois 1830). Galois treated his fields as polynomial algebras modulo an irreducible polynomial as we do here, and he knew that the multiplicative group of such an algebra is cyclic. Jordan (1870) proved that there exist fields of order p^n for all prime p, and Serret (1885) extended the theory by investigating the factorization of polynomials over $GF(p)$. It was E. H. Moore (1893) who proved the central result that all finite fields are Galois fields.

Herstein (1964) presents a very readable discussion of classical Galois theory. We have not even touched upon some of the very elegant aspects of the theory. Herstein also discusses applications of Galois theory to the solution of polynomial equations by radicals and to ruler-and-compass constructions. The brief monograph on Galois theory by Gaal (1971) includes an unusual number of good examples. Artin (1959) and Van der Waerden (1931) give deeper treatments of the material. The 1901 work by L. E. Dickson stands as a historic contribution because it was the first major text on the subject.

Exercises

8.3.1 Let $f(x)$ be a polynomial over a field of characteristic p. Prove that $f(x^p) = [f(x)]^p$. Prove that $f(x^{p^i}) = [f(x)]^{p^i}$.

8.3.2 Construct a representation of $GF(2^3)$ by using the irreducible polynomial $x^3 + x^2 + 1$ as the modulus. Find a field element that generates the multiplicative group and identify the other elements as powers of this one. Find the orders and minimum polynomials of each of the elements in this representation.

8.3.3 Show that the representation of $GF(2^3)$ constructed in Ex. 8.3.2 is isomorphic to the representation of $GF(2^3)$ generated by the irreducible polynomial $x^3 + x + 1$.

8.3.4 Prove that, in a field of characteristic p, the sum $\alpha + \alpha + \cdots + \alpha$ is 0 for nonzero α is and only if the number of terms in the sum is a multiple of p. HINT: Represent α as a polynomial with coefficients in $GF(p)$.

8.3.5 Let q be an integer not less than 2. Prove that $q^m - 1$ divides $q^n - 1$ if and only if m divides n.

8.3.6 Construct a representation of $GF(2^4)$, using the irreducible polynomial $x^4 + x + 1$ as a modulus. Use the fact that the other irreducible polynomials of degree 4 over $GF(2)$ are $x^4 + x^3 + 1$ and

$$x^4 + x^3 + x^2 + x + 1$$

to compute the minimum polynomials of each field element.

8.3.7 The *reciprocal polynomial* of a polynomial $f(x)$ of degree n is the polynomial $\overline{f(x)} = x^n f(1/x)$. Prove that, if $f(x)$ has a root α in the field K, then $\overline{f(x)}$ has the root α^{-1} in K. Prove that, if $f(x)$ is irreducible over $GF(2)$, then $f(x) = \overline{f(x)}$ if and only if the degree of $f(x)$ is 2 or less.

8.3.8 Prove that the mapping $f: K \rightarrow K$, such that $f(\alpha) = \alpha^p$, is a field automorphism if K has characteristic p.

8.3.9 Prove that the mapping $f: K \rightarrow K$, such that $f(\alpha) = \alpha^t$, is a field isomorphism if and only if t is relatively prime to $p^n - 1$, where p^n is the order of K.

8.3.10 Let α be an element of $GF(2^6)$ that generates the multiplicative group. Group the 63 nonzero field elements into sets such that the members of each set have the same minimum polynomial.

8.3.11 Prove that an irreducible polynomial is the minimum polynomial of each of its roots.

8.3.12 Find the values of $I_2(7)$, $I_2(8)$, and $I_3(n)$ for $1 \leq n \leq 6$.

8.4 The Representation of Finite Fields

In this section we investigate the representation of finite fields, with particular emphasis on different representations of the same field.

As a running example for this section, we treat the representation of $GF(2^4)$. We know that we can represent this field as an extension field of degree 4 over $GF(2)$. To do so, we construct a polynomial algebra modulo an irreducible polynomial over $GF(2)$ of degree 4. There are three irreducible polynomials from which we may choose, namely $x^4 + x^3 + x^2 + x + 1$, $x^4 + x + 1$, and $x^4 + x^3 + 1$. It does not matter which polynomial we select, because the representations determined by the various polynomials are isomorphic. Let us arbitrarily choose the polynomial $x^4 + x^3 + x^2 + x + 1$.

The addition table in the polynomial algebra is trivial to construct, because it is componentwise addition of the coefficients of polynomials in $GF(2)$. To construct the multiplication table, let us find some field element that generates the multiplicative group of the field. Table 8.4.1 shows the multiplicative order of each field element. We observe that the polynomial $1 + x$ represents a field element of order 15. We take this element to be α, and we show the powers of α in the third column of the table. The representation given in the table depicts the entire multiplicative and additive structure of the field, because the first column shows how to add field elements and the third column shows how to multiply them. For example, $\alpha^3 \cdot \alpha^5 = \alpha^8$, and

$$\alpha^3 + \alpha^5 = (1 + x + x^2 + x^3) + (1 + x^2 + x^3) = x = \alpha^{12}.$$

TABLE 8.4.1 A Representation of $GF(2^4)$ as the Polynomial Algebra Modulo $(x^4 + x^3 + x^2 + x + 1)$

Field Element	Order	Power of $\alpha = 1 + x$	Power of $\beta = 1 + x + x^2$
1	1	$\alpha^{15} = 1$	$\beta^{15} = 1$
x	5	α^{12}	β^6
$1 + x$	15	α	β^{13}
x^2	5	α^9	β^{12}
$1 + x^2$	15	α^2	β^{11}
$x + x^2$	15	α^{13}	β^4
$1 + x + x^2$	15	α^7	β
x^3	5	α^6	β^3
$1 + x^3$	15	α^8	β^{14}
$x + x^3$	15	α^{14}	β^2
$1 + x + x^3$	15	α^{11}	β^8
$x^2 + x^3$	3	α^{10}	β^{10}
$1 + x^2 + x^3$	3	α^5	β^5
$x + x^2 + x^3$	15	α^4	β^7
$1 + x + x^2 + x^3$	5	α^3	β^9

In this representation, the field element designated as α is arbitrary. Any field element of order 15 will do. Hence we can just as easily pick the polynomial $1 + x + x^2$ as a generator, which we shall call β. The fourth column of the table gives the powers of β, and from this we have a representation of the field in terms of powers of β. Clearly, the α representation and the β representation are isomorphic under the map $\beta^i = \alpha^{7i}$.

An interesting exercise is to group the field elements into sets such that the elements in a set have the same minimum polynomial. Recall from Section 8.3 that the d roots of an irreducible polynomial of degree d over $GF(p)$ are the d distinct pth powers of one of the roots. In our example $p = 2$, so that, if α is a root of an irreducible polynomial, then so are α^2, $\alpha^{2^2} = \alpha^4$, and $\alpha^{2^3} = \alpha^8$. Similarly, if α^3 is a root of an irreducible polynomial, then so are $(\alpha^3)^2 = \alpha^6$, $(\alpha^3)^{2^2} = \alpha^{12}$, and $(\alpha^3)^{2^3} = \alpha^{24} = \alpha^9$. Following this procedure, we find that the elements partition naturally into 5 sets, which we show here with elements as powers of α:

$$\{\alpha, \alpha^2, \alpha^4, \alpha^8\},$$
$$\{\alpha^3, \alpha^6, \alpha^9, \alpha^{12}\},$$
$$\{\alpha^7, \alpha^{11}, \alpha^{13}, \alpha^{14}\},$$
$$\{\alpha^5, \alpha^{10}\},$$
$$\{\alpha^{15} = 1\}.$$

From Section 8.3, we know that all of the irreducible factors of $x^{2^4-1} - 1 = x^{15} - 1$ must have degrees that divide 4. Using this and the tables of irreducible polynomials, we find that $x^{15} - 1$ factors into

$$(x + 1)(x^2 + x + 1)(x^4 + x^3 + x^2 + x + 1)(x^4 + x + 1)(x^4 + x^3 + 1).$$

Now we associate irreducible polynomials with powers of α. The root of $(x + 1)$ lies in $GF(2^1)$, so this must be $\alpha^{15} = 1$. The two roots of $(x^2 + x + 1)$ lie in $GF(2^2)$, so they satisfy the equation $x^3 - 1 = 0$. Note that α^5 and α^{10} have this property, while all the other elements except 1 do not have this property. Hence α^5 and α^{10} are roots of $x^2 + x + 1$.

The other sets of roots are all roots of fourth-degree polynomials. We have selected one of the polynomials as the modulus of our representation, and we have discovered that the polynomial x has order 5 and that x is a root of $x^4 + x^3 + x^2 + x + 1$ in this representation. Among the sets of roots, the only powers of α with order 5 are 3, 6, 9, and 12. Hence these powers of α are roots of $x^4 + x^3 + x^2 + x + 1$. Now we must match the two remaining sets of roots with $x^4 + x^3 + 1$ and $x^4 + x + 1$. To do so, we must see whether α is a root of $x^4 + x^3 + 1$ or of $x^4 + x + 1$. Inspection of the table shows that $\alpha^4 + \alpha^3 + 1 = 0$, and $\alpha^4 + \alpha + 1 = x^4 + x + 1 \neq 0$. Hence, α, α^2, α^4, and α^8 are the roots of $x^4 + x^3 + 1$, and the remaining powers of α are roots of $x^4 + x + 1$.

Now let us examine a different representation of the same field. For this representation, we choose the irreducible polynomial $x^4 + x + 1$ to be the modulus of the polynomial algebra. We obtain the representation shown in Table 8.4.2. Note that the polynomial x has order 15 in this representation, whereas it had order 5 in Table 8.4.1. Hence this representation is different, yet it is isomorphic to the representation in Table 8.4.1. To relate the representations of Tables 8.4.1 and 8.4.2, note that α and $(\alpha')^7$ are roots of $x^4 + x^3 + 1$. Hence we are tempted to have the field isomorphism map α into $(\alpha')^7$.

The mapping $f(\alpha^i) = (\alpha')^{7i}$ clearly is an isomorphism of the multiplicative structures, and we can show rigorously that it also is an isomorphism of the additive structures. For example, $\alpha^3 + \alpha^5 = \alpha^{12}$, and

$$f(\alpha^3) + f(\alpha^5) = (\alpha')^{21} + (\alpha')^{35} = (\alpha')^6 + (\alpha')^5 = (\alpha')^9 = (\alpha')^{84} = f(\alpha^{12}).$$

Because both the multiplicative and the additive structures are isomorphic, the fields are isomorphic.

Thus Tables 8.4.1 and 8.4.2 show how $GF(2^4)$ is an extension field of degree 4 over $GF(2)$. We can also construct a representation of $GF(2^4)$ as the algebra of polynomials modulo an irreducible polynomial of degree 2 over the coefficient field $GF(2^2)$. To see this, let us pick an irreducible polynomial over $GF(2^2)$ from Table 8.4.3 and generate the algebra of polynomials modulo this irreducible polynomial. Table 8.4.4 gives the field representation for the polynomial $x^2 + \beta x + 1$. If we

TABLE 8.4.2   A Representation of $GF(2^4)$ as the Polynomial Algebra Modulo $(x^4 + x + 1)$

Field Element	Order	Power of $\alpha' = x$	Minimum Polynomial
1	1	$(\alpha')^{15}$	$x + 1$
x	15	α'	$x^4 + x + 1$
$1 + x$	15	$(\alpha')^4$	$x^4 + x + 1$
x^2	15	$(\alpha')^2$	$x^4 + x + 1$
$1 + x^2$	15	$(\alpha')^8$	$x^4 + x + 1$
$x + x^2$	3	$(\alpha')^5$	$x^2 + x + 1$
$1 + x + x^2$	3	$(\alpha')^{10}$	$x^2 + x + 1$
x^3	5	$(\alpha')^3$	$x^4 + x^3 + x^2 + x + 1$
$1 + x^3$	15	$(\alpha')^{14}$	$x^4 + x^3 + 1$
$x + x^3$	5	$(\alpha')^9$	$x^4 + x^3 + x^2 + x + 1$
$1 + x + x^3$	15	$(\alpha')^7$	$x^4 + x^3 + 1$
$x^2 + x^3$	5	$(\alpha')^6$	$x^4 + x^3 + x^2 + x + 1$
$1 + x^2 + x^3$	15	$(\alpha')^{13}$	$x^4 + x^3 + 1$
$x + x^2 + x^3$	15	$(\alpha')^{11}$	$x^4 + x^3 + 1$
$1 + x + x^2 + x^3$	5	$(\alpha')^{12}$	$x^4 + x^3 + x^2 + x + 1$

let α be a root of this polynomial, then we can associate powers of α with each polynomial in the algebra, as indicated in the table. Note, for example, that the product $\alpha^2 \cdot \alpha^9$ is represented by the polynomial product $\beta x \cdot (1 + \beta x) \bmod (x^2 + \beta x + 1)$. We have

$$\beta x \cdot (1 + \beta x) \equiv \beta x + \delta x^2 \equiv \beta x + (\delta + x) \equiv \delta + \delta x \bmod (x^2 + \beta x + 1).$$

From Table 8.4.4, we observe that $\delta + \delta x$ is the polynomial representation of $\alpha^{11} = \alpha^2 \cdot \alpha^9$, as we expect. Table 8.4.4, like Tables 8.4.1 and 8.4.2, contains sufficient information to construct the complete tables for the arithmetic operations in this field.

TABLE 8.4.3 Irreducible Polynomials over $GF(2^2)$*

x
$x + 1$
$x + \beta$
$x + \delta$
$x^2 + x + \beta$
$x^2 + x + \delta$
$x^2 + \beta x + 1$
$x^2 + \beta x + \beta$
$x^2 + \delta x + 1$
$x^2 + \delta x + \delta$

+	0	1	β	δ
0	0	1	β	δ
1	1	0	δ	β
β	β	δ	0	1
δ	δ	β	1	0

$\cdot$	0	1	β	δ
0	0	0	0	0
1	0	1	β	δ
β	0	β	δ	1
δ	0	δ	1	β

FIG. 8.4.1 Arithmetic Operations in $GF(2^2)$

*See the tables of arithmetic operations given in Fig. 8.4.1.

TABLE 8.4.4 A Representation of $GF(2^4)$ as the Polynomial Algebra Modulo $(x^2 + \beta x + 1)$ with Coefficients in $GF(2^2)$

Field Element	Order	Power of $\alpha = 1 + x$
1	1	α^{15}
β	3	α^5
δ	3	α^{10}
x	5	α^{12}
$1 + x$	15	α
$\beta + x$	5	α^3
$\delta + x$	15	α^4
βx	15	α^2
$1 + \beta x$	5	α^9
$\beta + \beta x$	5	α^6
$\delta + \beta x$	15	α^8
δx	15	α^7
$1 + \delta x$	15	α^{13}
$\beta + \delta x$	15	α^{14}
$\delta + \delta x$	15	α^{11}

Exercises

8.4.1 Construct a representation of $GF(2^4)$ by using the irreducible polynomial $x^2 + x + \beta$ over $GF(2^2)$ as the modulus. Find the orders of all of the elements, and find an element that generates the multiplicative group. Show all nonzero elements as powers of the generator. Find the minimum polynomials over $GF(2)$ of each of the nonzero field elements.

8.4.2 In the representation of Table 8.4.4, find the minimum polynomials of each of the field elements over $GF(2^2)$.

8.4.3 Let $f(x)$ be an irreducible polynomial with coefficients in $GF(p^m)$, and let α in $GF(p^n)$ be a root of $f(x)$. Show that $GF(p^m)$ is a subfield of $GF(p^n)$ and that m divides n.

8.4.4 In Ex. 8.4.3, prove that the roots of $f(x)$ are $\alpha,\ \alpha^{p^m},\ \alpha^{p^{2m}},\ \dots,\ \alpha^{p^{n-m}}$.

8.4.5 Partition the elements of $GF(2^5)$ and $GF(2^6)$ into sets of elements such that the elements in a single set have the same minimum polynomial.

8.4.6 Let $f(x)$ be the minimum polynomial of α, and let $g(x)$ be the minimum polynomial of α^{-1}. Prove that $f(x)$ and $g(x)$ differ only by a reversal of coefficients.

8.4.7 Construct a polynomial representation of $GF(3^2)$ and identify the minimum polynomial of each of the field elements.

8.4.8 Construct a polynomial representation of $GF(5^2)$ and identify the minimum polynomial of each of the field elements.

8.4.9 Prove that, if α and β are roots of an irreducible polynomial of degree n over $GF(p)$, then in the multiplicative group of $GF(p)$ the order of α is equal to the order of β.

8.4.10 An irreducible polynomial of degree n over $GF(p)$ is said to be *primitive* if one of its roots is a generator of the multiplicative group of $GF(p^n)$. Prove that, if n is prime, then all irreducible polynomials of degree n are primitive.

8.4.11 Show that there exists an irreducible polynomial that is not primitive.

8.4.12 Find a formula for the number of primitive irreducible polynomials of degree n over $GF(p)$. HINT: See Ex. 2.5.12.

NINE | Linear Finite-State Machines

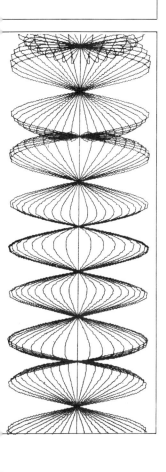

NINE | Linear Finite-State Machines

In this chapter we reexamine finite-state machines. In particular, we investigate the behavior of machines whose states can be represented as vectors in a vector space over a finite field, and whose next-state functions are linear. In the early part of the chapter we study autonomous linear machines—that is, machines without inputs. Such machines change state at regular clock intervals in a predetermined fashion. Therefore they can be viewed as counters or sequence generators. Our major result in this study is the observation that feedback shift-registers are standard building blocks for autonomous linear machines. We also show how to determine the cycle structure of the state graph for such machines without enumerating the states in the machine.

In the later part of this chapter we investigate applications of the theory of autonomous machines. It is somewhat surprising that the theory is applicable to Pólya enumeration of equivalence classes of switching functions. Although the two topics appear to have nothing in common, the mathematics used to compute the cycle structures of state graphs is identical to the mathematics required to compute cycle structures of groups in the Pólya sense.

We also apply the theory to the generation of very long sequences. These sequences have certain randomness properties, and they often are called *pseudorandom* sequences. We show how to construct a short and efficient computer program that produces random numbers by simulating a pseudorandom sequence generator. These sequences also have been applied in secure communication systems and in ranging experiments. One example of the latter application was a 1959 radar experiment for determining the distance between the earth and Venus.

Finally, we show how to construct encoders and decoders for Hamming single-error correcting codes, using feedback shift-registers. Encoders and decoders of this type presently are in use in several computer systems.

9.1 Linear Machines

In this section we set the context for later sections by defining the notion of linear machine and by exploring some of the elementary properties of the definition. We assume that the reader has some familiarity with linear algebra, so we only briefly review some of the concepts here. For good treatments of the requisite background material, see the books by Nering (1963) and DeRusso et al. (1965, Ch. 4).

We recall that, for a vector space V over a scalar field K, a *linear combination* of vectors $v_1, v_2, \ldots, v_n$ is a vector of the form

$$c_1 v_1 + c_2 v_2 + \cdots + c_n v_n,$$

where the constants c_i are scalars in K. A set of nonzero vectors $v_1, v_2, \ldots, v_n$ is said to be *dependent* if some linear combination of the vectors with at least one nonzero scalar coefficient is equal to the vector 0. The vectors are said to be *independent* if the only linear combination of the vectors that is equal to the vector 0 has all of the scalars c_i equal to 0. The set of all linear combinations of a set of n independent vectors is a vector space of *dimension n*. A set of independent vectors that generates a vector space V is said to be a *basis* for V, and all bases for V have the same number of elements. Thus an n-dimensional vector space is a vector space for which all bases contain n vectors. A function $f: V \to V'$ is a *linear function* from a vector space V into a vector space V' over the same scalar field as V if, for all c_1 and c_2 in K and all v_1 and v_2 in V,

$$f(c_1 v_1 + c_2 v_2) = c_1 f(v_1) + c_2 f(v_2).$$

After this brief summary of the background material, we are ready to begin the study of linear finite-state machines. Fig. 9.1.1 shows a typical example of such a machine; a formal definition follows.

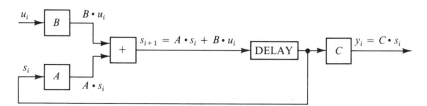

FIG. 9.1.1 Block Diagram of a Typical Linear Finite-State Machine

Definition   A machine M is a *linear finite-state machine* if

    (i) the state space S_M of M, the input space I_M, and the output space Y_M are each vector spaces over a finite field K (we let the dimensions of the spaces be n, p, and r, respectively); and

    (ii) for the state vector s_i, input vector u_i, and output vector y_i at time i, the next-state and output functions are of the forms

$$M(s_i, u_i) = s_{i+1} = A \cdot s_i + B \cdot u_i$$

and
$$\delta(s_i) = y_i = C \cdot s_i,$$

where A, B, and C are matrices over K; s_i, u_i, and y_i are column vectors; and the matrix operations are the usual matrix operations with arithmetic performed in the field K.

The dimensions of the matrices A, B, and C are $n \times n$, $n \times p$, and

$r \times n$, respectively, to be consistent with the dimensions of the three vector spaces.

In the preceding definition, the output function depends only on the present state of the machine, which is consistent with the Moore model of finite-state machines. The more usual practice is to use an output function derived from the Mealy model of finite machines, for which the output function depends both on the present state and the present input. Thus the output function for such a formulation is

$$y_i = C \cdot s_i + D \cdot u_i,$$

where D is an $r \times p$ matrix over K. We shall continue to use the simpler Moore model with a reminder that the two machine models are equivalent. As an example of a linear machine, consider the machine M whose matrices are shown in Fig. 9.1.2. M has a three-dimensional state space over $GF(2)$. The inputs and outputs of M are both two-dimensional vector spaces over $GF(2)$. The matrices A, B, and C are sufficient to specify the machine.

$$\begin{bmatrix} 0 & 0 & 1 \\ 1 & 0 & 1 \\ 0 & 1 & 0 \end{bmatrix} \quad \begin{bmatrix} 1 & 1 \\ 0 & 1 \\ 1 & 0 \end{bmatrix} \quad \begin{bmatrix} 1 & 0 & 0 \\ 0 & 1 & 1 \end{bmatrix}$$
$$A \qquad\qquad B \qquad\qquad C$$

FIG. 9.1.2   Matrices A, B, and C for Machine M

If we place M in the initial state $s_1 = \langle 0, 1, 0 \rangle^t$, and it receives an initial input $u_1 = \langle 0, 1 \rangle^t$, then the next state

$$s_2 = A \cdot s_1 + B \cdot u_1 = \langle 0, 0, 1 \rangle^t + \langle 1, 1, 0 \rangle^t = \langle 1, 1, 1 \rangle^t.$$

Table 9.1.1 shows the behavior of M for a short sequence of inputs.

This type of machine is called a linear machine because the next-state and output functions are linear functions. We prove some trivial lemmas concerning the linearity of the functions after introducing some definitions.

Definition The state $s = \langle 0, 0, \ldots, 0 \rangle^t$ of a linear machine is called the *0-state*, or *state 0*, and the input $u = \langle 0, 0, \ldots, 0 \rangle^t$ is called the *0-input*, or *input 0*.

Lemma The next-state function of a linear machine M is a linear function of the present state when the present input is the 0-input.

Proof: We must show that, for any two vectors s and t and for any two scalar field elements α and β,

$$M(\alpha s + \beta t, 0) = \alpha M(s, 0) + \beta M(t, 0),$$

where $M(s, u)$ is the next state when s is the present state and u is the present input. But, because $M(s, 0) = A \cdot s$, we have

$$M(\alpha s + \beta t, 0) = A \cdot (\alpha s + \beta t)$$
$$= \alpha A \cdot s + \beta A \cdot t = \alpha M(s, 0) + \beta M(t, 0). \qquad \square$$

The next two lemmas are similar to the preceding one, and we offer them here without proof.

Lemma The next-state function of a linear machine is a linear function of the present input when the present state is the 0-state.

Lemma The output function of a linear machine is a linear function of the present state.

One essential consequence of linearity is that we can easily analyze the behavior of sequential machines. Because we can determine $M(s + t, 0)$ from $M(s, 0)$ and $M(t, 0)$, we can determine the next state for any particular state under the 0-input by knowing only $M(e_i, 0)$ for some set of basis vectors $e_1, e_2, \ldots, e_n$ of the vector space of states. This follows because every state s in the state space is a linear combination of basis vectors.

When we consider the effect of a nonzero input, linearity allows us to use superposition to determine the behavior of the next state. That is, $M(s, u) = M(s, 0) + M(0, u)$ so that, if we know the next-state function for every state with the 0-input, and for the 0-state with every input, then the next-state function for nonzero state s and nonzero input u is just the sum (superposition) of $M(s, 0)$ and $M(0, u)$.

Now suppose that M starts in state s_1 and receives the input sequence $u_1 u_2 \ldots u_m$. This input sequence naturally decomposes into the sum of m input sequences $v_i = 0^{i-1} u_i 0^{m-i}$, where $1 \leq i \leq m$. That is, each v_i is a sequence of 0-inputs except for input u_i at time i. Does superposition hold to the extent that

TABLE 9.1.1 State Transitions and Outputs of a Linear Machine for a Short Sequence of Inputs*

Vector	Time				
	1	2	3	4	5
State	0	1	1	0	1
	1	1	0	0	1
	0	1	1	1	0 ·
Input	0	0	1	0	
	1	0	0	0	
Output	0	1	1	0	1
	1	0	1	1	1

*Each column shows the state, input, and output vectors of the machine at a particular time, with the time increasing by unit intervals from column to column moving left to right.

$$M(s_1, u_1 u_2 \ldots u_m) =$$
$$M(s_1, 0^m) + M(0, v_1) + M(0, v_2) + \cdots + M(0, v_m)?$$

In other words, is the behavior of M to a sequence of inputs equal to the superposition of its behavior to individual inputs? The following theorem answers this question affirmatively.

Theorem   For every input sequence $u_1 u_2 \ldots u_m$ and for every state s_1,

$$M(s_1, u_1 u_2 \ldots u_m) = M(s_1, 0^m) + \sum_{i=1}^{m} M(0, v_i),$$

where each v_i is defined as in the preceding paragraph and where M is a linear next-state function.

Proof: We prove this theorem by induction on the length of the input sequence.

Basis Step: For input sequences of length 1, a previous lemma gives us

$$M(s_1, u_1) = M(s_1, 0) + M(0, u_1) = M(s_1, 0) + M(0, v_1),$$

so the theorem holds for $m = 1$. For input sequences of length 2, we have

$$
\begin{aligned}
M(s_1, u_1 u_2) &= A \cdot (A \cdot s_1 + B \cdot u_1) + B \cdot u_2 \\
&= A^2 \cdot s_1 + A \cdot B \cdot u_1 + B \cdot u_2 \\
&= M(s_1, 00) + M(0, u_1 0) + M(0, 0 u_2) \\
&= M(s_1, 0^2) + M(0, v_1) + M(0, v_2),
\end{aligned}
$$

so the theorem holds for $m = 2$. In this argument, we have used the following equalities:

$$
\begin{aligned}
M(s_1, 00) &= A \cdot (A \cdot s_1 + B \cdot 0) + B \cdot 0 = A^2 \cdot s_1, \\
M(0, u_1 0) &= A \cdot (A \cdot 0 + B \cdot u_1) + B \cdot 0 = A \cdot B \cdot u_1, \\
M(0, 0 u_2) &= A \cdot (A \cdot 0 + B \cdot 0) + B \cdot u_2 = B \cdot u_2.
\end{aligned}
$$

Induction Step: We assume that the theorem holds for all sequences of length $m - 1$ or less and show that it holds for sequences of length m. From the linearity of the next-state function, we have

$$M(s_1, u_1 u_2 \ldots u_m) = A \cdot M(s_1, u_1 u_2 \ldots u_{m-1}) + B \cdot u_m.$$

By expansions similar to that given at the end of the basis step, we can easily show that $B \cdot u_m = M(0, u_m) = M(0, 0^i u_m)$ for all i. Thus we can rewrite the right side of the preceding equality as

$$M(s_1, u_1 u_2 \ldots u_m) = M\big(M(s_1, u_1 u_2 \ldots u_{m-1}), 0\big) + M(0, 0^{m-1} u_m)$$
$$= \Big(M(s_1, 0^m) + \sum_{i=1}^{m-1} M(0, v_i)\Big) + M(0, v_m),$$

where the second equality follows from the induction hypothesis. Thus the theorem is true for sequences of length m, and the proof is complete.   □

The proof of the preceding theorem indicates that $M(s_1, u_1 u_2 \ldots u_m)$ in matrix notation is

$$M(s_1, u_1 u_2 \ldots u_m) = M(s_1, 0^m) + \sum_{i=1}^{m} M(0, v_i)$$

$$= A^m \cdot s + \sum_{i=1}^{m} A^{m-i} \cdot B \cdot u_i.$$

Readers familiar with continuous linear systems will recognize this superposition as the discrete analog of convolution for such systems.

We see here that the response of a linear machine to the 0-input contributes to the response of the machine to a sequence of nonzero inputs. In the following sections of this chapter we investigate the autonomous behavior of linear machines. In essence, this investigation is a study of the behavior of linear machines in response to the 0-input. When we study linear machines as encoders and decoders in the latter part of this chapter, we return briefly to linear machines with nonzero inputs.

A number of texts on the subject of linear finite-state machines have appeared in recent years; they are good sources of material that cannot be covered in the brief survey here. A thorough basic text by Gill (1966) is self-contained and has an excellent summary of the mathematical background for the material. Kautz (1965) presents a collection of several landmark papers in the area, thereby providing an excellent view of the historical development of the subject. An advanced text by Harrison (1969) takes up such topics as synthesis and decomposition properties of linear machines.

Exercises

9.1.1 Let M be a machine with the next-state matrix A given in Fig. 9.1.3. Under the assumptions that arithmetic is in $GF(2)$ and that the input is the 0-input, determine the successors of each of the states of M.

$$A = \begin{bmatrix} 0 & 1 & 0 \\ 1 & 1 & 1 \\ 0 & 1 & 1 \end{bmatrix}$$

FIG. 9.1.3   Next-State Matrix A for Machine M in Ex. 9.1.1

9.1.2 Let M be the machine with the matrices A, B, and C given in Fig. 9.1.4. Assume that arithmetic is in $GF(2)$ and that M is started in state $\langle 0, 1, 0 \rangle^t$. Simulate the first seven state transitions of M when given the input $\langle 1, 1 \rangle^t$ followed by six instances of the 0-input.

$$A = \begin{bmatrix} 1 & 0 & 1 \\ 0 & 0 & 1 \\ 1 & 1 & 1 \end{bmatrix} \quad B = \begin{bmatrix} 1 & 0 \\ 1 & 1 \\ 0 & 1 \end{bmatrix} \quad C = \begin{bmatrix} 1 & 1 & 1 \\ 0 & 1 & 0 \end{bmatrix}$$

FIG. 9.1.4 Matrices A, B, and C for Machine M in Ex. 9.1.2

9.1.3 Let M be a linear machine over a finite field, and let M be stimulated by a sequence of 0-inputs. Prove that the sequence of states entered by M eventually becomes periodic. Prove that, if M is stimulated indefinitely by the repetition of one input symbol, then the sequence of states entered by M eventually becomes periodic so long as the input symbol is constant.

9.1.4 In Ex. 9.1.3, show that the period of the cycle of states attained by M cannot exceed $q^n - 1$ if the states of M are n-tuples over $GF(q)$. HINT: The 0-state always is on a cycle by itself. The other states may all be on one cycle.

9.1.5 Let M be an autonomous machine or, equivalently, a machine stimulated by the 0-input. Suppose that, when M is started in state $\langle 0, 1, 1 \rangle^t$, it goes through the sequence of states $\langle 1, 0, 1 \rangle^t$, $\langle 1, 1, 0 \rangle^t$, and $\langle 0, 1, 1 \rangle^t$. Find at least one A matrix for M that exhibits this behavior. Assume that the scalar field is $GF(2)$.

9.1.6 Let M be an autonomous machine, and assume that one cycle of states is the cycle $\langle 0, 1, 1 \rangle^t$, $\langle 1, 0, 1 \rangle^t$, $\langle 1, 1, 0 \rangle^t$, $\langle 0, 0, 1 \rangle^t$, $\langle 0, 1, 1 \rangle^t$. Let $GF(2)$ be the scalar field for M, and prove that M cannot be a linear machine. HINT: Show that the next-state function is nonlinear.

9.1.7 Let M be a linear machine without inputs, whose states are n-tuples over $GF(2)$. Construct an algorithm for finding the next-state matrix A for M when the successor states of a few of the states of M are specified. What is the least number of successor states that must be specified in order to determine A uniquely?

9.1.8 Prove that we can uniquely determine A in Ex. 9.1.7 if we are given the successor states of the basis for the vector space of states.

9.2 Autonomous Linear Machines   Autonomous machines are machines with no inputs. If an autonomous machine is linear, then its next-state function is $s_{i+1} = M(s_i) = A \cdot s_i$ or, equivalently, it is a linear machine for which the B matrix is a 0-matrix. In preparation for our analysis of the cycle structure associated with an autonomous linear machine, we show in this section that it is sufficient to analyze linear feedback shift-registers. Every autonomous linear machine has the same cycle structure as an autonomous machine composed entirely of feedback shift-registers. This result is a direct consequence of the well-known result from linear matrix theory that the companion matrix of a matrix has block submatrices on its diagonal and has zeros elsewhere. In this case, each submatrix corresponds to a feedback shift-register.

$$A = \begin{bmatrix} 0 & 1 & 0 & 1 \\ 1 & 1 & 0 & 0 \\ 1 & 1 & 1 & 0 \\ 1 & 0 & 1 & 0 \end{bmatrix}$$

FIG. 9.2.1   Matrix A for an Autonomous Machine M

As an example of an autonomous machine, consider the machine M whose A matrix over $GF(2)$ is shown in Fig. 9.2.1. The state space of M has $2^4 = 16$ states, and A is a mapping of this space into itself. Because A in this case maps the state space onto itself, A permutes the state space. For example, the state $\langle 0, 0, 1, 0 \rangle^t$ is mapped by A onto the state $\langle 0, 0, 1, 1 \rangle^t$. As a notational convenience, we denote each state in the state space by an integer between 0 and 15, such that each binary state vector is associated with the integer for which it is the binary representation. For example, state $2 = \langle 0, 0, 1, 0 \rangle^t$ is mapped by A onto state $3 = \langle 0, 0, 1, 1 \rangle^t$. Fig. 9.2.2 depicts the action of A on the states of

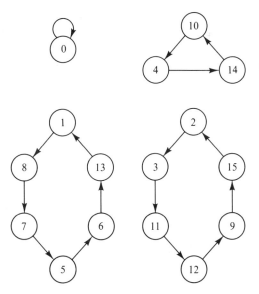

FIG. 9.2.2   State Graph of the Autonomous Machine M

M. Because *A* is a permutation, the state graph is a collection of disjoint cycles.

From Fig. 9.2.2 and the properties of permutations, it is quite evident that *A* is invertible. In fact, A^{-1} is the transformation obtained by reversing the arrows in Fig. 9.2.2. We can show that A^{-1} is a linear transformation by making use of some elementary linear algebra.

Definition   An $n \times n$ matrix *A* over a field *K* is said to be *nonsingular* if $det(A) \neq 0$, where the determinant is calculated in the arithmetic of the field *K*.

From linear algebra we know that the matrix A^{-1} exists if and only if *A* is nonsingular or, equivalently, if and only if the transformation *A* is a permutation of states in the *n*-dimensional vector space of states. In our example, *A* is nonsingular, so we know that there is a unique $n \times n$ matrix A^{-1} with the property $A \cdot A^{-1} = A^{-1} \cdot A = I$, where *I* is the $n \times n$ identity matrix. We can determine $A^{-1} \cdot x$ for any *x* by expanding *x* into a sum of basis vectors and then observing how A^{-1} maps each of the basis vectors. The natural basis to use for this calculation is the set of states $\{8, 4, 2, 1\}$. From Fig. 9.2.2 we see that A^{-1} maps this basis set into the states 1, 10, 15, and 13, respectively. Hence A^{-1} must be the matrix shown in Fig. 9.2.3, where the columns of A^{-1} are the images of states 8, 4, 2, and 1 (see Ex. 9.2.1).

$$A^{-1} = \begin{bmatrix} 0 & 1 & 1 & 1 \\ 0 & 0 & 1 & 1 \\ 0 & 1 & 1 & 0 \\ 1 & 0 & 1 & 1 \end{bmatrix}$$

FIG. 9.2.3   Matrix A^{-1} for the Autonomous Machine M

Autonomous linear machines that have singular next-state matrices are degenerate machines, in the sense that some subset of states must be transient. However, the nontransient states must lie on cycles much like the cycles of Fig. 9.2.2. In fact, the state graph of a singular machine must contain only disconnected cycles, possibly with transient states leading into them, as shown in Fig. 9.2.4. Hence a singular machine must eventually reach a cycle of states after a sufficiently long time, regardless of its initial state. Thus the difference between a singular and a nonsingular machine is the presence of transient states in the singular machine.

For this reason, the central results in autonomous-machine theory concern the nonsingular machines and the cycle structure of their state graphs. In the remainder of this section, we limit our attention solely to nonsingular machines, simply noting that the behavior of singular machines is a direct extension of the results here.

We have discovered thus far that the state diagrams of nonsingular

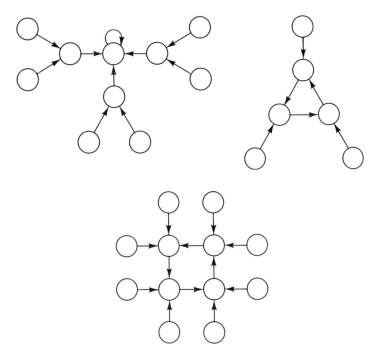

FIG. 9.2.4   State Graph of a Typical Singular Linear Machine

machines are collections of disjoint cycles. Because linearity guarantees that state 0 always is mapped onto state 0 by the matrix A, we know that state 0 is always on a 1-cycle. If we are working in $GF(q)$, the state space has q^n states, so that we must account for the $q^n - 1$ nonzero states in order to determine the complete cycle structure of the next-state function. The problem that we address in the remainder of this section is the determination of the lengths of the cycles of the next-state function. Clearly, we can answer this question directly by exhaustively examining the $q^n - 1$ nonzero states and finding their successor states. However, we shall see that this approach involves far more work than we need to do, and that such an approach does not help us to construct machines with specific cycle structures. We wish to find the cycle structure of the state space as some compact function of the A matrix.

To determine the cycle structure of a matrix, consider the conditions under which we obtain a cycle of states. Let s_1 be some starting state, and let $s_2 = A \cdot s_1$, $s_3 = A \cdot s_2$, and so forth until we reach $s_{r+1} = s_1$. Then the states $s_1, s_2, \ldots, s_r$ lie on a cycle of length r. Moreover, we have $s_{i+1} = A^i \cdot s_1$, so that $A^r \cdot s_1 = s_1$ or $(A^r - I) \cdot s_1 = 0$. The problem of finding the cycle structure reduces to the problem of finding for which r there exist states s such that the equation $(A^r - I) \cdot s = 0$ is satisfiable. Fortunately, we can easily answer this question through the use of linear algebra and the theory of finite fields.

Because we are looking for solutions to the equation $(A^r - I) \cdot s = 0$ for various r, we note that there exist similarity transformations of A that do not change the solvability of the equation for the transformed matrix. So first we show how similarity transformations preserve cycle structure; then we show that every matrix is equivalent in cycle structure to some canonical matrix whose cycle structure can be computed directly. We proceed by first examining similarity transformations of matrices.

Definition   If A is any $n \times n$ matrix over a field K, and if Q is a nonsingular $n \times n$ matrix over the same field, then the matrix $A' = Q \cdot A \cdot Q^{-1}$ is said to be *similar* to A.

We can easily show that similarity is an equivalence relation on $n \times n$ matrices: **(i)** it is reflexive because $I \cdot A \cdot I^{-1} = A$; **(ii)** it is symmetric because $(A' = Q \cdot A \cdot Q^{-1}) \Rightarrow (A = Q^{-1} \cdot A' \cdot Q)$; **(iii)** it is transitive because

$$((A' = Q \cdot A \cdot Q^{-1}) \wedge (A'' = P \cdot A' \cdot P^{-1})) \Rightarrow$$
$$(A'' = (P \cdot Q) \cdot A \cdot (P \cdot Q)^{-1}).$$

For the matrix A of Fig. 9.2.1, we can construct a similar matrix A' by using the matrices Q and Q^{-1} given in Fig. 9.2.5. We have selected Q as an arbitrary nonsingular matrix. The figure shows how we obtain A' under the similarity transformation $A' = Q \cdot A \cdot Q^{-1}$. The A matrix

$$
Q = \begin{bmatrix} 0 & 1 & 1 & 0 \\ 0 & 1 & 0 & 1 \\ 1 & 0 & 0 & 0 \\ 0 & 1 & 1 & 1 \end{bmatrix}
\qquad
Q^{-1} = \begin{bmatrix} 0 & 0 & 1 & 0 \\ 1 & 1 & 0 & 1 \\ 0 & 1 & 0 & 1 \\ 1 & 0 & 0 & 1 \end{bmatrix}
$$

$$A' = Q \cdot A \cdot Q^{-1}$$

$$
= Q \cdot \begin{bmatrix} 0 & 1 & 0 & 1 \\ 1 & 1 & 0 & 0 \\ 1 & 1 & 1 & 0 \\ 1 & 0 & 1 & 0 \end{bmatrix} \cdot Q^{-1}
$$

$$
= \begin{bmatrix} 0 & 1 & 0 & 1 \\ 1 & 0 & 0 & 0 \\ 0 & 1 & 0 & 0 \\ 0 & 0 & 1 & 0 \end{bmatrix}
$$

FIG. 9.2.5   Construction of the Matrix $A' = Q \cdot A \cdot Q^{-1}$

permutes the states in state space of the machine as shown in Fig. 9.2.6. Comparing Figs. 9.2.2 and 9.2.6, we observe that the cycle structures of A and A' are identical, except for the labeling of the states. This relationship is not a coincidence.

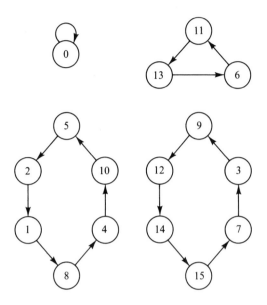

FIG. 9.2.6 State Graph of the Machine with Next-State Matrix $A' = Q \cdot A \cdot Q^{-1}$ (see Fig. 9.2.5)

Theorem If matrices A and A' are similar nonsingular state-transition matrices, then their state graphs have identical cycle structures and differ only in the labeling of the states.

    Proof: We prove the theorem by showing that there exists a one-to-one map f on the states such that, if $t = A \cdot s$, then $f(t) = A' \cdot f(s)$. Let $A' = Q \cdot A \cdot Q^{-1}$ for some nonsingular matrix Q. Let $f(s)$ be the map that takes s into $Q \cdot s$. Then if $t = A \cdot s$,

$$Q \cdot t = Q \cdot A \cdot s = Q \cdot Q^{-1} \cdot A' \cdot Q \cdot s = A' \cdot Q \cdot s,$$

so $f(t) = A' \cdot f(s)$. Then, if s is a state and t is its successor under A, f maps t into the successor of $f(s)$ under A'. Thus, for each predecessor-successor pair of states in A, there is a corresponding predecessor-successor pair of states in A'. But A is nonsingular, so that each state has a unique successor under A. Hence f is a one-to-one map, and it preserves the cycle structure of A. $\quad\square$

    For the next-state functions A and A' whose cycle sets are depicted in Figs. 9.2.2 and 9.2.6, the state-correspondence functions are shown in Table 9.2.1. By referring to Figs. 9.2.2 and 9.2.6, the reader can verify

TABLE 9.2.1   State-Correspondence Function for the Machines of Figs. 9.2.2 and 9.2.6

s (for Machine A)	$Q \cdot s$ (for Machine A')
0	0
1	5
2	9
3	12
4	13
5	8
6	4
7	1
8	2
9	7
10	11
11	14
12	15
13	10
14	6
15	3

that the two cycle sets are isomorphic under the mapping $f(s) = Q \cdot s$.

We have seen this theorem before in our treatment of group theory. Recall that h is the conjugate of g in a group G if there exists an x in G such that $h = x \cdot g \cdot x^{-1}$. We mention in Ex. 2.3.6 that conjugate elements of a group have the same cycle structure in all permutation representations of the group. Clearly, the set of nonsingular $n \times n$ matrices over a field K forms a group with matrix multiplication as the group operation. So, for every permutation representation of the group of matrices, similar matrices are represented by permutations with identical cycle structures. Each matrix is associated with a permutation of the state vectors, and it is not difficult to verify that these permutations constitute a representation of the group of matrices. Thus we can explain from our knowledge of group theory why similar linear machines have identical cycle structures.

From the preceding theorem, we see that the problem of finding the cycle structure induced by a matrix A reduces to the problem of finding the cycle structure for some matrix similar to A. If we are careful in our choice of the similar matrix, we can find the cycle structure by a simple computation. The theorem has even more significance with respect to the problem of synthesizing a machine with a prescribed cycle structure. The theorem allows us to obtain a particular cycle structure by constructing any machine in an equivalence class characterized by that structure, and it gives us the freedom to select the machine of least cost in an equivalence class. This latter property suggests that

$$
A = \begin{bmatrix}
a_1 & 1 & 0 & 0 & \cdots & 0 & 0 & 0 \\
a_2 & 0 & 1 & 0 & \cdots & 0 & 0 & 0 \\
a_3 & 0 & 0 & 1 & \cdots & 0 & 0 & 0 \\
\vdots & \vdots & \vdots & \vdots & \vdots\vdots\vdots & \vdots & \vdots & \vdots \\
a_{n-2} & 0 & 0 & 0 & \cdots & 0 & 1 & 0 \\
a_{n-1} & 0 & 0 & 0 & \cdots & 0 & 0 & 1 \\
a_n & 0 & 0 & 0 & \cdots & 0 & 0 & 0
\end{bmatrix}
$$

FIG. 9.2.7   General Form of the A Matrix of a Feedback Shift-Register

we should look for some standard machine in each equivalence class that is likely to be the least costly machine to synthesize. For this purpose we choose feedback shift-registers.

Definition   A linear machine is said to be a *feedback shift-register* if its A matrix has the general form shown in Fig. 9.2.7.

Matrices of the form shown in Fig. 9.2.7 lead to particularly simple physical realizations. Fig. 9.2.8 shows a network realization for A. The large squares in the figure represent memory cells, each of which can hold one field element. The field element s_i is stored in the ith memory cell. When such cells are connected in tandem as shown, we call the resulting network a shift register. In this network, the state of the network is the n-tuple of field elements stored in the register. The small squares between memory cells perform addition in the field, and the circular modules labeled $a_1, a_2, \ldots, a_n$ do multiplication in the field. In particular, a module with label a_i multiplies its input by the constant a_i, where a_i is a field element.

The shift register operates synchronously under the control of an external clock. At each clock tick, the contents of the memory cells are updated. Let s_i denote the field element in the ith cell, where the

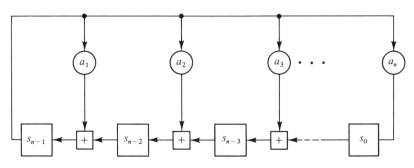

FIG. 9.2.8   A Linear Feedback Shift-Register

cells are numbered from 0 on the right to $n - 1$ on the left. At the clock tick, the element s_i in the ith cell is replaced by $a_{n-i} \cdot s_{n-1} + s_{i-1}$ if $i > 0$ or by $a_n \cdot s_{n-1}$ if $i = 0$. The change is assumed to occur instantaneously and simultaneously for all memory cells. Thus the register computes $A \cdot s$ where $s = \langle s_{n-1}, s_{n-2}, \ldots, s_0 \rangle^t$.

Data flows from right to left in the register, except for the data that flows from the leftmost cell. The reverse flow of data from the leftmost cell generally is called *feedback*, whence the name *feedback shift-register*.

Thus far we have assumed, for full generality, that the scalar field is any finite field. However, feedback shift-registers over $GF(2)$ are well-suited to computer applications. The adders shown in Fig. 9.2.8 do addition modulo 2, and they are simple and inexpensive to construct. The multipliers are particularly attractive to implement because they cost nothing at all. Multiplication by the field element 1 is merely a closed connection, and multiplication by the field element 0 is merely an open connection. Thus the multipliers shown in Fig. 9.2.8 are implemented by connecting the feedback line only to those stages for which the field element should be multiplied by 1. Multiplication in higher-order fields is considerably more difficult to implement.

Now we show that every linear autonomous machine is similar to a machine that is composed entirely of shift registers. Thus we can solve the problem of analyzing the state behavior of linear autonomous machines by investigating linear feedback shift-registers. To establish this assertion we draw upon matrix theory.

Definition The *characteristic polynomial* $\phi_A(\lambda)$ of an $n \times n$ matrix A is the polynomial $det(\lambda I - A)$.

It follows directly that the characteristic polynomial of a feedback shift-register is the polynomial

$$\phi_A(\lambda) = (\lambda^n - a_1 \lambda^{n-1} - a_2 \lambda^{n-2} - \cdots - a_n),$$

where $\langle a_1, a_2, \ldots, a_n \rangle^t$ is the first column of the A matrix of the shift register.

The next theorem states that every machine is equivalent to a machine that is composed solely of linear feedback shift-registers.

Theorem Every matrix A over a field K is similar to a matrix A' of the form shown in Fig. 9.2.9, where each of the submatrices A_i is a feedback shift-register matrix, and where the 0's are 0 submatrices of appropriate sizes. Moreover, the characteristic polynomial of each A_i either is irreducible over the field K or is a power of a polynomial that is irreducible over K. Finally, A' cannot be decomposed further into smaller block matrices on the diagonal.

We do not prove this theorem because it follows directly from well-known theorems of linear algebra. In particular, every matrix A is similar to a matrix known as its *companion matrix*, where the companion

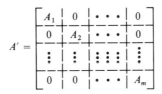

FIG. 9.2.9   General Form of the Matrix A'

matrix is in block-diagonal form. Moreover, each block factor is similar to a linear feedback shift-register, and the blocks are not further decomposable.

The correspondence between linear machines and feedback shift-registers first was noted by Zierler (1955) in a technical report. Elspas (1959) deserves credit for informing the technical community of this result in a well-written exposition on the theory of linear machines. We draw heavily upon the work of Elspas in the next section.

Exercises

9.2.1 Verify that the matrix A^{-1} of Fig. 9.2.3 is indeed the matrix inverse of A (Fig. 9.2.1).

9.2.2 Find the state graph for the machine M with the A matrix shown in Fig. 9.2.10, assuming that arithmetic is in $GF(2)$. Is A nonsingular?

9.2.3 Repeat Ex. 9.2.2 for the matrix A' shown in Fig. 9.2.10.

$$A = \begin{bmatrix} 0 & 1 & 1 & 1 & 1 \\ 1 & 1 & 0 & 0 & 0 \\ 1 & 0 & 1 & 0 & 1 \\ 0 & 1 & 0 & 1 & 0 \\ 0 & 0 & 1 & 0 & 0 \end{bmatrix} \qquad A' = \begin{bmatrix} 1 & 1 & 0 & 1 \\ 0 & 0 & 0 & 1 \\ 1 & 0 & 1 & 0 \\ 0 & 1 & 1 & 0 \end{bmatrix}$$

FIG. 9.2.10   Matrices A and A' for Exs. 9.2.2 and 9.2.3

$$A = \begin{bmatrix} 1 & 0 & 0 \\ 1 & 1 & 0 \\ 0 & 1 & 1 \end{bmatrix} \qquad Q = \begin{bmatrix} 1 & 1 & 0 \\ 1 & 0 & 1 \\ 0 & 1 & 0 \end{bmatrix}$$

FIG. 9.2.11   Matrices Q and A for Ex. 9.2.4

9.2.4 For the matrices Q and A shown in Fig. 9.2.11, compute $A' = Q^{-1} \cdot A \cdot Q$. Find the cycle structures for A and A'. Show the one-to-one correspondence of the states of the two matrices. Arithmetic is in $GF(2)$.

9.2.5 Find the cycle structure for the two matrices A and A' shown in Fig. 9.2.12. Show that A and A' are similar by finding a matrix Q with the property that $A' = Q^{-1} \cdot A \cdot Q$. All arithmetic is in $GF(2)$.

9.2.6 Find all of the nonsingular 2×2 matrices over $GF(2)$. Determine which of these matrices are inverses of each other, and find which of the matrices are similar to each other.

9.2.7 Repeat Ex. 9.2.6 for 2×2 matrices over $GF(3)$.

9.2.8 Show that the two matrices given in Fig. 9.2.13 are similar. Assume that the field is $GF(2)$, although the statement is true for every field.

$$A = \begin{bmatrix} 0 & 1 & 0 \\ 1 & 1 & 1 \\ 1 & 1 & 0 \end{bmatrix} \qquad A' = \begin{bmatrix} 0 & 1 & 1 \\ 1 & 1 & 1 \\ 0 & 1 & 0 \end{bmatrix}$$

FIG. 9.2.12   Matrices A and A' for Ex. 9.2.5

$$\begin{bmatrix} a_1 & 1 & 0 & \cdots & 0 \\ a_2 & 0 & 1 & \cdots & 0 \\ \vdots & \vdots & \vdots & \vdots & \vdots \\ a_{n-1} & 0 & 0 & \cdots & 1 \\ a_n & 0 & 0 & \cdots & 0 \end{bmatrix} \qquad \begin{bmatrix} 0 & 1 & 0 & \cdots & 0 \\ 0 & 0 & 1 & \cdots & 0 \\ \vdots & \vdots & \vdots & \vdots & \vdots \\ 0 & 0 & 0 & \cdots & 1 \\ a_n & a_{n-1} & a_{n-2} & \cdots & a_1 \end{bmatrix}$$

FIG. 9.2.13   Two Matrices for Ex. 9.2.8

9.3 The Cycle Structure of Linear Feedback Shift-Registers

We have shown that the study of the cycle structure of linear machines reduces to the study of feedback shift-registers. In this section we first examine shift registers for which the characteristic polynomial is irreducible; then we examine registers with a characteristic polynomial that is a power of an irreducible polynomial. These studies establish the behavior of all machines with A matrices that cannot be decomposed into smaller blocks. To obtain the cycle structure of an arbitrary linear machine, we derive the cycle structure of a composite machine from the cycle structure of its shift-register components.

To begin this analysis, consider a shift register M with an A matrix whose characteristic polynomial $\phi_A(x)$ is an irreducible polynomial over $GF(q)$, the field of matrix coefficients. We show that the cycle structure of this shift register has a 1-cycle for the 0-state and that all other states lie on cycles of equal length. The key theorem concerning the cycles of a feedback shift-register depends on a mapping from the states of M onto the ring of polynomials over $GF(q)$. The mapping enables us to construct particularly compact and revealing proofs of the theorems that follow.

Lemma

Let f be the bijection of the states of M onto polynomials of degree $n - 1$, such that f maps the state $s = \langle s_{n-1}, s_{n-2}, \ldots, s_1, s_0 \rangle^t$ into the polynomial $f(s) = s_{n-1}x^{n-1} + s_{n-2}x^{n-2} + \cdots + s_1x^1 + s_0$. Then f maps the state $A \cdot s$ into the polynomial $xf(s) \bmod \phi_A(x)$, where A is the linear next-state function of M, and where A is in canonical form.

Proof: From the matrix multiplication $A \cdot s$, we find that

$$f(A \cdot s) =$$
$$(a_1 s_{n-1} + s_{n-2})x^{n-1} + (a_2 s_{n-1} + s_{n-3})x^{n-2} + \cdots + a_n s_{n-1}.$$

Before reducing modulo $\phi_A(x)$, we have

$$xf(s) = s_{n-1}x^n + s_{n-2}x^{n-1} + \cdots + s_1x^2 + s_0x.$$

However,

$$s_{n-1}x^n \equiv a_1 s_{n-1}x^{n-1} + a_2 s_{n-2}x^{n-2} + \cdots + a_n s_{n-1} \bmod \phi_A(x),$$

when

$$\phi_A(x) = x^n - a_1 x^{n-1} - a_2 x^{n-2} - \cdots - a_n.$$

After substituting for $s_{n-1}x^n$, we find $xf(s) \bmod \phi_A(x) = f(A \cdot s)$. □

The lemma states that each shift of a shift register corresponds to multiplication by x in the algebra of polynomials modulo $\phi_A(x)$. Therefore we can use properties of the polynomial algebra derived previously to deduce new results about the cycle structure of the feedback shift-register. In particular, we can discover the structure of the cycles induced by A when $\phi_A(x)$ is irreducible.

Definition

For any monic polynomial $f(x)$ over $GF(q)$, the *period* of $f(x)$ is defined to be the least integer k such that $f(x)$ divides $x^k - 1$.

Theorem

Let $\phi_A(x)$ be the irreducible characteristic polynomial of a shift register with next-state matrix A. Then, if k is the period of $\phi_A(x)$, all nonzero states lie on cycles of length k.

Proof: We show first that no cycle has a length that exceeds k. From the preceding lemma, we have $xf(s) \equiv f(A \cdot s) \bmod \phi_A(x)$, so that

$$x^2 f(x) \equiv xf(A \cdot s) \equiv f(A^2 \cdot s) \bmod \phi_A(x)$$

and, more generally,

$$x^i f(s) \equiv f(A^i \cdot s) \bmod \phi_A(x)$$

for all i. If k is the period of $\phi_A(x)$, then $(x^k - 1)f(s)$ is a multiple of $\phi_A(x)$ for all states s, so that

$$f(A^k \cdot s) \equiv x^k f(s) \equiv f(s) \bmod \phi_A(x)$$

for all states s. Then $f(A^k \cdot s) = f(s)$, because both polynomials have degree less than $deg[\phi_A(x)]$. But $f(A^k \cdot s) = f(s)$ if and only if $A^k \cdot s = s$, because f is a one-to-one mapping. Therefore, the kth successor of every state s is s itself, so that every state lies on a cycle whose length does not exceed k.

To show that the nonzero states lie on cycles of length exactly k, note that, if the cycle containing state s has length $r < k$ where s is not the 0-state, then $A^r \cdot s = s$. Then $x^r f(s) \equiv f(s) \bmod \phi_A(x)$, and $(x^r - 1)f(s)$ is a multiple of $\phi_A(x)$. Because the degree of $f(s)$ is less than the degree of $\phi_A(x)$ and $\phi_A(x)$ is irreducible, $\phi_A(x)$ must divide $x^r - 1$ unless $f(s) = 0$. But $f(s) = 0$ if and only if s is the 0-state. If $f(s) \neq 0$ and $r < k$, we contradict the definition of period, for then the period of $\phi_A(x)$ is r and not k. In either case we reach a contradiction. Hence the length of the cycle containing s must be equal to the period. ☐

$$A_1 = \begin{bmatrix} 1 & 1 \\ 1 & 0 \end{bmatrix} \qquad A_2 = \begin{bmatrix} 0 & 1 & 0 \\ 1 & 0 & 1 \\ 1 & 0 & 0 \end{bmatrix} \qquad A_3 = \begin{bmatrix} 1 & 1 & 0 & 0 \\ 1 & 0 & 1 & 0 \\ 1 & 0 & 0 & 1 \\ 1 & 0 & 0 & 0 \end{bmatrix}$$

FIG. 9.3.1   The Matrices A_1, A_2, and A_3

Consider the matrices A_1 and A_2 shown in Fig. 9.3.1. The characteristic polynomials of both A_1 and A_2 are irreducible over $GF(2)$. In the first case the characteristic polynomial is $x^2 + x + 1$. In Chapter Eight we show that the roots of this polynomial lie in $GF(2^2)$ and therefore are roots of the polynomial $x^3 - 1$. Hence $x^2 + x + 1$ has period 3, and we find that its nonzero states lie on the 3-cycle (1 2 3). Similarly, A_2 has the characteristic polynomial $x^3 + x + 1$, which has its roots in $GF(2^3)$. Hence these roots satisfy $x^7 - 1 = 0$, and the polynomial must therefore have period 7. We find that the nonzero states of the shift register lie on the 7-cycle (1 2 4 3 6 7 5).

For each of these examples, all of the nonzero states fall on one cycle. This does not always happen. For example, the cycle structure associated with the matrix A_3 in Fig. 9.3.1 places all of the nonzero states on three 5-cycles. The characteristic polynomial of this matrix is

$$x^4 + x^3 + x^2 + x + 1.$$

In $GF(2)$, $x^5 - 1 = x^5 + 1$, and $x^5 + 1$ factors into

$$(x + 1)(x^4 + x^3 + x^2 + x + 1),$$

so that ϕ_A has period 5.

When we deal with n-stage shift-registers over the coefficient field $GF(q)$, then the size of the state set is q^n. The linearity of the next-state function guarantees that the 0-state is always a successor of itself, so that the 0-state always lies on a cycle by itself. Therefore, the longest possible cycle for a shift register is a cycle that contains all of the nonzero states, which is a cycle of length $q^n - 1$. The next lemma guarantees that we can construct registers that have cycles of this length for all finite fields $GF(q)$ and for all register lengths n.

Definition   An element α of the field $GF(q)$ is called a *primitive element* if it generates the multiplicative group of the field. The minimum polynomial of a primitive element is called a *primitive polynomial*.

Lemma   For every $n \geq 1$ there exists an n-stage feedback shift-register over $GF(q)$ that has one cycle of length 1 and one cycle of length $q^n - 1$. The latter cycle contains all of the nonzero states.

Proof: Let α be any primitive element of $GF(q^n)$. We show that its minimum polynomial $M_\alpha(x)$ has period $q^n - 1$ and, consequently, the shift register with $M_\alpha(x)$ as its characteristic polynomial satisfies the theorem. The roots of $M_\alpha(x)$ lie in $GF(q^n)$ and are the elements $\alpha, \alpha^q, \alpha^{q^2}, \ldots, \alpha^{q^{n-1}}$. The multiplicative order of α is $q^n - 1$ because it is primitive. All of the other roots also have multiplicative order $q^n - 1$ because q^i is relatively prime to $q^n - 1$ for all i, and thus α^{q^i} also generates the cyclic multiplicative group. $M_\alpha(x)$ does not divide $x^m - 1$ for m less than $q^n - 1$, because it has no roots of order less than $q^n - 1$. Therefore the period of $M_\alpha(x)$ is $q^n - 1$.   □

Thus we have shown that the length of sequences that can be generated by linear feedback shift-registers grows exponentially with the number of stages in the register. For $q = 2$ and $n = 50$,

$$q^n - 1 = 2^{50} - 1 \cong 1.25 \cdot 10^{15}.$$

We can easily construct a 50-stage shift-register that shifts every 100 nsec. For such a shift register to visit every state once without repetition would require just over three years. A 60-stage register would take over 3,000 years to complete a cycle! The fact that we can construct compact and inexpensive devices to generate such long cycles has had many practical implications, several of which are discussed later in this chapter.

Now we are ready to extend the results obtained thus far to the case for which the shift register has a characteristic polynomial that is a power of an irreducible polynomial. The next theorem shows that a shift register with characteristic polynomial $[\phi(x)]^i$ has all of the cycles produced by a register with characteristic polynomial $[\phi(x)]^{i-1}$. Each additional power of $\phi(x)$ simply adds more cycles to the state diagram without changing any of the cycles produced by lower powers of $\phi(x)$. In moving from a register with polynomial $[\phi(x)]^{i-1}$ to one with polynomial $[\phi(x)]^i$, all of the additional states lie on cycles of the same length, which happens to be equal to the period of $[\phi(x)]^i$.

Theorem Let A be a canonical feedback shift-register matrix over $GF(q)$ with the characteristic polynomial $\phi_A(x) = [\phi(x)]^m$, where $\phi(x)$ is irreducible of degree d. Let k_i be the period of $[\phi(x)]^i$ for each i such that $1 \leq i \leq m$. Then all of the nonzero states lie on μ_i distinct cycles of length k_i, where

$$\mu_i = (q^{di} - q^{d(i-1)})/k_i$$

for each i such that $1 \leq i \leq m$.

Proof: The shift register has q^n states, where $n = dm$ because dm is the degree of the characteristic polynomial of the register. We prove the theorem by showing that for each state s, if $f(s)$ is a multiple of $[\phi(x)]^{m-i}$ and is not a multiple of a higher power of $\phi(x)$, then s lies on a cycle of length k_i.

Suppose that, for some state s, $f(s) = a(x)[\phi(x)]^{m-i}$, where $a(x)$ is not divisible by $\phi(x)$. We claim that s lies on a cycle of length no greater than k_i. Because $[\phi(x)]^i$ has period k_i, $x^{k_i} - 1$ is a multiple of $[\phi(x)]^i$, and $(x^{k_i} - 1)f(s) = (x^{k_i} - 1)a(x)[\phi(x)]^{m-i}$ is a multiple of $\phi_A(x)$ because it contains $[\phi(x)]^m$ as a factor. Then

$$f(A^{k_i} \cdot s) \equiv x^{k_i}f(s) \equiv f(s) \bmod \phi_A(x),$$

so $A^{k_i} \cdot s = s$, and s lies on a cycle of length no greater than k_i.

Now we show that the cycle containing s has length exactly k_i. Suppose that the cycle has length $r < k_i$. Then $A^r \cdot s = s$, so that

$$x^r f(s) \equiv f(s) \bmod \phi_A(x),$$

and $(x^r - 1)f(s)$ is a multiple of $\phi_A(x)$. Because $\phi_A(x) = [\phi(x)]^m$ where $\phi(x)$ is irreducible, the product $(x^r - 1)f(s)$ must contain $\phi(x)$ as a factor m times. But $f(s)$ contains $\phi(x)$ precisely $m - i$ times, so that $[\phi(x)]^i$ must divide $x^r - 1$. Then, if s is nonzero, the period of $[\phi(x)]^i$ is r or less, and $r < k_i$. But this is a contradiction, and thus the cycle containing s has length k_i.

To compute the number of cycles of each length, we note that the number of states lying on cycles of length k_1 is the number of nonzero polynomials of degree less than n over $GF(q)$ that are multiples of $[\phi(x)]^{m-1}$. There are precisely $q^d - 1$ such polynomials, because there are $q^d - 1$ ways of picking the coefficients of $a(x)$ in the polynomial $f(s) = a(x)[\phi(x)]^{m-1}$ where $a(x)$ has degree less than d. Hence $q^d - 1$

states lie on cycles of length k_1, and thus there are $\mu_1 = (q^d - 1)/k_1$ cycles of length k_1.

Using this same argument, we can show that there are q^{di} polynomial multiples of $[\phi(x)]^{m-i}$ of degree less than n, and that of these $q^{d(i-1)}$ are multiples of $[\phi(x)]^{m-i+1}$. Hence $q^{di} - q^{d(i-1)}$ polynomials correspond to states that lie on cycles of length k_i. Thus there are

$$\mu_i = (q^{di} - q^{d(i-1)})/k_i$$

cycles of length k_i. □

To complete the analysis of this more general case, we need only determine the periods of polynomials of the form $[\phi(x)]^i$ where $\phi(x)$ is irreducible.

Lemma   Let $\phi(x)$ be irreducible over $GF(q)$ with period k. Then the period of $[\phi(x)]^i$ is kq^r, where r is the unique integer that satisfies $q^{r-1} < i \leq q^r$.

Proof: Because $\phi(x)$ is a divisor of $x^k - 1$, $[\phi(x)]^i$ is a divisor of $(x^k - 1)^i$. But $(x^k - 1)^i$ in turn divides $(x^k - 1)^{q^r} = x^{kq^r} - 1$, where the latter equality makes use of the relation $f(x)^{q^i} = f(x^{q^i})$ when $f(x)$ has coefficients in $GF(q)$. Hence the period of $[\phi(x)]^i$ divides kq^r. Because the period of an irreducible polynomial of degree d must be a divisor of $q^d - 1$, k must be a divisor of $q^d - 1$ and is therefore relatively prime to q^d and to q. Hence the period of $[\phi(x)]^i$ must be of the form kq^m where $m \leq r$. Because

$$(x^k - 1)^{q^m} = x^{kq^m} - 1,$$

$(x^k - 1)^i$ divides $x^{kq^m} - 1$ if and only if $kq^m \geq i$. By hypothesis, r is the least integer for which $kq^r \geq i$, so the period must be kq^r. □

The matrices A_4 and A_5 in Fig. 9.3.2 are canonical matrices with characteristic polynomials $(x^2 + x + 1)^2$ and $(x^2 + x + 1)^3$, respectively, where $x^2 + x + 1$ is irreducible over $GF(2)$. The period of $x^2 + x + 1$ is 3, so by the preceding lemma its square has period 6, and its cube has period 12. The predicted cycle structures for each matrix are listed in the figure, as are typical cycles and the corresponding polynomials.

We now have completed the study of the cycle structure of the individual block matrices that appear on the diagonal of a companion matrix. To complete our study of the cycles produced by autonomous linear machines, we must consider matrices with two or more block factors in their companion matrices.

Let us assume that the matrix A is equivalent to a matrix of the form shown in Fig. 9.3.3. Then the cycle structure of A is the same as that of the composite machine consisting of two disjoint registers, as shown in Fig. 9.3.4. The next lemma permits us to compute the cycle structure of the composite machine. In the lemma we use *gcd* to denote "greatest common divisor" and *lcm* to denote "least common multiple."

$$A_4 = \begin{bmatrix} 0 & 1 & 0 & 0 \\ 1 & 0 & 1 & 0 \\ 0 & 0 & 0 & 1 \\ 1 & 0 & 0 & 0 \end{bmatrix} \qquad A_5 = \begin{bmatrix} 1 & 1 & 0 & 0 & 0 & 0 \\ 0 & 0 & 1 & 0 & 0 & 0 \\ 1 & 0 & 0 & 1 & 0 & 0 \\ 0 & 0 & 0 & 0 & 1 & 0 \\ 1 & 0 & 0 & 0 & 0 & 1 \\ 1 & 0 & 0 & 0 & 0 & 0 \end{bmatrix}$$

i	k_i	$\mu_i(A_4)$	$\mu_i(A_5)$
0	1	1	1
1	3	1	1
2	6	2	2
3	12	—	4

$(A^i \cdot s)^t$	A_4 State	A_5 State
s^t	$0111 \cong x^2 + x + 1$	$000111 \cong x^2 + x + 1$
$(A \cdot s)^t$	$1110 \cong x(x^2 + x + 1)$	$001110 \cong x(x^2 + x + 1)$
$(A^2 \cdot s)^t$	$1001 \cong (x + 1)(x^2 + x + 1)$	$011100 \cong x^2(x^2 + x + 1)$
$(A^3 \cdot s)^t$	$0111 \cong x^2 + x + 1$	$111000 \cong x^3(x^2 + x + 1)$
$(A^4 \cdot s)^t$	——	$011011 \cong (x^2 + 1)(x^2 + x + 1)$
$(A^5 \cdot s)^t$	——	$110110 \cong x(x^2 + 1)(x^2 + x + 1)$
$(A^6 \cdot s)^t$	——	$000111 \cong x^2 + x + 1$

FIG. 9.3.2   Two Irreducible Matrices and Their Cycle Structures

$$A = \left[\begin{array}{c|c} A_1 & 0 \\ \hline 0 & A_2 \end{array} \right]$$

FIG. 9.3.3   General Block-Diagonal Form for the Matrix A

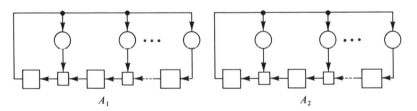

FIG. 9.3.4   A Composite Shift-Register, with Each Register Corresponding to a Block Factor of a Canonical Matrix with Two Blocks (Fig. 9.3.3)

Lemma   For each cycle of length k_1 induced by A_1 and for each cycle of length k_2 induced by A_2, the composite machine with matrix A will have $gcd(k_1, k_2)$ cycles of length $lcm(k_1, k_2)$.

Proof: When A_1 is started in a state on a k_1-cycle and A_2 is started in a state on a k_2-cycle, the composite machine will return to the composite starting state after exactly $lcm(k_1, k_2)$ steps. Hence the cycles produced by this combination must be of length $lcm(k_1, k_2)$. There is a total of $k_1 k_2$ composite states of the form $\langle s_1, s_2 \rangle$, where s_1 is a state from A_1 on the k_1-cycle and s_2 is a state from A_2 on the k_2-cycle. Because $gcd(k_1, k_2) = k_1 k_2 / lcm(k_1, k_2)$, the composite states must lie on $gcd(k_1, k_2)$ distinct cycles.   ☐

The matrix A_6 of Fig. 9.3.5 is in block-diagonal form, and its component registers have cycle sets $[1(1) + 1(3)]$ and $[1(1) + 3(5)]$, respectively, where the notation $[\mu_1(k_1) + \cdots + \mu_m(k_m)]$ indicates μ_i cycles of length k_i for each i. The lemma indicates that the composite cycle structure can be computed by "multiplying" the cycle representations together, using the relation

$$\mu_1(k_1) \times \mu_2(k_2) = gcd(k_1, k_2)\mu_1 u_2 [lcm(k_1, k_2)].$$

Thus we have

$$[1(1) + 1(3)] \times [1(1) + 3(5)]$$
$$= [1(1) \times 1(1) + 1(1) \times 3(5) + 1(3) \times 1(1) + 1(3) \times 3(5)]$$
$$= [1(1) + 3(5) + 1(3) + 3(15)]$$

The reader can easily verify that A_6 has the cycle structure predicted by this computation.

The following examples of cycle-set multiplication should help to clarify the details of the multiplication procedure. The reader should verify that these examples are correct by observing the ways that typical cycles of the composite machine are constructed from cycles of the individual machines. Note also that the cycles contain $2^8 = 256$ states.

$$[1(1) + 3(5)][1(1) + 3(5)] = [1(1) + 3(5) + 3(5) + 45(5)]$$
$$= [1(1) + 51(5)].$$

$$[1(1) + 1(15)][1(1) + 3(5)] = [1(1) + 1(15) + 3(5) + 15(15)]$$
$$= [1(1) + 3(5) + 16(15)].$$

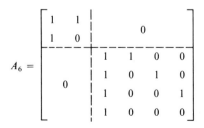

FIG. 9.3.5   The Block-Diagonal Matrix A_6

In the proof of the preceding lemma, we did not require that the decomposition of A into A_1 and A_2 be its companion-matrix decomposition. In fact, the lemma holds even if A_1 or A_2 in turn have two or more block-diagonal factors. Consequently, we conclude that, if the complete block factorization of a matrix is known to have the matrices $A_1, A_2, \ldots, A_m$ on the diagonal, then we can analyze the cycle structure of the composite matrix by forming the m-fold cycle-structure product of A_1 through A_m. The reasoning here is that, after we have computed the cycle structure of the product of A_{m-1} and A_m, we can replace the two matrices by their composite matrix A'_{m-1}, and we can then proceed to compute the cycle structure for the factors A_{m-2} and A'_{m-1}, and so forth.

With this last observation we have completed our analysis of the cycle structure of linear machines. We conclude this section with a summary of its major points. The cycle structure of a linear machine is completely determined by the block factors of its canonical matrix. A block factor whose characteristic polynomial is irreducible produces a single cycle of length 1 and other cycles, all of which have lengths equal to the period of the characteristic polynomial. When the characteristic polynomial is a power of an irreducible polynomial, each of the cycles has length equal to the period of some power of the irreducible polynomial, and the number of cycles of each length can be determined by formula. A machine with two or more block factors produces cycles whose lengths can be determined from the cycle lengths of the individual machines by using cycle-set arithmetic operations.

The material in this section is derived largely from the work of Elspas (1959) and Zierler (1955, 1959).

Exercises

9.3.1 Determine the periods of the following polynomials. The field is $GF(2)$.
(a) $x^2 + x + 1$;
(b) $x^3 + x^2 + x + 1$;
(c) $x^4 + x^3 + x^2 + x + 1$;
(d) $x^n + \cdots + x + 1$.

9.3.2 Determine the periods of the following polynomials. The field is $GF(2)$.
(a) $x^4 + x^2 + 1$;
(b) $x^6 + x^4 + x^2 + 1$;
(c) $x^{2n} + \cdots + x^2 + 1$.

9.3.3 Find the cycle structures for the feedback shift-registers with the following (irreducible) characteristic polynomials. The field is $GF(2)$.
(a) $x^4 + x + 1$;
(b) $x^4 + x^3 + 1$;
(c) $x^5 + x^2 + 1$;
(d) $x^6 + x^4 + x^2 + x + 1$.

9.3.4   Find the cycle structures for the feedback shift-registers with the follow-ing (irreducible) characteristic polynomials. The field is $GF(3)$.
(a) $x^2 + 2x + 1$;
(b) $x^2 + x + 1$;
(c) $x^3 + 2x + 1$;
(d) $x^3 + 2x^2 + 1$;
(e) $x^3 + x^2 + 1$.

9.3.5   Prove that the irreducible polynomial $f(x)$ of degree d over $GF(p)$ has period m if and only if each of its roots in $GF(p^d)$ has order m in the multiplicative group of the field.

9.3.6   Find the cycle structures associated with the shift registers whose char-acteristic polynomials are given below. Each register has one block factor in its canonical matrix; the characteristic polynomial of that factor is a power of an irreducible polynomial. The field is $GF(2)$.
(a) $(x^2 + x + 1)^4$;
(b) $(x^3 + x + 1)^4$;
(c) $(x^4 + x + 1)^4$;
(d) $(x^5 + x + 1)^3$;
(e) $(x^6 + x^4 + x^2 + x + 1)^3$.

9.3.7   Find the cycle structures for the shift registers whose polynomials are given in factored form below. Assume that the characteristic polynomial of each block factor in its canonical matrix is irreducible, and that the field is $GF(2)$.
(a) $(x^2 + x + 1)^4$;
(b) $(x^3 + x + 1)^4$;
(c) $(x^2 + x + 1)^2(x^3 + x + 1)^2$;
(d) $(x^3 + x + 1)(x^4 + x^3 + x^2 + x + 1)$.

9.3.8   For each $n = 2, 3, \ldots, 10$, find a shift register that exhibits at least one cycle of length n.

9.3.9   Prove that a shift register of length n with an irreducible characteristic polynomial must produce cycles of lengths that divide $p^n - 1$, where the field is $GF(p)$.

9.3.10   Suppose that an n-stage shift-register shifts every 250 nanoseconds. How long can such a register run without repeating a state when $n = 16$, 24, 32, and 36? How many stages are required to run for a year without repeating?

9.3.11   A register with characteristic polynomial $x^3 + x + 1$ over $GF(2)$ pro-duces the sequence 1001011 of period 7 on the output of its first stage. When this sequence is added component by component to a shift of itself, we obtain yet another shift of itself. This is called the *cycle-and-add property*. Thus $1001011 \oplus 1011100 = 0010111$. Prove that every maxi-mum-length sequence (a sequence of length $2^n - 1$) that can be produced by a linear n-stage shift-register has the cycle-and-add property.

9.4 A Reprise:
Pólya Theory

We return now to the problem of enumerating equivalence classes of switching functions, and we discover that the cycle-set arithmetic of Section 9.3 is precisely what we need to compute cycle index polynomials. In particular, we investigate the problem of counting the classes of functions of the form $f: \{0,1\}^n \to \{0,1\}$ under the action of S_n on the n variables of the functions. For example, $\pi = (1\,2\ldots n)$ changes $f(x_1, x_2, \ldots, x_n)$ into

$$f(x_{\pi(1)}, x_{\pi(2)}, \ldots, x_{\pi(n)}) = f(x_2, x_3, \ldots, x_n, x_1).$$

We mention in the earlier discussion that π acting on the subscripts of the n variables has the same effect as some permutation π' acting on the domain $\{0,1\}^n$. To enumerate the equivalence classes, we must construct the cycle index polynomial for the group acting on the domain of the functions, and this polynomial is quite different from the cycle index polynomial for the group acting on the variables of the functions. In Chapter Three we imply that we do the construction by exhaustive enumeration. In this section we show how to do that construction by using cycle-set arithmetic.

To construct the desired cycle index polynomial, we consider each permutation π in S_n and examine a corresponding linear machine. The cycles of states produced by the linear machine are in one-to-one correspondence with the cycles of domain elements under the action of π. Thus we can determine the appropriate cycle index polynomial by calculating the cycle structures of linear machines. The linear machines in question are composed entirely of circulating shift-registers, which are particularly easy to analyze.

The first portion of this section discusses the cycle structure of circulating shift-registers, and the latter part relates this topic directly to the Pólya enumeration.

Fig. 9.4.1 is a diagram of a *circulating shift-register*. This type of register is a feedback shift-register with no intermediate feedback taps.

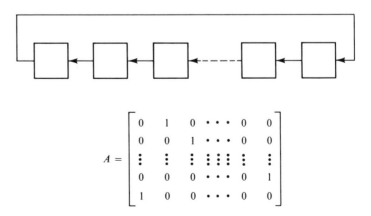

$$A = \begin{bmatrix} 0 & 1 & 0 & \cdots & 0 & 0 \\ 0 & 0 & 1 & \cdots & 0 & 0 \\ \vdots & \vdots & \vdots & \vdots\vdots\vdots & \vdots & \vdots \\ 0 & 0 & 0 & \cdots & 0 & 1 \\ 1 & 0 & 0 & \cdots & 0 & 0 \end{bmatrix}$$

FIG. 9.4.1   A Circulating Shift-Register and Its Next-State Matrix A

The next-state matrix associated with the register also is shown in the figure. From the figure, it is clear that the characteristic polynomial for this class of registers is $x^n - 1$. To determine the cycle structure for shift registers with the characteristic polynomial $x^n - 1$, the techniques of Section 9.3 require us to factor $x^n - 1$ into its irreducible factors. However, because circulating shift-registers are simple to analyze, we can compute their cycle structures more directly.

An n-stage circulating register produces cycles of length d if and only if d is a divisor of n. We reach this result by first introducing the notion of the period of an n-tuple and then showing that the period of a state vector s is equal to the length of the cycle in the state diagram that contains s.

Definition An n-tuple $s = \langle s_1, s_2, \ldots, s_n \rangle$ is said to have *period* d if d divides n, if $s_i = s_{i+d}$ for $1 \leq i \leq n - d$, and if d is the least integer for which these properties hold.

The n-tuple $\langle 0, 0, \ldots, 0 \rangle$ has period 1 and the 6-tuple $\langle 0, 1, 0, 1, 0, 1 \rangle$ has period 2. An n-tuple of period d consists of n/d identical copies of a d-tuple, concatenated together. Thus the 2-tuple $\langle 0, 1 \rangle$ is repeated three times to form the 6-tuple $\langle 0, 1, 0, 1, 0, 1 \rangle$ of period 2. The 5-tuple $\langle 0, 1, 0, 1, 0 \rangle$ has period 5 (not 2) because it does not contain an integral number of repetitions of $\langle 0, 1 \rangle$, although it does satisfy $s_i = s_{i+2}$ for $1 \leq i \leq n - 2$.

Lemma A circulating shift-register of length n produces cycles of length d if and only if d divides n. All states on a d-cycle of the register have period d.

Proof: We first assume that a state s lies on a cycle of length d, and we show that d divides n and that s has period d. In this proof we make use of some group theory from Chapter Two. A single shift of the state $s = \langle s_1, s_2, \ldots, s_n \rangle$ produces the new state $\langle s_2, s_3, \ldots, s_n, s_1 \rangle$. Thus a single shift cyclically permutes the components of s. This permutation generates a cyclic group, C_n. Let π denote the cyclical permutation $(1 \, 2 \ldots n)$ of the subscripts.

The set of all permutations in C_n that leave s invariant is a subgroup of C_n because it is finite and closed under function composition. But the order of every subgroup of C_n is a divisor of n, so that the subgroup has order m for some m that divides n. Let $d = n/m$. Then π^d generates the subgroup that leaves s invariant. Thus from state s, the shift register reaches state s again for the first time after d shifts, and therefore s lies on a cycle of length d. Moreover, because the state of the register after d shifts is

$$\langle s_{d+1}, s_{d+2}, \ldots, s_n, s_1, \ldots, s_d \rangle = s = \langle s_1, s_2, \ldots, s_n \rangle,$$

we have $s_i = s_{i+d}$ for $1 \leq i \leq n - d$. Hence d divides n, and s has period d.

To prove that a register produces d-cycles for each d that divides n, we note that the definition of period guarantees that a state with period

TABLE 9.4.1 Cycles Produced by a 6-Stage Circulating Register

Cycle Length	Representative State
1	000000
1	111111
2	010101
3	001001
3	011011
6	000001
6	000011
6	000101
6	000111
6	001011
6	010011
6	001111
6	010111
6	011111

64 States in Register

d is invariant under d shifts. Therefore all states with period d lie on d-cycles. □

As an example illustrating the preceding lemma, consider the cycles produced by a 6-stage register over $GF(2)$. Table 9.4.1 shows the cycle lengths and a typical state from each cycle. Note that the states on d-cycles have period d. Note also that the total number of states enumerated in Table 9.4.1 is $2^6 = 64$, which is the total number of 6-tuples over $GF(2)$.

Now we turn to an interesting result that makes use of material introduced in a different context. The number of cycles of length n produced by an n-stage register is equal to the number of irreducible polynomials of degree n.

Definition   Let $N_p(n)$ be the number of n-cycles produced by an n-stage circulating shift-register over $GF(p)$.

Recall from Chapter Eight that $I_p(n)$ denotes the number of irreducible polynomials of degree n over $GF(p)$.

Theorem $I_p(n) = N_p(n)$ for all prime p and integer $n \geq 1$.

Proof: The number of states of an n-stage register over $GF(p)$ is p^n. The sum of the cycle lengths therefore is equal to p^n.

For each d that divides n, an n-stage register produces $N_p(d)$ d-cycles. To see this, suppose that state $s = \langle s_1, s_2, \ldots, s_d \rangle$ has period d and therefore lies on a d-cycle of a d-stage register. Then the n-tuple

$$\langle s_1, \ldots, s_d, s_1, \ldots, s_d, \ldots, s_1, \ldots, s_d \rangle,$$

which has n/d copies of s, is a state of period d and lies on a d-cycle of the n-stage register. Similarly, every n-tuple that lies on a d-cycle of the n-stage register has period d and corresponds to a state on a d-cycle of the d-stage register.

Now, because each d-cycle contains d distinct states and, by definition, there are $N_p(d)$ such cycles, the total number of states of the n-stage register is given by the formula

$$p^n = \sum_{d|n} dN_p(d).$$

This expression is identical to the formula satisfied by $I_p(n)$ (see Section 8.3). Moreover, $I_p(n)$ is determined uniquely by the formula, so that $N_p(n)$ must be also, and the two functions are identical. □

Recall from the discussion in Section 8.3 that the values of $I_2(n)$ for $n = 1, 2, 3$, and 6 are $I_2(n) = 2, 1, 2$, and 9, respectively, which is in agreement with the cycle enumeration in Table 9.4.1.

We have now completed our discussion of the analysis of cycles of circulating shift-registers, and we next relate our results to Pólya enumeration. To set the context for the enumeration, let us review the problem that we wish to solve. Consider the set of functions from $\{0, 1\}^n$ into $\{0, 1\}$. Two functions $f_1(x_1, x_2, \ldots, x_n)$ and $f_2(x_1, x_2, \ldots, x_n)$ are said to be equivalent if there exists a permutation π of the subscripts such that

$$f_1(x_1, x_2, \ldots, x_n) = f_2(x_{\pi(1)}, x_{\pi(2)}, \ldots, x_{\pi(n)}).$$

We wish to count the number of equivalence classes of functions under the action of the permutations in the group S_n acting on the variables of the functions.

The solution provided by Pólya theory requires that we compute the cycle index polynomial of a permutation group S_n' acting on the domain of the functions. Thus we compute

$$Z_{S_n'}(x_1, x_2, \ldots, x_m) = \frac{1}{|S_n'|} \sum_{\pi' \in S_n'} p(\pi'),$$

where $p(\pi')$ is the product $x_1^{e_1} x_2^{e_2} \ldots x_m^{e_m}$ if π' has e_i cycles of length i for each i. Then the number of equivalence classes is

$$Z_{S_n'}(|R|, |R|, \ldots, |R|),$$

where R is the range of the functions.

In Chapter Three we assume that $p(\pi')$ for each group element π' is computed by exhaustive enumeration. However, exhaustive enumeration is practical only for small groups. For groups of large order we require more powerful tools, and our investigation of shift registers provides the requisite tools.

Recall from Section 3.3 that a permutation of S_n acting on the n variables of a switching function is associated with a permutation of

Before		After	
x_1	x_2	x_2	x_1
0	0	0	0
0	1	1	0
1	0	0	1
1	1	1	1

FIG. 9.4.2   The Domain $\{0, 1\}$ Before and After Applying the Permutation (1 2) to the Subscripts of the Variables

elements of the domain $\{0, 1\}^n$. For $n = 2$, as an example, the domain of 2-tuples is shown in the left part of Fig. 9.4.2. The permutation (1 2) in S_2 interchanges the columns of the table. This process is equivalent to the permutation of the rows of the table that leaves $\langle 0, 0 \rangle$ and $\langle 1, 1 \rangle$ fixed while interchanging $\langle 0, 1 \rangle$ and $\langle 1, 0 \rangle$. The right part of the figure shows the domain after application of the permutation (1 2) to the subscripts of the variables, and it is obvious that the same configuration would be obtained by interchanging the middle rows of the table on the left.

For comparison purposes, Fig. 9.4.3 shows the state diagram of a 2-stage circulating shift-register over $GF(2)$. Observe that the states are permuted among themselves just as the rows of the table are permuted in Fig. 9.4.2. This example illustrates the relation of the theory of circulating shift-registers to our enumeration problem.

In order to construct a cycle index polynomial for S_n', we must solve the following problem. Given a permutation of the columns of a table of the form of Fig. 9.4.2, we must find a permutation of the rows of the table that places the table in the same configuration as does the

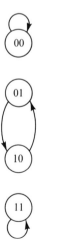

FIG. 9.4.3   State Diagram of a 2-Stage Circulating Shift-Register over $GF(2)$

column permutation. The row permutation is a permutation of the domain of the switching functions, whereas the column permutation is a permutation of the variables of the switching functions. Fig. 9.4.3 illustrates that, when the column permutation is a single cycle, say of length d, then the equivalent row permutation has the same cycle structure as the state diagram of a d-stage circulating shift-register. We state this result formally in the next lemma, and then we investigate permutations of variables that have two or more cycles.

The following definition makes precise the notion of associated permutation, which is useful in stating the lemma.

Definition Let π be a permutation of the n variables of the set of functions from $\{0,1\}^n$ into $\{0,1\}$. With each permutation π of the variables, we associate the permutation π' of the domain $\{0,1\}^n$ of the functions, such that

$$\pi'(\langle s_1, s_2, \ldots, s_n \rangle) = \langle s_{\pi(1)}, s_{\pi(2)}, \ldots, s_{\pi(n)} \rangle.$$

We say that π' is the *associated permutation* of π.

Note that π acts on the set $\{1, 2, \ldots, n\}$ and that π' acts on $\{0,1\}^n$. The degree of π is n, whereas the degree of π' is 2^n.

Lemma Let π be a permutation of the n variables of a switching function from $\{0,1\}^n$ into $\{0,1\}$, and let π permute the variables in a single cycle of length n. Then its associated permutation π' has $N_2(d)$ cycles of length d for each d that divides n.

Proof: The cyles of π' are in one-to-one correspondence with the cycles of the states of an n-stage circulating shift-register. The stages of the shift register are in one-to-one correspondence with the components of an n-tuple under the action of π. Because the complete proof of this lemma is a special case of the proof of the next theorem, we do not give the details here. $\square$

The previous example (Fig. 9.4.2) illustrates this lemma. We have not yet considered the cycle structure of a permutation π' on the domain of the switching functions when π has two or more cycles. It so happens that the cycles of π' are in one-to-one correspondence with the cycles of states of a linear machine composed entirely of circulating shift-registers. In fact, the circulating shift-registers in this machine are in one-to-one correspondence with the cycles of π.

Before we deal formally with this assertion, it is worthwhile to investigate a typical example. Fig. 9.4.4 shows a composite machine that

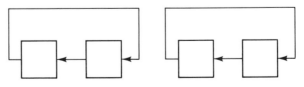

FIG. 9.4.4   A Composite Machine That Represents (1 2)(3 4)

corresponds to the permutation (1 2)(3 4) of the subscripts of the four variables for the domain $\{0, 1\}^4$. The composite machine has two 2-stage shift-registers, one for the cycle (1 2) and one for (3 4). The state diagram for the machine is shown in Fig. 9.4.5. In Fig. 9.4.6, we show the table for the domain before and after the permutation (1 2)(3 4) has been applied to the columns of the table. Observe that the associated permutation π' on the rows of the table is isomorphic to the permutation of the states illustrated in Fig. 9.4.5. We state this result formally in the next theorem.

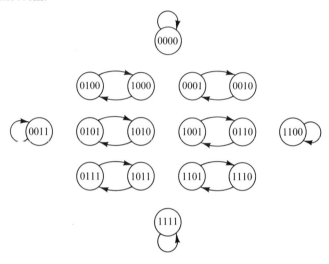

FIG. 9.4.5   **State Diagram of the Composite Machine Shown in Fig. 9.4.4**

Theorem   Let π be an arbitrary permutation of the n variables of the set of switching functions from $\{0, 1\}^n$ into $\{0, 1\}$. For each d-cycle in π, construct a d-stage circulating shift-register, and form a composite linear machine from the shift registers for the various cycles of π. Then the cycle structure of the state diagram of the composite machine is identical to the cycle structure of the permutation π' associated with π.

   Proof: To prove this theorem, it is sufficient to show that there exists an isomorphism F between the states of the linear machine and the n-tuples in $\{0, 1\}^n$ such that, if A is the next-state matrix of the composite machine, then for each state s we have $\pi'(F(s)) = F(A \cdot s)$. That is, F is a one-to-one mapping and, if state t is a successor of state s in the machine, then $F(t)$ is the image of $F(s)$ under π'. Let F be the function that maps the column state vector $s = \langle s_1, s_2, \ldots, s_n \rangle^t$ onto the n-tuple of variables $x = \langle x_1, x_2, \ldots, x_n \rangle$ such that $s_i = x_i$ for $1 \le i \le n$. Clearly, F is a bijection.

   To construct the next-state function for the machine corresponding to π, we set $a_{i,j} = 1$ in A if and only if π maps i into j; otherwise we set $a_{i,j} = 0$. Then A operating on the state s permutes the components of s

Before				After			
x_1	x_2	x_3	x_4	x_2	x_1	x_4	x_3
0	0	0	0	0	0	0	0
0	0	0	1	0	0	1	0
0	0	1	0	0	0	0	1
0	0	1	1	0	0	1	1
0	1	0	0	1	0	0	0
0	1	0	1	1	0	1	0
0	1	1	0	1	0	0	1
0	1	1	1	1	0	1	1
1	0	0	0	0	1	0	0
1	0	0	1	0	1	1	0
1	0	1	0	0	1	0	1
1	0	1	1	0	1	1	1
1	1	0	0	1	1	0	0
1	1	0	1	1	1	1	0
1	1	1	0	1	1	0	1
1	1	1	1	1	1	1	1

FIG. 9.4.6   The Domain $\{0, 1\}^4$ Before and After the Application of $(1\,2)(3\,4)$

just as π permutes the components of $x = \langle x_1, x_2, \ldots, x_n \rangle$. Observe that the machine whose next-state function is A has a d-stage circulating shift-register for each d-cycle in π. In fact, if π has the cycle $(i_1\, i_2 \ldots i_d)$, then A permutes the components of s according to the cycle

$$(s_{i_1}\, s_{i_2} \ldots s_{i_d}),$$

and the corresponding portion of the linear machine is a d-stage circulating shift-register that shifts these components of s.

To complete the proof we must show that, if $t = A \cdot s$, then $F(t) = \pi'(F(s))$. We have $\pi'(F(s)) = \pi'(x_1, x_2, \ldots, x_n)$, and π' carries the n-tuple $\langle x_1, x_2, \ldots, x_n \rangle$ into the n-tuple $\langle x_{\pi(1)}, x_{\pi(2)}, \ldots, x_{\pi(n)} \rangle$. But

$$F(t) = F(A \cdot s) = F(\langle s_{\pi(1)}, s_{\pi(2)}, \ldots, s_{\pi(n)} \rangle^t) = \langle x_{\pi(1)}, x_{\pi(2)}, \ldots, x_{\pi(n)} \rangle.$$

Thus $F(t) = \pi'(F(s))$, which is what we set out to prove.   □

In the case where π has just one cycle, this theorem reduces to the preceding lemma.

Fig. 9.4.7 shows a nontrivial example that illustrates the proof of the theorem. Here we have $\pi = (1\,3\,4)(2\,5)$. In the A matrix we observe, for example, that the first row has a single 1 in the third coordinate because $\pi(1) = 3$. The machine corresponding to the A matrix consists of two independent circulating shift-registers, one with three stages

$$A = \begin{bmatrix} 0 & 0 & 0 & 1 & 0 \\ 0 & 0 & 1 & 0 & 0 \\ 0 & 0 & 0 & 0 & 1 \\ 1 & 0 & 0 & 0 & 0 \\ 0 & 1 & 0 & 0 & 0 \end{bmatrix}$$

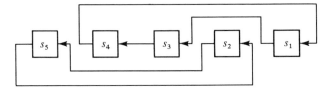

FIG. 9.4.7   Matrix for the Permutation (1 3 4)(2 5) and the Composite Machine Described by This Matrix

and one with two stages. By exhaustive enumeration, we can find that the cycle structure of π' is $[4(1) + 2(2) + 4(3) + 2(6)]$. We can compute this structure directly, however, because the cycle structures of the 3-stage and 2-stage registers are $[2(1) + 2(3)]$ and $[2(1) + 1(2)]$, respectively, and the product of these cycles is

$$[2(1) + 2(3)] \times [2(1) + 1(2)] = [4(1) + 2(2) + 4(3) + 2(6)].$$

At this point we can find the cycle structure for each π' in S_n' from the cycle structure of each π in S_n. Thus we can compute the cycle index polynomial of S_n' if we know the cycle structure of each permutation in S_n. Note that we still must learn how to find the cycle structures of the permutations in S_n. Fortunately, we can solve this problem by combinatorial techniques. We offer the solution without proof in the next lemma.

Lemma   For each n-tuple $\langle i_1, i_2, \ldots, i_n \rangle$ such that

$$\prod_{j=1}^{n} j i_j = n,$$

the number of permutations in S_n with i_1 cycles of length 1, i_2 cycles of length 2, . . . , and i_n cycles of length n is

$$n! / \prod_{j=1}^{n} (i_j)! j^{i_j}.$$

For example, the number of permutations in S_5 that have two 1-cycles and one 3-cycle can be computed by substituting $n = 5$, $i_1 = 2$, $i_3 = 1$, and $i_2 = i_4 = i_5 = 0$ in the formula given in the lemma. We obtain

$$5!/(2! \cdot 1^2 \cdot 1! \cdot 3^1) = 120/6 = 20.$$

We can verify this answer by an enumeration. There are ten ways to select two items from a set of five items to assign to the 1-cycles. For each such selection, we can arrange the remaining three items on a 3-cycle in two distinct ways. If the three elements are a, b, and c, then the two distinct 3-cycles of these elements are $(a\,b\,c)$ and $(a\,c\,b)$. Thus we have $10 \cdot 2 = 20$ permutations with the given cycle structure, which agrees with the result obtained from the formula.

With this last result we are able to state an algorithm for enumerating the equivalence classes of switching functions under the action of S_n'. The method is to compute the cycle structure of S_n using the preceding lemma, and then to use the shift-register theory to construct $Z_{S_n'}$. We complete the enumeration by computing the number of classes as $Z_{S_n'}(2, 2, \ldots, 2)$.

We now state the full algorithm.

Equivalence-Class Enumeration Algorithm.

Step 1. Compute the cycle structure of S_n.

Step 2. Choose any permutation π in S_n not already chosen. If none are left, terminate. For this permutation, each cycle of length i is analogous to an i-stage circulating shift-register. Record the cycle structure of the state diagram for each such register. An i-stage register has $N_2(d)$ cycles of length d for each d that divides i.

Step 3. Compute the cycle structure of the state diagram of the composite machine that is derived from the cycles of π. To do this, combine the cycle sets of the shift-register components according to the cycle-structure cross-product operation described in Section 9.3. Recall that, if one component has k_i cycles of length i and another component has k_j cycles of length j, then the composite machine has $gcd(i,j) \cdot k_i \cdot k_j$ cycles of length $lcm(i,j)$. After this computation is done, return to Step 2.

An example of the algorithm will clarify the details. Here we show the computation of $Z_{S_2'}(x_1, x_2)$. The group S_2 consists of the two permutations $e = (1)(2)$ and $(1\,2)$. The permutation in S_2' associated with $(1)(2)$ has the same cycle structure as the state diagram of a composite machine with two 1-stage circulating registers. Each 1-stage circulating register has two 1-cycles, so that the cycle structure of the composite machine is

$$[2(1)] \times [2(1)] = [4(1)]$$

in cycle-structure notation. The permutation $(1\,2)$ in S_2 is associated with a permutation in S_2' that has the same cycle structure as the state diagram of a 2-stage circulating register. This structure is $[2(1) + 1(2)]$ in cycle-structure notation. Thus we have

$$Z_{S_2'}(x_1, x_2) = \frac{1}{2}(x_1^4 + x_1^2 x_2),$$

TABLE 9.4.2   The Calculation of $Z_{S'_3}(x_1, x_2, x_3)$

Permutation	Shift-Register Cycle Structure	Term in $Z_{S'_3}$
(1)(2)(3)	$[2(1)] \times [2(1)] \times [2(1)]$ $= [8(1)]$	x_1^8
(1 2)(3), (1 3)(2), (2 3)(1)	$[2(1) + 1(2)] \times [2(1)]$ $= [4(1) + 2(2)]$	$3x_1^4 x_2^2$
(1 2 3), (1 3 2),	$[2(1) + 2(3)]$	$2x_1^2 x_3^2$

$$Z_{S'_3}(x_1, x_2, x_3) = \frac{1}{6}(x_1^8 + 3x_1^4 x_2^2 + 2x_1^2 x_3^2)$$

which is in agreement with the results of Chapter Three. The total number of equivalence classes is $Z_{S'_2}(2, 2) = 12$.

As an additional example, Table 9.4.2 shows the construction of $Z_{S'_3}$. Recall that S_3 contains the six permutations (1)(2)(3), (1 2)(3), (1)(2 3), (1 3)(2), (1 2 3), and (1 3 2). Because (1 2 3) and (1 3 2) have the same cycle structure and differ only in the labeling of the cycle s, they give rise to identical terms in $Z_{S'_3}$. Similarly, (1 2)(3), (1)(2 3), and (1 3)(2) give rise to identical terms. Consequently, Table 9.4.2 shows the computation for just one permutation for each distinct cycle structure. Table 9.4.3 lists Z_{S_n}, $Z_{S'_n}$, and the number of equivalence classes of the n-variable switching functions under $Z_{S'_n}$ for n up to 5.

TABLE 9.4.3   Cycle Index Polynomials and Equivalence-Class Enumerations for S_n Acting on the Variables of the Functions from $\{0,1\}^n$ into $\{0,1\}$

n	Z_{S_n}	$Z_{S'_n}$	Number of Equivalence Classes
2	$\frac{1}{2}(x_1^2 + x_2)$	$\frac{1}{2}(x_1^4 + x_1^2 x_2)$	12
3	$\frac{1}{6}(x_1^3 + 3x_1 x_2 + 2x_3)$	$\frac{1}{6}(x_1^8 + 3x_1^4 x_2^2 + 2x_1^2 x_3^2)$	80
4	$\frac{1}{24}(x_1^4 + 6x_1^2 x_2 + 3x_2^2 + 8x_1 x_3 + 6x_4)$	$\frac{1}{24}(x_1^{16} + 6x_1^8 x_2^4 + 3x_1^4 x_2^6 + 8x_1^4 x_3^4 + 6x_1^2 x_2 x_4^3)$	3,984
5	$\frac{1}{120}(x_1^5 + 10x_1^3 x_2 + 15x_1 x_2^2 + 20x_1^2 x_3$ $+ 20x_2 x_3 + 30x_1 x_4 + 24x_5)$	$\frac{1}{120}(x_1^{32} + 10x_1^{16} x_2^8 + 15x_1^8 x_2^{12} + 20x_1^8 x_3^8$ $+ 20x_1^4 x_2^2 x_3^4 x_6^2 + 30x_1^4 x_2^2 x_4^6 + 24x_1^2 x_5^6)$	37,333,248

Summary   We have applied cycle-set arithmetic to the problem of enumerating the equivalence classes of switching functions under the action of the symmetric group on the variables of the functions. To perform the enumeration, we construct the cycle index polynomial of S_n' acting on n-tuples in $\{0, 1\}^n$. The cycles of a permutation in S_n' are in one-to-one correspondence with the cycles of states of a composite feedback shift-register, and we can determine the latter cycles from cycle-set arithmetic.

The equivalence-class partition of the switching functions under S_n is of great practical significance because we can use one device to implement an entire equivalence class of functions, simply by varying the connections of function variables to device inputs. Hellerman (1963) published a catalog of the minimal-cost logic-network implementations for 3-variable functions in each of the 80 equivalence classes under S_3. Although the 4-variable functions lie in over 3,000 equivalence classes, it is within the capability of modern computers to find the minimal-cost realizations for each of these classes. This computation is substantially smaller than a computation over the 65,536 4-variable switching functions. Moreover, the catalog for the equivalence classes of 4-variable switching functions is sufficiently small to be held within the auxiliary memory of a modern computer system, where it can be used to automate some of the steps in the design of computers. Unfortunately, the number of equivalence classes grows as

$$2^{2^n}/n! \cong 2^{2^n - n(\log_2(n) - 1)},$$

which grows astronomically with n. For $n = 6$, the number of classes is of the order of 10^{13}, which is far too large for our techniques to have practical significance.

The enumeration techniques applied in this section have been adapted to other groups. A group that has received much investigation is the group of permutations and complementations of the variables of the set of switching functions. For this group, as for S_n', the determination of the cycle index polynomial can be done with cycle-set arithmetic. For a full discussion, see the paper by Harrison (1971).

The correspondence between the cycles of circulating registers and the cycle index polynomial of $Z_{S_n'}$ was first noticed by Harrison (1963). The information in Table 9.4.3 is reproduced from the work of Harrison (1971).

Exercises

9.4.1   Calculate the cycle structure for an n-stage circulating shift-register for $n = 2$ and $n = 3$. Enumerate the cycles exhaustively to check your calculation.

9.4.2   Find all 2-cycles and 3-cycles for a 6-stage circulating register, and verify that these are in one-to-one correspondence with the cycles enumerated in Ex. 9.4.1.

9.4.3 Verify the calculation of $Z_{S_4'}$ and $Z_{S_5'}$ in Table 9.4.3. Calculate the number of equivalence classes for each of these groups and compare with the numbers given in the table.

9.4.4 How many permutations in S_6 consist of two 3-cycles? How many consist of a 1-cycle, a 2-cycle, and a 3-cycle?

9.4.5 Calculate the term in $Z_{S_6'}$ that corresponds to the permutation $(1\,2\,3)(4\,5)(6)$.

9.4.6 Calculate the term in $Z_{S_6'}$ that corresponds to the permutation $(1\,2)(3\,4\,5\,6)$.

9.4.7 Find the cycle structure for the 4-stage shift-register whose next-state matrix is given in Fig. 9.4.8.

$$A = \begin{bmatrix} 0 & 1 & 0 & 0 \\ 0 & 0 & 1 & 0 \\ 0 & 0 & 0 & 1 \\ -1 & 0 & 0 & 0 \end{bmatrix}$$

FIG. 9.4.8   Next-State Matrix for Shift Register in Ex. 9.4.7

9.4.8 The shift-register in Ex. 9.4.7 differs from a circulating shift-register in that the feedback line multiplies an element by -1. A register of this type is called a *negacyclic* shift-register. Show that the cycles produced by an n-stage negacyclic register are of length d for each d that divides $2n$ and does not also divide n. Find a method for computing the number of cycles of length d; this method should be similar to the method for computing $I_p(d)$.

9.4.9 Consider the cycles produced by the register whose A matrix is given in Fig. 9.4.9. Show that the cycle structure is identical to that of a circulating register if an even number of -1's appear in the matrix, and is otherwise identical to that of a negacyclic register.

9.4.10 From the results of Exs. 9.4.8 and 9.4.9, develop a method for enumerating the equivalence classes under the action of the group of permutations and complementations of the variables.

$$A = \begin{bmatrix} 0 & \pm 1 & 0 & \cdots & 0 & 0 \\ 0 & 0 & \pm 1 & \cdots & 0 & 0 \\ \vdots & \vdots & \vdots & \vdots\vdots\vdots & \vdots & \vdots \\ 0 & 0 & 0 & \cdots & 0 & \pm 1 \\ \pm 1 & 0 & 0 & \cdots & 0 & 0 \end{bmatrix}$$

FIG. 9.4.9   Next-State Matrix for Register in Ex. 9.4.9

9.4.11 Enumerate the equivalence classes of 3-variable switching functions under the action of S_2 on the first two variables.

9.4.12 Enumerate the equivalence classes of 4-variable switching functions under the action of S_3 on the first three variables.

9.4.13 Enumerate the equivalence classes of 4-variable switching functions under the action of S_2 on the first two variables and of S_2 acting independently on the last two variables.

9.5 Primitive Shift-Registers and Maximum-Length Sequences

For each $n \geq 1$, there is an n-stage linear feedback shift-register over $GF(p^n)$ that has all of the nonzero states on a cycle of length $p^n - 1$. In this section, we investigate several applications of this class of shift registers.

Our first application is as a source of random numbers. Certain computer applications, such as simulation, require a sequence of randomly selected integers. The successive digits produced by some linear feedback shift-registers satisfy the specifications for the random integers, and therefore these linear shift-registers have been used in random-number-generator subroutines.

The second application of long shift-register sequences is for security in data-transmission systems where the data is of a sensitive nature. In such systems, the sender adds a shift-register sequence to his data before transmission, and the receiver subtracts the same shift-register sequence from the received sequence. The decoded sequence then is identical to the original transmitted data, but an uninformed observer who receives the sum sequence without knowing the shift-register sequence cannot decode the data. Thus the data-transmission system is secure from casual eavesdroppers.

The final application that we mention here is to radar range-measuring experiments. Shift-register sequences have been used to modulate radar pulses transmitted to distant bodies in the solar system. The modulation has the useful property that the reflected pulses can be discerned more easily from background noise that unmodulated radar pulses can be. Thus, range measuring can be done more accurately with shift-register sequences than without them. The distance from the earth to Venus has been measured by such techniques with greater accuracy than has been obtained with other less sophisticated techniques.

We begin our discussion with the definition of a primitive shift-register.

Definition An n-stage *primitive shift-register* is an n-stage linear feedback shift-register over $GF(p)$ that has all of the nonzero states on a single cycle of length $p^n - 1$. The characteristic polynomial of such a shift register is primitive and irreducible.

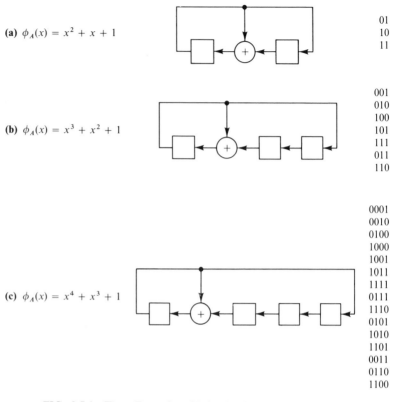

(a) $\phi_A(x) = x^2 + x + 1$

01
10
11

(b) $\phi_A(x) = x^3 + x^2 + 1$

001
010
100
101
111
011
110

(c) $\phi_A(x) = x^4 + x^3 + 1$

0001
0010
0100
1000
1001
1011
1111
0111
1110
0101
1010
1101
0011
0110
1100

FIG. 9.5.1   Three Examples of Primitive Shift-Registers and the Maximum-Length Cycles Associated with Them

We know from Section 9.3 that there is an n-stage primitive shift-register for each prime p and each $n \geq 1$.

To find examples of primitive shift-registers over $GF(2)$, consider the factorization of polynomials of the form $x^{2^n-1} - 1$ over $GF(2)$. Recall that $x^3 - 1$ factors into $(x + 1)(x^2 + x + 1)$. The roots of the irreducible polynomial $x^2 + x + 1$ each have multiplicative order 3 in $GF(2^2)$, so that $x^2 + x + 1$ is the characteristic polynomial of a 2-stage primitive shift-register. The shift register is shown in Fig. 9.5.1(a), together with a list of the states on its maximum-length cycle.

To construct a 3-stage primitive shift-register, observe that $x^7 - 1$ has two factors of degree 3, both with period 7. Thus we can use either $x^3 + x + 1$ or $x^3 + x^2 + 1$ as feedback polynomial, for each of these is primitive over $GF(2)$. The 3-stage register for $x^3 + x^2 + 1$ and its maximum-length cycle are shown in Fig. 9.5.1(b).

To construct a 4-stage primitive shift-register, we find that $x^{15} - 1$ has three irreducible factors over $GF(2)$ of degree 4, namely

$$x^4 + x^3 + x^2 + x + 1, \quad x^4 + x^3 + 1, \text{ and } x^4 + x + 1.$$

The latter two have period 15 and therefore are primitive, whereas the first one has period 5 and does not lead to a maximum-length cycle. Fig. 9.5.1(c) shows the 4-stage register for $x^4 + x^3 + 1$ and the states on its maximum-length cycle.

Table 9.5.1 lists primitive irreducible polynomials of several different degrees. The data in the table are drawn from a table produced by a rather sophisticated computer search (Watson 1962). The polynomials in each case have the smallest possible number of nonzero coefficients, so that the corresponding shift-registers have a minimum number of modulo-2 adders.

We now turn to a discussion of the application of primitive shift-registers to the generation of random numbers.

Computer simulations and other applications require a program to behave probabilistically, and the probabilistic behavior usually is produced by selecting a random number. In some sense this procedure is like casting a die. Algorithms of this type often are called *Monte Carlo algorithms*, because of the analogy with games of chance. Most computer-center libraries have a random-number-generator subroutine

TABLE 9.5.1 Primitive Irreducible Polynomials over
$GF(2)$ of Selected Degrees

n	$f(x)$
8	$x^8 + x^4 + x^3 + x^2 + 1$
11	$x^{11} + x^2 + 1$
12	$x^{12} + x^6 + x^4 + x + 1$
15	$x^{15} + x + 1$
16	$x^{16} + x^5 + x^3 + x^2 + 1$
23	$x^{23} + x^5 + 1$
24	$x^{24} + x^4 + x^3 + x + 1$
31	$x^{31} + x^3 + 1$
32	$x^{32} + x^7 + x^5 + x^3 + x^2 + x + 1$
35	$x^{35} + x^2 + 1$
36	$x^{36} + x^6 + x^5 + x^4 + x^2 + x + 1$
40	$x^{40} + x^5 + x^4 + x^3 + 1$
47	$x^{47} + x^5 + 1$
48	$x^{48} + x^7 + x^5 + x^4 + x^2 + x + 1$
59	$x^{59} + x^6 + x^5 + x^4 + x^3 + x + 1$
60	$x^{60} + x + 1$
63	$x^{63} + x + 1$
64	$x^{64} + x^4 + x^3 + x + 1$
70	$x^{70} + x^5 + x^3 + x + 1$
80	$x^{80} + x^7 + x^5 + x^3 + x^2 + x + 1$
90	$x^{90} + x^5 + x^3 + x^2 + 1$
100	$x^{100} + x^8 + x^7 + x^2 + 1$

and in some cases these generators are based upon primitive shift-registers.

The sequence of nonzero states on a maximum-length cycle of a primitive shift-register enjoys certain randomness properties. For example, each nonzero state appears exactly once during a cycle of $p^n - 1$ states, so that the probability of a given state appearing in a register is $1/(p^n - 1)$ and is the same for all states. If we examine the contents of just a few of the n stages, say of k stages, then we also obtain very good uniformity. Each nonzero k-tuple appears in the first k stages exactly p^{n-k} times, and the zero k-tuple appears in the first k stages exactly $p^{n-k} - 1$ times. The difference between the two expressions arises because the 0-state does not appear on the cycle. Thus the probability of finding a particular k-tuple in the first k stages is given by

$$1/(p^{n-k}) \qquad \text{if the } k\text{-tuple is nonzero,}$$

or by $\qquad 1/(p^{n-k} - 1) \quad$ if the k-tuple is zero.

The probability distribution for k-tuples is virtually uniform when $n - k$ is large. When $k = 1$, $n = 32$, and $p = 2$, we find that the differences in the probabilities is less than one part in 10^9. If the leading digit is taken as a sign digit, as it is in most computers, our results indicate that the 32-bit integers produced by a primitive shift-register have nearly equal probabilities of being positive or of being negative.

Because the random-number-generator subroutine is invoked frequently—perhaps a million or more times—during the execution of a typical Monte Carlo program, it is mandatory that a random-number-generator subroutine be very fast. The simulation of a linear feedback shift-register is extremely easy, and only a few microseconds are required to produce a random number in this way. We next describe a typical algorithm for simulating a feedback shift-register. It is relatively simple to translate the algorithm into the machine language of most computers, and the timing is such that each step requires one to two microseconds. We assume that the computer has n-bit registers that can hold n-tuples of the form $x = \langle x_0, x_1, \ldots, x_{n-1} \rangle$. The algorithm generates a sequence of binary $(n - 1)$-tuples. The polynomial

$$f(x) = x_0 + f_1 x + \cdots + f_{n-1} x^{n-1}$$

is a previously selected polynomial known to be primitive over $GF(2)$ and is the generator polynomial for the algorithm. We assume that the state vector $s = \langle s_0, s_1, \ldots, s_{n-2} \rangle$ is initialized to any nonzero state before this algorithm is called for the first time.

Random-Number-Generator Algorithm:

Step 1. Multiply s by two. (This can be done with an end-off shift or by adding the state to itself, whichever is faster.)

Step 2. If $s_{n-1} = 0$, then exit; otherwise go to Step 3. (On most computers this can be done by testing if $s \geq 0$.)

Step 3. Replace s by $\langle s_0 +_2 f_0, s_1 +_2 f_1, \ldots, s_{n-2} +_2 f_{n-2}, 0 \rangle$ and exit. (On most computers, this can be done in a single operation by using the EXCLUSIVE OR instruction.)

The algorithm simulates an $(n - 1)$-stage shift-register in an obvious way. Step 1 does the shift and places the feedback digit in s_{n-1}. If the digit is 0, then the state of the register is correct, and the algorithm terminates in Step 2. If $s_{n-1} \neq 0$, then the effect of feedback is simulated by Step 3.

For most computers, the algorithm can be modified easily to simulate an n-stage register by modifying Step 2. The add or shift operation of n-tuples in Step 1 can cause an overflow, and the overflow occurs if and only if the feedback digit for an n-stage register is 1. Hence, Step 2 is changed from a test of the sign bit to a test of the overflow indicator. For this modification, we use a primitive irreducible polynomial $f(x)$ of degree n, and the coefficient f_{n-1} of this polynomial should be added to s_{n-1} in Step 3.

Earlier we indicated some of the randomness properties of the states on a maximum-length cycle. Unfortunately, however, maximum-length cycles have characteristics that make them unacceptable for general use. For example, Step 1 always doubles the state vector s. Hence, about half of the time, successive states differ by a factor of 2. Thus the successor of a state is to some extent predictable or, equivalently, the sequence of states has high *serial correlation*. The factor-of-2 correlation is by no means the only serial-correlation problem, but it does indicate that serial correlation exists. Obviously, we must insist that a random sequence be reasonably free from serial correlation; otherwise, the highly nonrandom sequence $1, 2, 3, \ldots, 2^n - 1$ would be acceptable. This sequence has all of the nonzero states on one cycle, so it shares this randomness property with the maximum-length state cycle of a primitive shift-register. Yet this sequence clearly is unacceptable as a random sequence because of the serial correlation. Thus we must examine a sequence not only for the frequency of occurrence of each number, but also for the order in which they occur.

For this reason, the linear shift-register is less desirable than other methods for generating random numbers. However, it is not difficult to construct a good random number generator by simulating *two* linear feedback shift-registers. Next we give an algorithm for this type of generator. This algorithm requires about three times as much time as the preceding one, and it requires some additional storage registers. In this algorithm, both $f(x)$ and $g(x)$ are primitive irreducible polynomials of degree n and m, respectively. We obtain the best results when $2^n - 1$ and $2^m - 1$ are relatively prime. This is the case, for example, when $m = 31$ and $n = 32$, or when $m = 35$ and $n = 36$. The variable y is a vector of 2^k states, each of which is an n-tuple. The notation $y[i]$ denotes the ith element of the vector y. We assume that y is initialized with 2^k arbitrary nonzero n-tuples.

Random-Number-Generator Algorithm (*with Two Shift Registers*):

Step 1. Generate a random number x using the polynomial $g(x)$ as a feedback polynomial.

Step 2. Select a k-tuple of digits from x. (The k coordinate positions may be fixed ahead of time.) Treat this k-tuple as an integer i in the range $0 \leq i \leq 2^k - 1$.

Step 3. Output the integer $y[i]$ as the next random integer.

Step 4. Generate a random number z using the polynomial $f(x)$.

Step 5. Set $y[i]$ to z and exit.

When $g(x) = x^2 + x + 1$ and $f(x) = x^3 + x + 1$, the random integers selected in Steps 1 and 4 have periods 3 and 7, respectively. Hence the pairs repeat after 21 times. The integer $y[i]$ in Step 3 repeats with period 21, except possibly for a transient initial sequence when the generator is first started. In general, the period of the composite generator is the least common multiple of the individual periods and is independent of k. However, the sequence produced by the generator does depend on k. The shift register for $g(x)$ essentially scrambles the sequence produced by $f(x)$, and it does so by displacing digits in the $f(x)$ sequence by as much as 2^k positions.

In general, serial correlation tends to decrease as k increases. Specific applications may be sensitive to serial correlations over short intervals, and k should be set high enough to produce uncorrelated sequences for these intervals. Knuth (1969) discusses various methods for testing generators for serial correlation, and such methods can be used to determine if a random number generator meets the criteria for a specific application. The analytic calculation of the serial correlation of a composite primitive shift-register remains an open problem.

The second of the three applications we discuss in this section is the use of primitive shift-registers in secure communication systems. Fig. 9.5.2 illustrates the problem that we wish to solve here. A message must be sent from point A to point B, and an unfriendly eavesdropper listens to all messages en route from A to B. We wish to find a simple cryptographic system to encode messages between A and B, so that the eavesdropper will not understand what he hears, yet we require that the encoding and decoding of the message be relatively simple operations. We can conjure up various cloak-and-dagger plots in which such cryptographic systems are especially useful. Indeed, the advances in the state of the art in such systems have been stimulated by diplomatic and military needs. Kahn (1967) describes numerous secure communication techniques of this type in his rather revealing and entertaining peek into the annals of cryptography. Of late, data communication of a nonmilitary nature has grown substantially, and with it has grown the need for protecting data from unauthorized access. The data communication we refer to here is largely from computer to computer, or from computer terminal to computer. Several secure communication

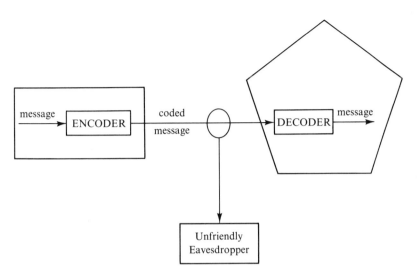

FIG. 9.5.2 A Secure Communication System

systems to protect this form of data transmission have been invented and marketed. We describe here a system based upon the use of primitive shift-registers.

First we discuss the notion of maximum-length sequence. Then we show a secure communication system that uses such sequences.

Definition   A *maximum-length sequence* of length $p^n - 1$ is a sequence of elements of $GF(p)$ that appears in a given stage of an n-stage primitive shift-register over $GF(p)$ as it cycles through the nonzero states on its maximum-length cycle.

We usually select the stage at either end of a primitive shift-register as a source of digits for a maximum-length sequence. The successive digits in a maximum-length sequence have very good randomness properties, including low serial correlation, even though the successive states of a primitive shift-register generally have high serial correlation. A typical maximum-length sequence is 1110100, which is generated by the shift register in Fig. 9.5.1(*b*). Observe that each of the three stages of the shift register generates this sequence or a cyclic shift of this sequence.

Now we show how to use maximum-length sequences in a secure communication system. Fig. 9.5.3 shows a crude block diagram of the system. We see that the sender has at his disposal a primitive n-stage shift-register. He transmits digits to the receiver at unit-time intervals, such that at time i the transmitted digit v_i is the sum

$$v_i = d_i +_p r_i,$$

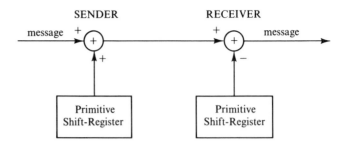

FIG. 9.5.3   A Secure Communication System Based on a Maximum-Length Sequence

where d_i is the message digit at time i, and r_i is the ith digit of the maximum-length sequence produced by the primitive shift-register. In most applications of interest, the field is $GF(2)$ and $p = 2$, so that addition is modulo-2 addition.

The receiver reverses the encoding process of the sender. The receiver has an identical primitive shift-register, which is synchronized with the sender's register. Thus the receiver computes

$$v_i +_p (-r_i) = (d_i +_p v_i) +_p (-v_i)$$
$$= d_i +_p (r_i +_p (-r_i))$$
$$= d_i.$$

The receiver subtracts the random digit r_i from the received sequence and thus obtains the original message digit. When the field is $GF(2)$, the transmitter and receiver are identical, because addition and sub-traction modulo 2 are identical operations.

The eavesdropper listens to the sequence $v_i, v_{i+1}, v_{i+2}, \ldots$, which appears to be quite random when the sequence r_i is random. The shift-register sequence generators must be primitive, because otherwise all of the nonzero register states do not appear on one cycle, and the sequences produced by such nonprimitive registers do not have the requisite randomness properties.

Secure communication systems of this type are in common use. Some commercial systems are slightly more sophisticated than this simple model because the linear feedback shift-register is replaced by some other sequence generator that makes the system more difficult to crack for an eavesdropper who knows what kind of coding is being used. For example, the composite primitive shift-register described in connection with random number generation is one possible replacement for the single shift-register.

Our third application for primitive shift-registers is that of range measuring. Fig. 9.5.4 is a diagram of a typical radar ranging system. A pulse train is transmitted from the radar to a distant body such as the moon or a planet, and the signal reflected from the body returns to the receiver on earth sometime later. Because the pulse train travels at a

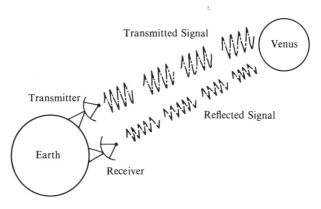

FIG. 9.5.4 A Radar Ranging System

known speed (the speed of light), we can compute the distance of the body from the earth by measuring the time-delay between pulse transmission and pulse reception. The difficulty in performing this measurement arises because the received signal is quite weak and is virtually hidden by normal background noise.

Fig. 9.5.5 illustrates the problem of exactly determining the time-delay. The figure shows several examples of received pulses, with background noise progressively more dominant. To calculate the time-delay of the pulse, we must be able to determine the position of the leading edge of the pulse. When the noise is small compared to pulse amplitude, the leading edge is quite discernible. As the background noise increases, the leading edge of the pulse becomes less and less discernible and, in fact, the entire pulse more or less disappears in the noise.

Very accurate ranging has been done by modulating the transmitted pulse with a maximum-length sequence. For example, if the maximum-length sequence is the sequence 1110100 produced by the last stage of the register in Fig. 9.5.1(*b*), then the radar transmits a short pulse during each of the first three time intervals, followed by a time interval in which no pulse is sent, followed by an interval with a pulse, and finally by two intervals with no pulses. We choose the sequence to be sufficiently long so that its period exceeds the uncertainty in our prior knowledge of the delay-time of the received sequence. The sequence normally is transmitted repeatedly for many periods, so that many different synchronization points are available in the received sequence.

In our model, we assume that the transmitted signal is a binary sequence, and that the received sequence also is binary but is a corrupted version of the transmitted sequence. In actual practice, the ranging receivers operate on the continuous wave form of the received sequence, thus attaining greater accuracy than that attained with the discrete version of the received sequence. The difference between the continuous and discrete versions of the received sequence is unimportant for our discussion.

Thus, in our model, we assume that we know the transit time of a signal to within one period of a maximum-length sequence, and we wish to determine the transit time to within the transmission time for one digit. Consider, for example, what happens as we repeatedly transmit the sequence 1110100. In order to compute the transmit time to the precision of one digit-time, we must determine where the sequence 1110100 lies in the received sequence. Hence, if we receive . . . 11101001110100 . . . , we compare the original sequence 1110100 to the seven subsequences 1110100, 1101001, 1010011, . . . , 0111010 and look for the best match. Each of the subsequences is a full period of 1110100 but is delayed by 0 to 6 digit-times. In this case we have assumed no background noise, and the first of the seven subsequences is an exact match. Thus we know to within one digit-time when 1110100 was received.

In a noisy environment, the received sequence may have several bits in error. We wish to find the sequence 1110100 in the received sequence, even if errors occur. Thus our model of this problem is similar to the model we use in Chapter Five for error-correcting codes. Indeed, we make use of several results from Chapter Five to show that maximum-length sequences have optimal properties for this application.

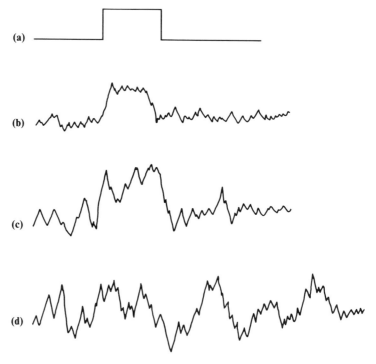

FIG. 9.5.5   A Pulse Waveform Corrupted by Various Levels of Background Noise

Let us restate our problem in terms of error-correcting codes. Consider a code that has the m-tuple $v = \langle v_0, v_1, \ldots, v_{m-1} \rangle$ as a code vector, and that has as the other code vectors the $m - 1$ cyclic shifts of v. We must find a v such that v is as different as possible from each of the other code vectors in the code.

From Section 5.2, we recall that the Hamming distance is a measure of the difference between two code vectors; for two code vectors v and w, the distance $H(v, w)$ is equal to the number of coordinate positions in which v and w differ. Recall also that, for random errors, the maximum-likelihood decoding rule requires us to decode a received word as the nearest code vector (in the sense of Hamming distance) to the received word. For the ranging problem, we let v be a maximum-length sequence, and we transmit v repeatedly. The received signal is a corrupted version of the periodic sequence. We essentially receive corrupted versions of v and of each of its $m - 1$ cyclic shifts, and we must determine which of the corrupted shifts of v is the original v. A maximum-likelihood decoder for this problem compares each of the corrupted received shifts of v to v and chooses the nearest one to v as the one that truly is v.

TABLE 9.5.2 Hamming Distances Between 1110100 and Length-7 Subsequences of 11101011110000

Subsequence	Hamming Distance to 1110100
1110101	1
1101011	5
1010111	3
0101111	5
1011110	3
0111100	2
1111000	2

As an example, suppose that we transmit two periods of 1110100, and we receive the corrupted signal 11101011110000, which is in error in digits 7 and 12. To determine where 1110100 falls within the first seven positions, we compare 1110100 to each of the first seven subsequences of length 7. We obtain the Hamming distances shown in Table 9.5.2. Because the first subsequence is the nearest to 1110100, we choose it as the starting position. By using the maximum-likelihood rule, we minimize the probability of an incorrect decision. Thus we can determine the correct time-delay, even when the received sequence is corrupted by some errors. In practice, the receiver uses the entire received sequence, possibly spanning several periods, to determine the correct phase of the transmitted signal in the received sequence.

In the remainder of this section, we show that a maximum-length sequence of length $m = 2^n - 1$ can be received with as many as $2^{n-2} - 1$ errors and still be decoded properly. Moreover, this is the best possible error-correcting capability for any sequence of length $2^n - 1$.

First we investigate several properties of maximum-length sequences, and we prove that the Hamming distance between a sequence of length $2^n - 1$ and a cyclic shift of itself is 2^{n-1} for all cyclic shifts other than the identity shift. From this observation we derive the error-correcting capability.

We can establish the Hamming-distance property from an interesting characteristic of maximum-length sequences that is sometimes called the *cycle-and-add property*, which is set forth in the next lemma.

Lemma Let $v = \langle v_0, v_1, \ldots, v_{m-1} \rangle$ be a maximum-length sequence over $GF(2)$. Let $u = \langle u_0, u_1, \ldots, u_{m-1} \rangle$ be any cyclic shift of v other than v itself. Then $u + v = \langle u_0 + v_0, u_1 + v_1, \ldots, u_{m-1} + v_{m-1} \rangle$ is also a cyclic shift of v, where the addition of sequence components is in $GF(2)$.

Proof: We base this proof on the linearity of the sequential machine that produces the sequence v. Let $m = 2^n - 1$ so that v is a maximum-length sequence generated by an n-stage primitive shift-register. Assume that v is produced by starting the shift register in state s. From Section 9.1 we have

$$v_0 = C \cdot s,$$
$$v_1 = C \cdot A \cdot s,$$
$$\ldots$$
$$v_i = C \cdot A^i \cdot s,$$

where v_i is the ith component of v, and where A and C are the next-state and output matrices, respectively, for the shift register. Because the sequence u is a cyclic shift of v, say a shift of j positions, it is produced by starting the register in state $t = A^j \cdot s$. That is,

$$u_0 = C \cdot t = C \cdot A^j \cdot s,$$
$$u_1 = C \cdot A \cdot t = C \cdot A^{j+1} \cdot s,$$
$$\ldots$$
$$u_i = C \cdot A^{j+i} \cdot s,$$

And because $C \cdot A^i \cdot s + C \cdot A^i \cdot t = C \cdot A^i \cdot (s + t)$ for all states s and t and for all integers i, we have

$$u_0 + v_0 = C \cdot s + C \cdot t = C \cdot (s + t),$$
$$u_1 + v_1 = C \cdot A \cdot s + C \cdot A \cdot t = C \cdot A \cdot (s + t),$$
$$\ldots$$
$$u_i + v_i = C \cdot A^i \cdot s + C \cdot A^i \cdot t = C \cdot A^i \cdot (s + t),$$

and the sequence $u + v = \langle u_0 + v_0, u_1 + v_1, \ldots, u_{m-1} + v_{m-1} \rangle$ is the sequence produced from the same shift register started in state $s + t$. Because s is distinct from t, the n-tuple $s + t$ must be nonzero, and therefore it must lie on the cycle of nonzero states of the primitive shift-register. Thus, for some integer k such that $1 \leq k \leq m - 1$, we have $A^k \cdot s = s + t$, and the shift register reaches state $s + t$ after k shifts. Hence the sequence $u + v$ is produced by starting the register in state $A^k \cdot s$, and thus $u + v$ must be a cyclic shift of k positions of the sequence v. ◻

As an example of the preceding lemma, consider the sequence 1110100 that we have investigated previously. Observe that

$$1110100 + 0100111 = 1010011$$

and $1110100 + 1001110 = 0111010$, so that the sum in each case is a cyclic shift of 1110100.

Now we can calculate the distance between a maximum-length sequence v and a shift of itself, u. In fact, from the preceding lemma

$$H(u, v) = H(u + v, 0) = H(v, 0)$$

because $u + v$ is a cyclic shift of v and is the same distance from 0 as is v. Recall that $W(v) = H(v, 0)$ is the Hamming weight of v, and it is this quantity that determines the error-correcting capability of a ranging system. The next lemma shows that the Hamming weight of a maximum-length sequence is approximately half of the length of the sequence.

Lemma The weight of a binary maximum-length sequence of length $m = 2^n - 1$ is 2^{n-1}.

Proof: Let the states of a primitive-shift-register generator be n-tuples of the form $s = \langle s_0, s_1, \ldots, s_{n-1} \rangle$. Without loss of generality, let us select the coordinate s_0 as the output digit. As the shift register moves through the nonzero states, every nonzero n-tuple appears exactly once during a period. There are 2^{n-1} n-tuples s such that $s_0 = 1$, and there are $2^{n-1} - 1$ nonzero n-tuples such that $s_0 = 0$. Hence the output sequence has 2^{n-1} ones and $(2^{n-1} - 1)$ zeros. Then its weight is 2^{n-1}, the number of ones in the sequence. ◻

Again referring to the sequence 1110100, we note that the sequence has four 1's and three 0's, in agreement with the lemma.

The crucial theorem on the error-correcting capability of maximum-length sequences is quite simple to prove from the preceding lemmas.

Theorem Any combination of $2^{n-2} - 1$ or fewer errors per period can be corrected in a ranging system with a maximum-length sequence of length $2^n - 1$.

Proof: The distance from a maximum-length sequence v of length $2^n - 1$ to any of its cyclic shifts is exactly 2^{n-1} because of the cycle-and-add property. The maximum-likelihood decoding rule for codes of minimum distance $2d + 1$ corrects any combination of d or fewer random errors, but leads to decoding errors if $d + 1$ or more errors occur. Setting $2d + 1 = 2^{n-1}$ gives $d = 2^{n-2} - 1/2$. Because d must be integral, we must lower this to $2^{n-2} - 1$. ◻

Maximum-length sequences of length $2^3 - 1 = 7$ can correct one error per period. No error correction is possible for maximum-length sequences of length $2^2 - 1 = 3$. For large n, the error-correction capability is approximately $m/4$, where m is the length of the sequence.

Now we show that we cannot find sequences of length $m = 2^n - 1$ with better error-correcting properties. First, we show that a good sequence v must have Hamming weight 2^{n-1} or $2^{n-1} - 1$. The average Hamming distance between v and a cyclic shift of v, where the average is computed over the $m - 1$ nonzero cyclic shifts of v, is given by

$$\text{Average distance} = \frac{1}{m-1} \sum_{k=1}^{m-1} \sum_{i=0}^{m-1} (v_i +_2 v_{i+k}). \tag{9.5.1}$$

Although the addition of the terms in parentheses is modulo-2 addition, the summations are ordinary arithmetic. The subscript of v_{i+k} is computed modulo m. The inner summation over i computes the distance between v and the sequence formed when v is shifted cyclically by k coordinates. The outer summation varies k from 1 to $m - 1$ to include all of the nonzero cyclic shifts. Because the term $(v_i +_2 v_{i+k})$ is nonzero if and only if either v_i or v_{i+k} (but not both) is nonzero, we can rewrite Eq. 9.5.1 in the following form:

$$\text{Average distance} = \frac{1}{m-1} \sum_{i=0}^{m-1} \sum_{k=1}^{m-1} (v_i +_2 v_{i+k})$$

$$= \frac{1}{m-1} \left(\sum_{\substack{i=0, \\ v_i=1}}^{m-1} \sum_{\substack{k=1, \\ v_{i+k}=0}}^{m-1} 1 + \sum_{\substack{i=0, \\ v_i=0}}^{m-1} \sum_{\substack{k=1, \\ v_{i+k}=1}}^{m-1} 1 \right)$$

$$= \frac{1}{m-1} \left(W(v)(m - W(v)) + (m - W(v))W(v) \right)$$

$$= \left(2W(v)(m - W(v)) \right)/(m-1), \tag{9.5.2}$$

where $W(v)$ is the Hamming weight of v. By differentiating Eq. 9.5.2 with respect to $W(v)$, we find that Eq. 9.5.2 has a maximum when $W(v) = m/2$. Because $W(v)$ is integral and m is odd, the maximum is attained when $W(v) = (m \pm 1)/2$. For these cases, $W(v) = 2^{n-1}$ and $W(v) = 2^{n-1} - 1$. When we substitute either of these quantities into Eq. 9.5.2, we obtain

$$\text{Average distance} = \left(2^n \cdot (2^{n-1} - 1) \right)/(2 \cdot (2^{n-1} - 1))$$

$$= 2^{n-1}.$$

Because this is the average distance, and the minimum distance cannot exceed the average distance, we conclude that the minimum distance between v and a cyclic shift of itself is no more than 2^{n-1}. But this is precisely the minimum distance for maximum-length sequences. Consequently, the cyclic shifts of a maximum-length sequence differ from the sequence itself by the greatest distance possible. This tells us that the class of maximum-length sequences has optimal properties for ranging.

Two different ranging experiments on the planet Venus were conducted in 1959 and 1961, each using maximum-length sequences of length $2^{13} - 1 = 8,191$. These experiments produced more accurate measures of the distance from the earth to Venus than ever had been made previously.

In the first part of this section, we note that primitive shift-registers have been used in random number generators, although other competitive techniques also are in frequent use. The book by Knuth (1969) is the definitive work on random number generators. Tausworthe has made an extensive study of the serial-correlation properties of maximum-length sequences. He has found generally that, although sequences of *n*-tuples from an *n*-stage register show serial correlation, the sequence of digits from a particular stage has very low serial correlation (Titsworth 1962 and Tausworthe 1965). The composite random number generator described in this section was suggested by MacLaren and Marsaglia (1965). Maisel and Gnugnoli (1972) discuss random number generators for computer simulations.

The article by Shannon (1949) is an interesting study of the mathematical theory underlying secure communication systems. Relatively little material is published in this field, perhaps because secure communication systems are more secure when less is known about them. Earlier in the section we mention the book by Kahn (1967) as a source for some of the nonmathematical aspects of the history of cryptography.

Radar ranging experiments are discussed in some detail by Golomb et al. (1964), Tausworthe (1964), and Victor et al. (1961).

Golomb (1967) gives substantial information about shift-register sequences in general.

Exercises

9.5.1 Write a computer program to find all of the primitive irreducible polynomials of degree 5 and 6 over $GF(2)$.

9.5.2 Prove that the set of *m* cyclic shifts of a maximum-length sequence of length $m = 2^n - 1$, together with the 0 sequence, forms a representation of $GF(2^n)$. Addition of field elements is done by adding the corresponding sequences component by component. If α is a root of the characteristic polynomial of the generator polynomial, then multiplication of a field element β by α^i is done by cyclically shifting the β sequence *i* times.

9.5.3 Let *u*, *v*, and *w* be three different cyclic shifts of a maximum-length sequence such that $u = v + w$. Also, let *v* and *w* differ by *i* cyclic shifts. Show that there do not exist two other cyclic shifts v' and w' such that $u = v' + w'$ and such that v' and w' also differ by *i* cyclic shifts. HINT: Use the representation of Ex. 9.5.2. Express *u*, *v*, and *w* as powers of α, and find *u* as a product of two field elements rather than as a sum of two field elements.

9.5.4 Consider a maximum-length sequence of length $2^n - 1$, and let the sequence be the sequence of digits that appears in a particular stage, say the first stage, of an n-stage primitive shift-register. Prove that each of the nonzero n-tuples over $GF(2)$ appears as a subsequence of n consecutive digits.

9.5.5 Suppose that an observer sees a subsequence of length $2n - 1$ of a maximum-length sequence of length $2^n - 1$ over $GF(2)$. Also suppose that the observer knows that the sequence is generated by an n-stage shift-register. Prove that he then can uniquely determine the entire sequence.

9.5.6 Consider a maximum-length sequence over $GF(2)$ of length $2^n - 1$. Prove that each nonzero binary $(n - 1)$-tuple occurs exactly twice as a subsequence of $n - 1$ consecutive digits of the maximum-length sequence. Prove that each nonzero binary k-tuple occurs exactly 2^{n-k} times as a subsequence of k consecutive digits.

9.5.7 A particular nonzero $(n - 1)$-tuple appears twice in a maximum-length sequence of length $2^n - 1$, once beginning at position i and once beginning at j, where $j > i$. We define the distance between these positions to be the minimum of $i - j$ and $j - i$, where both subtractions are performed modulo $2^n - 1$. Thus the distance between two occurrences of an $(n - 1)$-tuple is an integer in the interval $1 \le x \le 2^{n-1} - 1$. Prove that the distance between the occurrences of a particular nonzero $(n - 1)$-tuple in a maximum-length sequence is unique for that $(n - 1)$-tuple. That is, if u and v are distinct nonzero $(n - 1)$-tuples, then the two occurrences of u will be separated by a different distance than the distance between the two occurrences of v.

9.6 Shift-Register Decoders and Encoders

In this section we return to the topic of Hamming codes, and we discover that we can both encode and decode these codes by using primitive shift-registers. In fact, to implement a data transmission for a specific Hamming code, we need three copies of a primitive shift-register. One register is used for encoding, one is used for calculating the parity checks, and the third is used to perform error correction.

Recall from Section 5.2 that a Hamming single-error correcting code of length $n = 2^r - 1$ is the set of all n-tuples in the null space of an $r \times n$ matrix H that has each of the nonzero binary r-tuples as its columns. Fig. 9.6.1 shows a suitable H matrix for a Hamming code of length 7. A binary 7-tuple v is in the code if and only if $v \cdot H^t = 0$. Thus, 1011000 and 0110001 are both code vectors, but 1100000 and 1000001 are not.

Recall that we show the minimum distance of the code to be 3 by showing that the code has minimum weight 3. In particular, because no column of H is all 0's, every nonzero code vector must have two or more 1's. Moreover, because all columns of H are distinct, no two columns sum to 0. Therefore, every nonzero code vector has three or more 1's.

Now we show an equivalent definition of Hamming code derived from a primitive element of $GF(2^r)$. This definition leads naturally to the use of primitive shift-registers as encoders and decoders.

Let α be a primitive element of $GF(2^r)$, and let H be the $1 \times 2^r - 1$ matrix over $GF(2^r)$ given by

$$H = \begin{bmatrix} 1 & \alpha & \alpha^2 & \alpha^3 & \alpha^4 & \ldots & \alpha^{2^r - 2} \end{bmatrix}. \tag{9.6.1}$$

Consider the null space of H over $GF(2)$. This is the set of all n-tuples over $GF(2)$ such that $v \cdot H^t = 0$, where $n = 2^r - 1$. It so happens that the null space of H is a Hamming single-error correcting code.

$$H = \begin{bmatrix} 1 & 0 & 0 & 1 & 1 & 1 & 0 \\ 0 & 1 & 0 & 0 & 1 & 1 & 1 \\ 0 & 0 & 1 & 1 & 1 & 0 & 1 \end{bmatrix}$$

FIG. 9.6.1   An H Matrix for a Hamming Code of Length 7

Theorem   Let H be the matrix described in Eq. 9.6.1. The null space of H is a single-error correcting code.

Proof: To prove the theorem it is sufficient to show that the minimum weight of a nonzero vector is at least 3.

No vector in the null space has weight 1 because each of the $2^r - 1$ distinct powers of α is nonzero. Similarly, no vector in the null space has weight 2, because the existence of such a vector would imply that $\alpha^i + \alpha^j = 0$ for some i and j. But the field elements form a group under addition, so that each field element has a unique additive inverse. In fields of characteristic 2, each element is its own additive inverse. Hence $\alpha^i + \alpha^i = 2\alpha^i = 0$, and we cannot find an α^j with j distinct from i such that $\alpha^i + \alpha^j = 0$. Thus the minimum weight of a vector in the null space is at least 3, and the null space is a Hamming single-error correcting code.   ☐

As an example of this definition of Hamming code, consider a Hamming code of length 7. For this we use $GF(2^3)$, and we choose α to be a root of the primitive polynomial $x^3 + x^2 + 1$. Table 9.6.1 gives a representation of $GF(2^3)$ as the polynomial algebra modulo $x^3 + x^2 + 1$, where the polynomial x is chosen to be α.

Table 9.6.1 leads to an interesting observation concerning the H matrices of Fig. 9.6.1 and Eq. 9.6.1. When we replace each field element in Eq. 9.6.1 by a binary 3-tuple whose components are the coefficients of the polynomial representation of the field element, we obtain the H matrix of Fig. 9.6.1. For example, α^5 is represented by $1 + x$, and this

TABLE 9.6.1   The Representation of $GF(2^3)$ as a Polynomial Algebra Modulo $x^3 + x^2 + 1$

Polynomial	Field Element
1	$\alpha^7 = 1$
x	α
$1 + x$	α^5
x^2	α^2
$1 \qquad + x^2$	α^3
$x + x^2$	α^6
$1 + x + x^2$	α^4

in turn is represented by the column vector $\langle 1, 1, 0 \rangle^t$. Then α^5 is the sixth column in Eq. 9.6.1, and $\langle 1, 1, 0 \rangle^t$ is the sixth column in Fig. 9.6.1.

This example shows clearly that every H matrix of the form shown in Eq. 9.6.1 is equivalent to an $r \times 2^r - 1$ matrix over $GF(2)$ whose columns are nonzero and distinct. By definition, such a matrix is a parity-check matrix for a Hamming code.

The next theorem shows that the code vectors for the code derived from H (Eq. 9.6.1) are in one-to-one correspondence with the polynomial multiples of $M_\alpha(x)$, the minimum polynomial of α. This theorem establishes the method of encoding and decoding the Hamming code. To encode, we treat the information digits as a polynomial over $GF(2)$, and from it we construct another polynomial that is a multiple of $M_\alpha(x)$. This polynomial is a code word. To decode, we divide a received polynomial by $M_\alpha(x)$ to determine if the received polynomial is a multiple of $M_\alpha(x)$ and extract the information digits. Both the encoder and the polynomial-divider networks are primitive shift-registers.

Theorem   Let $v = \langle v_0, v_1, \ldots, v_{n-1} \rangle$ be an n-tuple over $GF(2)$ in the null space of H, where H is of the form given in Eq. 9.6.1. Then the polynomial

$$v(x) = \sum_{i=0}^{n-1} v_i x^i$$

is a multiple of $M_\alpha(x)$, the minimum polynomial of α over $GF(2)$. Conversely, every polynomial $v(x)$ of degree less than n that is a multiple of $M_\alpha(x)$ corresponds to an n-tuple $\langle v_0, v_1, \ldots, v_{n-1} \rangle$ in the null space of H.

   Proof: Let v be an n-tuple in the null space of H. Then

$$\sum_{i=0}^{n-1} v_i \alpha^i = 0.$$

Equivalently, $v(\alpha) = 0$, where $v(x)$ is the polynomial derived from the vector v. From Section 8.3 we know that, if α is a root of $v(x)$, then $v(x)$ is a multiple of $M_\alpha(x)$. This proves the first part of the theorem.

To prove the second part, let $v(x) = u(x)M_\alpha(x)$ for some polynomial $u(x)$, and let the degree of $v(x)$ be less than n. Then $v(\alpha) = 0$, because $M_\alpha(\alpha) = 0$. Then

$$\sum_{i=0}^{n-1} v_i \alpha^i = 0,$$

and the n-tuple v is in the null space of H. $\qquad\square$

Now we are prepared to encode and decode Hamming codes with primitive shift-registers. We treat the encoding problem first. For this problem, we wish to find a one-to-one mapping between the set of binary k-tuples, where $k = n - r = 2^r - 1 - r$, and the set of binary n-tuples in the null space of a matrix H of the form in Eq. 9.6.1. For our purposes, it is convenient to deal exclusively with polynomials over $GF(2)$ instead of n-tuples over $GF(2)$. Thus, in the remainder of this section, we associate the n-tuple $v = \langle v_0, v_1, \ldots, v_{n-1} \rangle$ with the polynomial $v(x) = \Sigma_i v_i x^i$, as we have done in the preceding proofs.

Let $u(x)$ be a polynomial of degree less than k over $GF(2)$ that we wish to encode as a polynomial $v(x)$ of degree less than n. We say that $u(x)$ is the *information polynomial* and that $v(x)$ is the *code polynomial* for $u(x)$. From the preceding theorem, $v(x)$ must be a multiple of $M_\alpha(x)$. Because α is a primitive element of $GF(2^r)$, we know from Section 8.3 that $M_\alpha(x)$ is a polynomial of degree r. (The roots of an irreducible polynomial of degree r over $GF(2)$ belong to $GF(2^r)$ and to no smaller field.) From this we have $r \leq deg[v(x)] \leq n - 1$. These are precisely 2^{n-r} polynomials over $GF(2)$ with degrees in this interval, and there are precisely $2^k = 2^{n-r}$ distinct information polynomials over $GF(2)$ for this code. Therefore we can construct a one-to-one map between information polynomials and code polynomials.

We choose a construction technique that encodes $u(x)$ as a polynomial $v(x)$ whose k high-order coefficients are the coefficients of $u(x)$, and whose r low-order coefficients are essentially check digits. This choice makes decoding especially simple in the absence of errors. Thus, the code polynomial for $u(x)$ is $v(x) = u(x)x^r + c(x)$, where $c(x)$ is a *check polynomial* and has degree less than r. Because the code for $u(x)$ must be a multiple of $M_\alpha(x)$, we have

$$u(x)x^r + c(x) \equiv 0 \bmod M_\alpha(x). \qquad (9.6.2)$$

From Euclid's division algorithm, we have

$$u(x)x^r = q(x)M_\alpha(x) + \left[(u(x)x^r) \bmod M_\alpha(x) \right] \qquad (9.6.3)$$

for some polynomial $q(x)$. Here we have expressed the remainder polynomial as $u(x)x^r \bmod M_\alpha(x)$. Recall that the remainder has degree less than r, and that both it and $q(x)$ are unique for a given $u(x)$ and $M_\alpha(x)$. From Eq. 9.6.3, we have

$$u(x)x^r - \left[u(x)x^r \bmod M_\alpha(x) \right] \equiv 0 \bmod M_\alpha(x). \qquad (9.6.4)$$

Because we assume that the coefficient field is $GF(2)$, we can change the minus sign in Eq. 9.6.4 to a plus sign. Comparing Eqs. 9.6.2 and 9.6.4, we see that $c(x) = [u(x)x^r \bmod M_\alpha(x)]$. Thus, to encode a polynomial $u(x)$, we must perform the following algorithm.

Encoding Algorithm.
Step 1. The $k = n - r$ high-order coefficients of $v(x)$ are equal to $u(x)x^r$.
Step 2. The r low-order coefficients of $v(x)$ are equal to
$$-[u(x)x^r \bmod M_\alpha(x)].$$
(For $GF(2)$, ignore the minus sign.)

Step 1 requires no special computation because no polynomial coefficients are modified. The nontrivial step is the second step, which we can perform with a primitive shift-register.

Fig. 9.6.2 shows a primitive shift-register divider. This register computes $[u(x)x^3 \bmod M_\alpha(x)]$ for $M_\alpha(x) = x^3 + x^2 + 1$. The three stages of the register are associated with the coefficients of 1, x, and x^2 as indicated in the figure. Because each shift of the register moves a coefficient to the next higher power of x, a shift corresponds to multiplication by x. When the coefficient of x^2 is multiplied by x, it is fed back to earlier stages as coefficients of x^2 and 1. Thus $x^3 = x^2 + 1$ in the register, and indeed $x^3 \equiv (x^2 + 1) \bmod (x^3 + x^2 + 1)$.

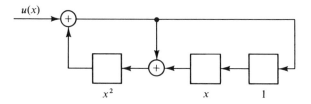

FIG. 9.6.2   A Register That Computes $[u(x)x^3 \bmod (x^3 + x^2 + 1)]$

To compute $[u(x)x^3 \bmod (x^3 + x^2 + 1)]$ with the register shown in Fig. 9.6.2, we clear the register, and then enter $u(x)$ one coefficient at a time, starting with the highest-order coefficient. The register shifts after each coefficient of $u(x)$ is presented to it. When all coefficients have been presented to the register and the last shift is completed, the register contains $[u(x)x^3 \bmod (x^3 + x^2 + 1)]$.

As an example, let $u(x) = x^3 + x$. The digits presented to the shift-register are $u_3 = 1$, $u_2 = 0$, $u_1 = 1$, and $u_0 = 0$, in that order. The states of the register after each shift are $x^2 + 1$, $x^2 + x + 1$, $x^2 + x$, and 1, respectively. From polynomial division, we find that
$$(x^3 + x)x^3 \equiv 1 \bmod (x^3 + x^2 + 1),$$
which agrees with the shift-register calculation. The proof of the next theorem shows why the shift-register calculates $[u(x)x^r \bmod M_\alpha(x)]$.

Theorem   Let $f(x)$ be a primitive polynomial of degree r over $GF(2)$. Construct a shift-register as shown in Fig. 9.6.2, such that the feedback line is connected to stage i if and only if the coefficient of x^i in $f(x)$ is 1. Let $u(x)$ be any polynomial over $GF(2)$. Then this shift register computes $[u(x)x^r \bmod f(x)]$ when $u(x)$ is entered one digit at a time, highest-order digit first, provided that the register initially is in state 0.

  Proof: The proof is by induction on the degree of $u(x)$.

  Basis step. Let $u(x)$ have degree 0. If $u(x) = 1$, then the shift-register contains $x^r \bmod f(x)$ after a shift; if $u(x) = 0$, it contains 0 after a shift. In either case, the register contains $[u(x)x^r \bmod f(x)]$.

  Induction step. We assume that the theorem is true for all polynomials $u(x)$ of degree n or less, and we show that it is true for all $u(x)$ of degree $n + 1$. Let $u(x)$ be a polynomial of degree $n + 1$ and let

$$w(x) = u_1 + u_2 x + \cdots + u_{n+1} x^n.$$

Then $u(x) = w(x)x + u_0$. This yields the following congruences:

$$\begin{aligned} u(x)x^r &\equiv [(w(x)x^{r+1} + u_0 x^r) \bmod f(x)] \\ &\equiv [([w(x)x^r \bmod f(x)]x + u_0 x^r) \bmod f(x)]. \end{aligned} \quad (9.6.5)$$

By the induction hypothesis, we can compute $w(x)x^r \bmod f(x)$ by entering the $n + 1$ coefficients of $w(x)$ to the shift register. Because a shift of the register is equivalent to a multiplication by x followed by a reduction modulo $f(x)$, when the register contains $w(x)x^r \bmod f(x)$ before a shift, it contains

$$w(x)x^{r+1} \bmod f(x) = [[w(x)x^r \bmod f(x)]x \bmod f(x)]$$

after the shift. Then from Eq. 9.6.5 we find that we can compute $u(x)x^r \bmod f(x)$ by entering the coefficient u_0 when the register contains $w(x)x^r \bmod f(x)$. The latter relation follows because the state of the register after a shift is the superposition of $[u_0 x^r \bmod f(x)]$ and $[w(x)x^{r+1} \bmod f(x)]$, which is what we require in Eq. 9.6.5. Thus we can compute $[u(x)x^r \bmod f(x)]$ by entering the coefficients of $u(x)$ to a register that is initially clear. $\square$

  In our previous example, we computed $(x^6 + x^4) \bmod f(x)$ for $f(x) = x^3 + x^2 + 1$. It is easy to verify by direct computation that the first few values of $x^i \bmod (x^3 + x^2 + 1)$ are those shown in Table 9.6.2. We have observed previously that the successive states of the register

TABLE 9.6.2   Some Values of $x^i \bmod (x^3 + x^2 + 1)$

i	$x^i \bmod f(x)$
3	$x^2 + 1$
4	$x^2 + x + 1$
5	$x + 1$
6	$x^2 + x$

while computing $(x^6 + x^4) \bmod f(x)$ are the states $x^2 + 1$, $x^2 + x + 1$, $x^2 + x$, and 1. Note that these states are $x^3 \bmod f(x)$, $x^4 \bmod f(x)$, $(x^5 + x^3) \bmod f(x)$, and $(x^6 + x^4) \bmod f(x)$, respectively, in agreement with the assertions in the proof of the theorem.

Now that we have shown how to encode an information polynomial $u(x)$ as a code polynomial $v(x)$, we turn to the problem of decoding $v(x)$ to obtain $u(x)$. Decoding turns out to be essentially identical to encoding. Because the leading k coefficients of $v(x)$ are identical to the k coefficients of $u(x)$, no special decoding is required to obtain the coefficients of $u(x)$ when no errors occur. However, if one or more errors occur, some special action is required.

Suppose that $v(x)$ is a code polynomial, and that it is received as $\hat{v}(x)$, where $\hat{v}(x)$ differs from $v(x)$ by one more errors. Formally, we write

$$\hat{v}(x) = v(x) + \varepsilon(x),$$

where $\varepsilon(x)$ is the *error polynomial* and has 1's in the error positions and 0's elsewhere. To determine if $\hat{v}(x)$ is a code polynomial, it is sufficient to determine if it is a multiple of $M_\alpha(x)$. That is, we compute $\hat{v}(x) \bmod M_\alpha(x)$, and we say that no error occurred if and only if $\hat{v}(x) \bmod M_\alpha(x) = 0$. Note that there are some error polynomials $\varepsilon(x)$ of weight 3 or more that change one code polynomial $v(x)$ into a different code polynomial $\hat{v}(x)$. When this happens, we cannot decode correctly because the error has exceeded the limits of the protection offered by the code.

We can quite easily make the calculation of $\hat{v}(x) \bmod M_\alpha(x)$. In fact, this is the same calculation that we used to encode $u(x)$, and we can use the register shown in Fig. 9.6.2 to find out if detectable errors have occurred. Recall that this register computes $[\hat{v}(x)x^r \bmod M_\alpha(x)]$ when we enter $\hat{v}(x)$ sequentially. But $\hat{v}(x)x^r$ is a multiple of $M_\alpha(x)$ if and only if $\hat{v}(x)$ is, because $M_\alpha(x)$ does not contain x as a factor. Hence we compute $[\hat{v}(x)x^r \bmod M_\alpha(x)]$ with the register shown in Fig. 9.6.2, and we say that no detectable error has occurred if the final state of the register is 0; otherwise we say that a detectable error has occurred. If the reader prefers to compute $\hat{v}(x) \bmod M_\alpha(x)$, he can verify that the register shown in Fig. 9.6.3 performs this operation. The two registers differ only in the position at which the input polynomial is entered.

Even though we choose to use a single-error correcting code, we need

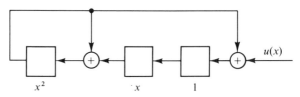

FIG. 9.6.3 A Register That Computes $[u(x) \bmod (x^3 + x^2 + 1)]$

not correct errors. We may prefer to signal the existence of an error when one or more are detected. In some systems this is a reasonable and valid method of operation. For example, under some circumstances in both computers and communication systems, it is possible to repeat an operation after error detection, with a high probability that the repeated operation is done without error.

If we use a Hamming code for error detection only, then two copies of the shift-register in Fig. 9.6.2 are sufficient for both encoding and decoding. For some computer systems, encoding and decoding are never done concurrently, so that a single shift-register suffices as both encoder and decoder.

If we choose to correct single errors, then we must increase the complexity of the decoder. Fortunately, the necessary increase in complexity is not a very large one. One method of error correction requires one other copy of the shift-register used for encoding and decoding.

Because the received polynomial is $\hat{v}(x) = v(x) + \varepsilon(x)$ and because $v(\alpha) = 0$, we have

$$\hat{v}(\alpha) = v(\alpha) + \varepsilon(\alpha) = \varepsilon(\alpha).$$

For error correction, we assume that $\varepsilon(x)$ has weight 1. That is, $\varepsilon(x) = x^i$ for some i, and $\hat{v}(x)$ is in error in position i. Thus

$$\hat{v}(\alpha) = \varepsilon(\alpha) = \alpha^i.$$

From this relation we can construct the following error-correction algorithm.

Algorithm for Error Correction.
Step 1. Compute $\hat{v}(\alpha)$.
Step 2. If $\hat{v}(\alpha) = 0$, indicate that no error has occurred, and terminate.
Step 3. If $\hat{v}(\alpha) = \alpha^i$, then set $v(x) = \hat{v}(x) + x^i$, and terminate.

This algorithm involves two nontrivial tasks. First, we must evaluate the polynomial $\hat{v}(x)$ for $x = \alpha$, which requires the ability to do arithmetic in $GF(2^r)$, the field containing α. Second, given the field element α^i, we must find i. In most computer systems, $r \le 6$, so that the second operation requires only a table of size $2^6 = 64$ or less. It is rather interesting to discover that the shift register in Fig. 9.6.3 can perform both Step 1 and Step 3. In fact, the decoding operation we describe previously leaves the polynomial representation of $\hat{v}(\alpha) = \alpha^i$ in this shift register while evaluating $\hat{v}(x) \bmod M_\alpha(x)$. We prove this somewhat surprising fact in the next lemma. After establishing this result, the only computation left to describe is the computation of i from α^i.

The following lemma establishes that the decoder in Fig. 9.6.3 computes $\hat{v}(\alpha)$.

Lemma Consider the representation of $GF(2^r)$ as a polynomial algebra modulo $M_\alpha(x)$, where α is a primitive element in the field. Then, for any polynomial $f(x)$ over $GF(2)$, the polynomial representation of $f(\alpha)$ is the polynomial $f(x) \bmod M_\alpha(x)$.

Proof: To prove this lemma, we use the Euclidian algorithm to find $q(x)$ and $c(x)$ such that $f(x) = q(x)M_\alpha(x) + c(x)$, where $0 \le deg[c(x)] < r$. Then $f(x) \bmod M_\alpha(x) = c(x)$, and $f(\alpha) = c(\alpha)$. In the representation of $GF(2^r)$ as a polynomial algebra modulo $M_\alpha(x)$, the element α is represented by the polynomial x. Then α^i is represented by x^i for each i in the interval $1 \le i \le r - 1$. Because

$$c(x) = \sum_{i=0}^{r-1} c_i x^i = \Sigma_i c_i x^i,$$

the polynomial representation of the field element $c(\alpha) = \Sigma_i c_i \alpha^i$ is the polynomial $\Sigma_i c_i x^i = c(x)$. Thus the polynomial representation of $f(\alpha)$ is $c(x)$, which is $f(x) \bmod M_\alpha(x)$. ☐

Because the register in Fig. 9.6.3 computes $\hat{v}(x) \bmod M_\alpha(x)$, the lemma indicates that it also computes the polynomial representation of α^i, when $\hat{v}(x)$ is in error in coordinate position i.

As an example of the lemma, let us return to the Hamming code defined by the polynomial $x^3 + x^2 + 1$. Let $u(x)$ be the information polynomial $x^2 + x + 1$, and we find that the code polynomial for $u(x)$ is

$$v(x) = u(x)x^3 + [u(x)x^3 \bmod M_\alpha(x)].$$

Because $[u(x)x^3 \bmod M_\alpha(x)] = 1$, the code polynomial $v(x)$ is

$$x^5 + x^4 + x^3 + 1.$$

Now suppose that an error occurs in the coefficient of x^4. In this case, the received polynomial $\hat{v}(x)$ is $x^5 + x^3 + 1$. We find

$$\hat{v}(x) \bmod M_\alpha(x) = x^2 + x + 1,$$

and Table 9.6.1 gives this as the representation of α^4, which indicates an error in $\hat{v}_4$.

The only decoding operation we have yet to describe is the computation of the error position i from the value of $\alpha^i = \hat{v}(a)$. We now show how to do this computation with a shift-register identical to the register in Fig. 9.6.3. The complete decoder, showing both primitive shift-registers, is illustrated in Fig. 9.6.4. The upper register computes $\hat{v}(\alpha) = \alpha^i$, and as each digit in $\hat{v}(x)$ enters the register, the digit is simultaneously stored in an n-stage buffer. When a complete code polynomial has been received and $\hat{v}(\alpha)$ has been computed in the upper register, the contents of the upper register are transferred to the lower register for the error-correction operation. In this way, the lower register can perform the error correction while the upper register decodes the next code polynomial. The lower register shifts once for each received digit, thereby multiplying the register contents by α. With each shift of the register, one digit of $\hat{v}(x)$ leaves the buffer, with the highest-order co-

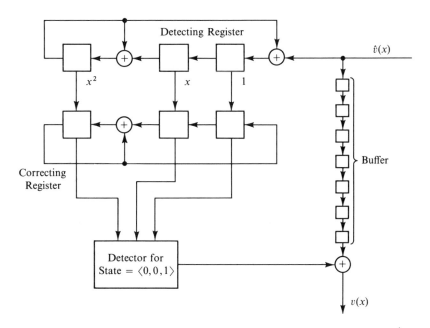

FIG. 9.6.4   An Error-Correcting Decoder for the Hamming Code Defined by
$M_\alpha(x) = x^3 + x^2 + 1$

efficient leaving first. If the initial state of the corrector register is α^i, then the register contains α^{i+j} after j shifts, at which time the coefficient of $\hat{v}_{n-j}$ leaves the buffer. The incorrect digit $\hat{v}_i$ leaves the buffer when $i = n - j$, and at this time the shift register contains $\alpha^{i+(n-i)} = \alpha^n = 1$. Thus we can detect when the incorrect digit is leaving the register by testing for the polynomial 1 in the error-corrector shift-register. The network shown in the figure does just this, and produces a signal that changes the incorrect digit as it leaves the n-stage buffer.

As an example of error correction, consider how the network in Fig. 9.6.4 corrects $\hat{v}(x) = x^5 + x^3 + 1$, which we investigated previously. The initial state of the error corrector is $x^2 + x + 1$, which is the representation of α^4. The new states of the register and the corresponding components of $\hat{v}(x)$ after each shift are shown in Table 9.6.3. When the polynomial 1 appears, the coefficient $\hat{v}_4$ is changed. Because

TABLE 9.6.3   States of the Corrector Register and the Corresponding Components of $\hat{v}(x)$ After Each Shift

State	Coefficient
$x + 1$	$\hat{v}_6$
$x^2 + x$	$\hat{v}_5$
1	$\hat{v}_4$

$\hat{v}(\alpha) = \alpha^4$, the error occurred in the coefficient $\hat{v}_4$ and the error correction is performed properly.

Summary We have seen how to implement Hamming codes with primitive shift-registers. We use a primitive shift-register in three ways in the system: one in the encoder and two in the decoder. It is particularly pleasing to find that encoding and decoding can be done relatively simply with shift registers.

The shift-register networks for the Hamming code were first described by Peterson (1961). Encoders for multiple-error correcting codes are identical to the Hamming encoder; only the feedback polynomial of the shift-register encoder is different. Similarly, we can do error detection for multiple-error correcting codes in the fashion indicated here. However, the operation of multiple-error correction is a good deal more complex than that of single-error correction. Berlekamp (1968) and Peterson and Weldon (1972) describe in detail several approaches to multiple-error correction.

Exercises

9.6.1 Prove that the shift register in Fig. 9.6.3 computes $u(x) \bmod M_\alpha(x)$. Prove that, if $u(x)$ is added to the input of the stage holding x^i, then the shift register computes $[u(x)x^i \bmod M_\alpha(x)]$.

$$H = \begin{bmatrix} 1 & \alpha & \alpha^2 & \alpha^3 & \alpha^4 & \cdots & \alpha^{2^r-2} \\ 1 & 1 & 1 & 1 & 1 & \cdots & 1 \end{bmatrix}$$

FIG. 9.6.5 Matrix H for Ex. 9.6.2

9.6.2 Consider the null space over $GF(2)$ of the matrix H shown in Fig. 9.6.5, where α is a primitive element of $GF(2^r)$. Prove that the null space is an error-correcting code that can correct any single error or any two adjacent errors.

9.6.3 Prove that the code vectors in Ex. 9.6.2 are in one-to-one correspondence with the polynomial multiples of $(x + 1)M_\alpha(x)$.

9.6.4 Find shift-register encoders and decoders for the code of Ex. 9.6.2 for $r = 3$ and $r = 4$. Assume that the decoders do error detection but do not do error correction.

9.6.5 Let $\hat{v}(x)$ be a received vector for the code of Ex. 9.6.2. Show how to compute $\hat{v}(\alpha)$ and $\hat{v}(1)$. For error correction, what action should a decoder take when $\hat{v}(1) = 1$? What action should the decoder take when $\hat{v}(1) = 0$? Show how to construct an error-correcting decoder for the code, using shift registers and a buffer in the style of Fig. 9.6.4. HINT: For double adjacent errors, $\hat{v}(\alpha) = \alpha^i + \alpha^{i+1} = \alpha^i(1 + \alpha) = \alpha^{i+j}$, where $\alpha^j = 1 + \alpha$. Construct a shift register that starts with α^{i+j} and multiplies by α until it reaches α^j.

9.6.6 Fig. 9.6.3 shows a primitive shift register with characteristic polynomial $M_\alpha(x)$. This shift register multiplies the polynomial representation of a field element by α. Construct a linear feedback shift-register that multiplies by α^2. Use $M_\alpha(x) = x^3 + x^2 + 1$ and $x^4 + x^3 + 1$.

9.6.7 Decode the following received code vectors for the code derived from $M_\alpha(x) = x^3 + x^2 + 1$. The high-order coefficients are on the right.
  (a) 0010111;
  (b) 0100111;
  (c) 1111100;
  (d) 1000001.

9.6.8 Decode the following received code vectors for the code derived from $M_\alpha(x) = x^4 + x^3 + 1$. The high-order coefficients are on the right.
  (a) 111101011001000;
  (b) 111100010011010;
  (c) 111100001110001;
  (d) 100000000000001.

9.6.9 Let $v(x)$ be a code polynomial in the code derived from the polynomial $M_\alpha(x)$. Show that every cyclic shift of $v(x)$ also is a code polynomial. That is, $\left[v(x)x^i \bmod (x^{2^r-1} - 1) \right]$ is a code vector if v is.

9.6.10 Ex. 9.6.9 indicates that code vectors lie on cycles. Find the cycle structure of the code vectors for the code derived from $M_\alpha(x) = x^3 + x^2 + 1$.

9.6.11 Find the cycle structure of the shift register over $GF(2)$ whose characteristic polynomial is
$$(x^7 - 1)/(x^3 + x^2 + 1) = (x^3 + x + 1)(x + 1).$$
Compare this result to that of Ex. 9.6.10.

9.6.12 Consider a register whose characteristic polynomial is $(x^{2^r-1} - 1)/M_\alpha(x)$ over $GF(2)$. Prove that every cycle produced by this register is in one-to-one correspondence with a cycle of code vectors in the code derived from $M_\alpha(x)$, and moreover that corresponding cycles have the same length.

TEN Boolean Algebra with Applications to Computer Design

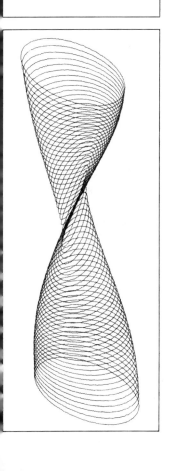

TEN | Boolean Algebra with Applications to Computer Design

Of all the algebraic structures discussed in this text, boolean algebra has had the greatest impact on computer design. We mention in Chapter One that Shannon (1938) and Shestakoff (1941) were the first to recognize the use of boolean algebra in this area. Research has progressed substantially since their time, to the point where the theory of computer design, sometimes known as *switching theory*, is a discipline in its own right, quite apart from its applications.

In this chapter we present the mathematical foundations of switching theory, and we discuss some of the applications of the mathematics to the design of computers. In the first section we investigate algebraic structures known as lattices, of which boolean algebras form one class. We discuss the various types of lattices and their relations to each other, but we omit the deeper properties of lattices.

In the second section we concentrate on boolean algebras, and we derive two major theoretical results. One is that every boolean algebra is isomorphic to an algebra of sets under union and intersection. The other is that every element of a boolean algebra has a normal-form representation as a sum of generators of the algebra. The latter result enables us to show that every switching function can be implemented by a network that contains only three different types of devices. This result also establishes a basis for techniques of logical design.

10.1 Lattices

In this section we investigate lattices and some algebraic structures derived from lattices. A lattice essentially is a partially ordered set in which each pair of elements has a greatest lower bound and least upper bound. Important classes of lattices are characterized by other properties such as distributivity, universal bounds, and complementation. The principle object of this section is to develop the background for boolean algebras, which are lattices having all three of these lattice properties.

Here we define various lattices as algebraic structures and show how they are related to each other. In Section 10.2 we concentrate on boolean algebras and relate the theory to computer design. We discover there that, for each n, the n-variable switching functions form a boolean algebra. Moreover, boolean algebra is a powerful tool for the design of networks to realize switching functions.

We begin our survey of lattices with a brief review of partially ordered sets. Later we find that a lattice is a partially ordered set with two compositions, + and ·, which are defined in terms of the partial-ordering relation.

Recall from Section 1.4 that a partial ordering on a set S is a relation on S that is reflexive, antisymmetric, and transitive. If $\geq$ is a partial ordering on the set S, then we say that the algebraic structure $\langle S, \geq \rangle$ is a *partially ordered set*. We use the symbol $\geq$ to indicate a partial ordering because the relation $\geq$ is a partial ordering on **Z**. Fig. 10.1.1 shows some graphs of partially ordered sets. Following the convention of Section 1.4, we omit edges in the graphs that are self-loops or that are implied by transitivity. To further simplify the graphs, we adopt the additional convention of drawing all edges so that they are directed downward; thus we are able to omit the arrowheads. In a graph for the partial-ordering relation $\geq$, an edge directed downward from node x to node y denotes $x \geq y$.

What property is common to all of the graphs in Fig. 10.1.1? We claim that each of these graphs is acyclic and, moreover, that the graph of every partial ordering is acyclic. Recall that a graph is *acyclic* if it contains no cycles other than self-loops, where the direction of the edges is taken into account in making this determination. For example, the graph shown in Fig. 10.1.1(*a*) is acyclic in its present form, but it becomes a cyclic graph if we reverse the edge between the top and bottom nodes of the graph.

The absence of cycles truly is the property that characterizes the graph of a partial ordering. Recall from Section 1.4 that the graph of every

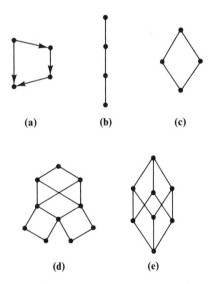

(a)   (b)   (c)

(d)   (e)

FIG. 10.1.1   Graphs of Partial Orderings

partial ordering is acyclic. Conversely, every acyclic graph is a graph of some partial ordering. This result follows because we can construct a partial ordering from an acyclic graph by including in the partial ordering all of the ordered pairs associated with the edges of the graph, as well as all other ordered pairs required for transitivity and reflexitivity.

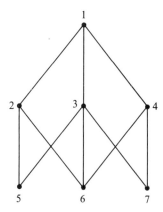

FIG. 10.1.2   Graph of a Partially Ordered Set

Now we take a closer look at partially ordered sets, and we find that some have special properties that lead to useful applications. In particular, we find that upper and lower bounds are of central importance. In the context of set theory, the union and intersection of two sets are upper and lower bounds, respectively, for the pair of sets. In the context of computer design, the upper and lower bounds for a pair of binary-valued signals are switching functions known as AND and OR, respectively. Here we formalize the notion of bounds, and from this formalization we define the algebraic structure of lattices.

Definition   Let $\langle X, \geq \rangle$ be a partially ordered set. Then y in X is said to be an *upper bound* for x in X if $y \geq x$. Similarly, x is said to be a *lower bound* for y.

Definition   For x, y, and z in a partially ordered set $\langle X, \geq \rangle$, we say that z is a *least upper bound* for x and y if $z \geq x$ and $z \geq y$ and if

$$((w \geq x) \wedge (w \geq y)) \Rightarrow (w \geq z)$$

for all w in X. We define *greatest lower bound* similarly.

Fig. 10.1.2 provides examples that clarify these definitions. The least upper bound of nodes 5 and 7 is node 3; node 1 is another upper bound but not a least upper bound. Node 4 is an upper bound for node 6, but not for node 5. The greatest lower bound for nodes 2 and 4 (in fact, the only lower bound for them) is node 6. Nodes 6 and 7 have no lower bound, and thus they have no greatest lower bound. Nodes 6 and 7 have three upper bounds (nodes 1, 3, and 4), but none of these is a least upper bound.

Note that, if a node has a least upper bound, then the least upper bound is unique. This follows because, if y and z are least upper bounds for x, then $y \geq z$ and $z \geq y$, so $y = z$.

From Fig. 10.1.2 we conclude that greatest lower and least upper bounds need not exist for every pair of elements in a partially ordered set. We are quite interested, however, in those algebraic structures for which these bounds do indeed exist everywhere. Clearly they exist for every pair in a totally ordered set, but there also exist partially ordered sets for which they exist everywhere. The partially ordered sets with this property are lattices.

Definition The algebraic structure $\langle X, \geq, +, \cdot \rangle$ is a *lattice* if

    (i) $\langle X, \geq \rangle$ is a partially ordered set,

    (ii) $+$ is a composition on X such that $x + y$ is the least upper bound for x and y, and

    (iii) $\cdot$ is a composition on X such that $x \cdot y$ is the greatest lower bound of x and y.

In this definition, note that we require the least upper and greatest lower bounds to be compositions defined for each pair of elements. This is equivalent to saying that the least upper and greatest lower bounds exist and are unique for each pair of elements in the lattice.

Fig. 10.1.3 shows some examples of graphs of lattices. Fig. 10.1.3(a) is the graph of the partial ordering of the integer divisors of 24 under the relation $x \mid y$. In this case the operation $x + y$ is $lcm(x, y)$ and $x \cdot y$ is $gcd(x, y)$. Fig. 10.1.3(b) shows the subsets of the set $\{a, b, c\}$ partially ordered by the relation $T \subseteq S$. Here $+$ is union and $\cdot$ is intersection. Fig. 10.1.3(c) shows the subgroups of the permutation group

$$\{e, (1\,2)(3\,4), (1\,3)(2\,4), (1\,4)(2\,3)\}$$

partially ordered by the relation "H is a subgroup of G." In this case $\cdot$ is group intersection, because the intersection of two subgroups is again a subgroup. The operation $+$ is not union, however, because the union of two subgroups is not necessarily a group (because it need not be closed under the group operation). Here we define the least upper bound of two subgroups H_1 and H_2 to be the smallest subgroup H containing both H_1 and H_2. From these examples, the reader can verify that the compositions $+$ and $\cdot$ are well defined for each of the partial-ordering relations $\mid$, $\subseteq$, and "is a subgroup of."

The compositions $+$ and $\cdot$ in a lattice automatically are associative and commutative, and every lattice element is an idempotent for both compositions.

Lemma Let $\langle X, \geq, +, \cdot \rangle$ be a lattice. Then

    (i) $+$ and $\cdot$ are associative;

    (ii) $+$ and $\cdot$ are commutative; and

    (iii) $x + x = x \cdot x = x$ for all x in X.

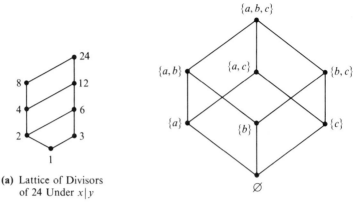

(a) Lattice of Divisors of 24 Under $x \mid y$

(b) Lattice of Subsets of $\{a, b, c\}$ Under $T \subseteq S$

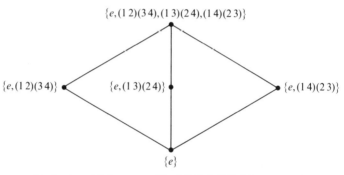

(c) Lattice of Subgroups of $\{e, (1\,2)(3\,4), (1\,3)(2\,4), (1\,4)(2\,3)\}$ Under "H Is a Subgroup of G"

FIG. 10.1.3   Examples of Graphs of Lattices

Proof: The proof is somewhat tedious. Here we prove only **(i)** as an example, and we leave the remainder of the proof as an exercise (Ex. 10.1.1).

To prove **(i)**, we must show that $x + (y + z) = (x + y) + z$ for all x, y, and z in X. We do this by showing that both $x + (y + z)$ and $(x + y) + z$ are least upper bounds for $(x + y)$ and $(y + z)$, and that therefore they are the same bound. We show here only that $x + (y + z)$ is a least upper bound; the other part of the proof follows similarly.

The method of proof involves the repeated use of the following fact: if w is an upper bound for x and y, then $w \geq x + y$, because $x + y$ is the least upper bound for x and y.

Because $x + (y + z) \geq x, y$, the preceding argument gives us $x + (y + z) \geq (x + y)$. Because $x + w \geq w$ for all w, we have

$$x + (y + z) \geq (y + z).$$

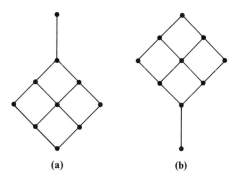

FIG. 10.1.4   A Lattice and Its Dual Lattice

Thus it follows from the preceding argument that

$$x + (y + z) \geq (x + y) + (y + z).$$

However, $(x + y) \geq x$, so that

$$(x + y) + (y + z) \geq x + (y + z) \geq (x + y) + (y + z).$$

Then we have $x + (y + z) = (x + y) + (y + z)$, which is what we set out to prove. A similar proof holds for the associativity of $\cdot$.   □

It is interesting to observe the similar characteristics of $+$ and $\cdot$ in a lattice. In working through the details of the proof of the preceding lemma, the reader will discover that the proofs of the stated properties for $+$ are identical to those for $\cdot$. This is not just a coincidence, but rather it is an essential characteristic that we use frequently.

Indeed, the operations $+$ and $\cdot$ are interchangeable, provided that we suitably modify the partial ordering $\geq$. Fig. 10.1.4 illustrates this point. The lattice in Fig. 10.1.4(*a*) reappears upside down in Fig. 10.1.4(*b*). It should be obvious that the composition $+$ on the lattice in Fig. 10.1.4(*a*) is isomorphic to the composition $\cdot$ on the lattice in Fig. 10.1.4(*b*). Clearly the partial ordering relation $\geq$ is not the same for the two lattices. In fact, if $\geq$ is the partial ordering for the first lattice, then $\leq$ is the partial ordering for the second.

We present the next definition to formalize this observation.

Definition   Let $\langle X, \geq \rangle$ be a partially ordered set. Then $\langle X, \leq \rangle$ is the partially ordered set obtained from $\langle X, \geq \rangle$ by reversing the order of the pairs in the relation $\geq$.

Note that changing $\geq$ to $\leq$ is equivalent to turning the graph of the lattice upside down. Next we show that every statement that holds for a lattice with $\geq$, $+$, and $\cdot$ also holds if we change $\geq$ to $\leq$ and interchange $+$ and $\cdot$. This result is called the *Principle of Duality*.

Theorem   If $\langle X, \geq, +, \cdot \rangle$ is a lattice, then $\langle X, \leq, \cdot, + \rangle$ is an isomorphic lattice.
       Proof: See Ex. 10.1.2.   □

Corollary   (*Principle of Duality*). Every statement about the lattice $\langle X, \geq, +, \cdot \rangle$ also is true of its *dual lattice* $\langle X, \leq, \cdot, + \rangle$.

As an example of the use of the Principle of Duality, consider the problem of proving that $\cdot$ is an associative composition for every lattice when we are given that $+$ is associative for every lattice. By duality, the composition $\cdot$ on a lattice is isomorphic to the composition $+$ on

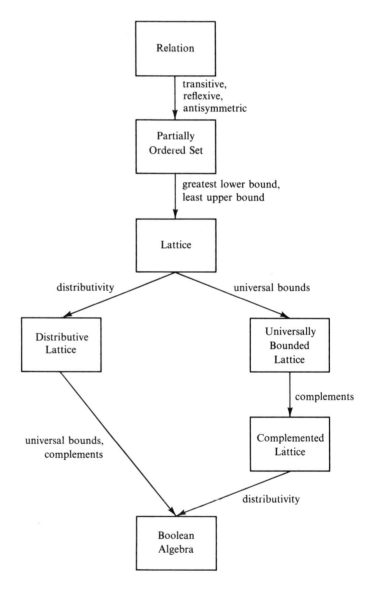

FIG. 10.1.5   Structures Derived from Lattices

the dual lattice. Because + is associative on the dual lattice, · is associative on the first lattice.

Because the major application of lattices to computer science is through boolean algebra, we quickly introduce three different properties by which we characterize some lattices, and we note that a boolean algebra is a lattice that possesses all three of these properties. The properties are distributivity, universal bounds, and complementation. Fig. 10.1.5 shows how we characterize lattices with these properties. Note that the graph in Fig. 10.1.5 is itself a lattice.

Definition   A lattice X is a *distributive lattice* if, for all x, y, and z in X,

$$x + (y \cdot z) = (x + y) \cdot (x + z),$$
$$x \cdot (y + z) = (x \cdot y) + (x \cdot z).$$

Fig. 10.1.6 shows examples of graphs for both distributive and non-distributive lattices. Each of the two graphs in Fig. 10.1.6(a) fails to satisfy $a \cdot (b + c) = (a \cdot b) + (a \cdot c)$. We claim that each of the graphs in Fig. 10.1.6(b) satisfies distributivity, but this claim is rather tedious to check. Fig. 10.1.6(a) helps us to perform this check, because a distributive lattice cannot contain either of the nondistributive lattices as a sublattice. It is rather surprising to find that every nondistributive lattice contains at least one of these two nondistributive lattices as a sublattice. Hence these two graphs in some sense characterize the nondistributive lattices.

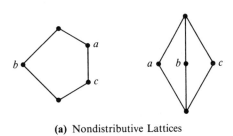

(a) Nondistributive Lattices

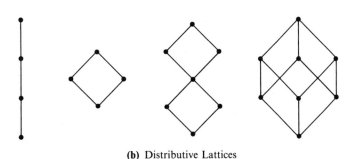

(b) Distributive Lattices

FIG. 10.1.6   Examples of Graphs for Nondistributive and Distributive Lattices

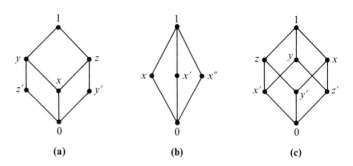

FIG. 10.1.7 Examples of Complements in Lattices

Definition A lattice X is a *universally bounded lattice* (or *bounded lattice*) if there exist elements 1 and 0 in X such that $1 \geq x$ and $x \geq 0$ for all x in X. We call the elements 1 and 0 the *universal upper bound* and the *universal lower bound*, respectively.

The existence of universal bounds is something that is quite easy to spot in graphs of lattices, for we need merely look for a topmost and a bottommost node. However, every finite lattice must have universal bounds, so this particular property automatically is satisfied in the applications of interest to us. Note that, for any finite or infinite lattice, if a universal upper bound exists, then it must be unique. By duality, the same observation holds as well for a universal lower bound.

Definition In a bounded lattice X, we say that an element x' is the *complement* of the element x if $x \cdot x' = 0$ and $x + x' = 1$.

Definition A bounded lattice X is a *complemented lattice* if, for each x in X, there is an x' in X such that x' is the complement of x.

The complementation property certainly does not hold for all bounded lattices. Fig. 10.1.7(a) shows a lattice in which the element x has no complement. Fig. 10.1.7(b) shows a lattice in which x has two complements, and Fig. 10.1.7(c) shows a lattice in which each element has a unique complement. Note that the latter lattice is distributive as well as complemented, so that it is a boolean algebra.

An element in a general lattice does not necessarily have a unique complement. However, if an element in a distributive lattice has a complement, then that complement is unique.

Lemma Let X be a bounded distributive lattice, and let x in X be an element with a complement. Then x has a unique complement in X.

Proof: Suppose that x has two complements, y_1 and y_2. Then $x \cdot y_1 = x \cdot y_2 = 0$ and $x + y_1 = x + y_2 = 1$. From these properties and distributivity, we have

$$y_1 = y_1 \cdot 1 = y_1 \cdot (x + y_2) = y_1 \cdot x + y_1 \cdot y_2 = 0 + y_1 \cdot y_2 = y_1 \cdot y_2.$$

Similarly, $y_2 = y_1 \cdot y_2$. Then $y_1 = y_2$. □

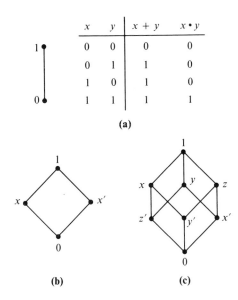

x	y	$x + y$	$x \cdot y$
0	0	0	0
0	1	1	0
1	0	1	0
1	1	1	1

(a)

(b)   **(c)**

FIG. 10.1.8   **Examples of Graphs for Boolean Algebras**

The lattice in Fig. 10.1.7(a) is a distributive lattice in which all of the elements except x have complements. These complements are unique.

Definition   A *boolean algebra* (or *boolean lattice*) is a complemented distributive lattice.

The graph in Fig. 10.1.7(c) is the graph of a boolean algebra. We sometimes denote the algebraic structure of a boolean algebra as a structure of the form $\langle X, \geq, +, \cdot, ', 1, 0 \rangle$, where the first four components give the structure of the underlying lattice, $'$ denotes complementation, and 1 and 0 are the universal bounds.

Fig. 10.1.8 shows some simple examples of boolean algebras. The simplest nondegenerate algebra is the algebra on the set $\{0, 1\}$. When we treat 0 and 1 as the truth values *false* and *true*, respectively, we see that the compositions $+$ and $\cdot$ in the algebra are isomorphic to the logical connectives $\vee$ ("or") and $\wedge$ ("and"), respectively. Of course, this isomorphism is well known, and it is the basis for the name *logic design* given to the practice of switching-circuit design.

One boolean algebra that we use frequently in this text is set algebra. Consider, for example, the subsets of a set X under union and intersection. The lattice for this algebra is $\langle \mathscr{P}(X), \supseteq, \cup, \cap \rangle$, where $\cup$ and $\cap$ are the operations $+$ and $\cdot$, respectively. The universal bounds 0 and 1 are the empty set $\varnothing$ and the set X. Complementation is set complementation, which takes each subset Y into $\overline{Y}$, its complement in X. Thus set algebra is the algebraic structure $\langle \mathscr{P}(X), \supseteq, \cup, \cap, \bar{\ }, \varnothing, X \rangle$. The dual boolean algebra is this algebra with $\supseteq$ replaced by $\subseteq$, with $\cup$ and

$\cap$ interchanged, and with $\emptyset$ and X interchanged. Hence the dual boolean algebra is the structure $\langle \mathscr{P}(X), \subseteq, \cap, \cup, ^-, X, \emptyset \rangle$. Many authors prefer to omit the partial-ordering relation from the algebraic structure notation, but we choose to include it here to show more clearly the lattice substructure of a boolean algebra. In the next section we discuss several interesting characteristics of boolean algebras.

Summary   In this section we introduce the notion of lattice and look at a few special types of lattices. A lattice is a partially ordered set in which each pair of elements has a unique least upper bound and a unique greatest lower bound. The operations of $+$ and $\cdot$ (that is, of least upper bound and greatest lower bound) in a lattice are associative and commutative, and every element is an idempotent with respect to both operations.

A lattice in which $+$ distributes over $\cdot$ and $\cdot$ distributes over $+$ is a distributive lattice. Not every lattice is distributive. Similarly, not every lattice contains a least or a greatest element; those that do are called universally bounded lattices. Every finite lattice is universally bounded. The complement of an element x in a universally bounded lattice is an element x' such that $x \cdot x' = 0$ and $x + x' = 1$, so that complementation is analogous to, but not identical to, the notion of inversion in semigroups. Complements are unique in a distributive lattice, but need not be unique in a nondistributive lattice.

A boolean algebra is a distributive lattice in which each element has a complement. Set algebra is a primary example of this structure, but another example of major interest is the algebra of switching functions, which we treat in the next section.

Birkhoff and MacLane (1962) give a more detailed treatment of lattices and boolean algebras. Texts by Hohn (1960) and Whitesitt (1961) discuss applications of boolean algebra. The work by Halmos (1967) is an advanced monograph on the set-theoretical aspects of boolean algebra.

Exercises

10.1.1   Prove that the following assertions hold for every lattice:
(a) $+$ and $\cdot$ are commutative;
(b) $x + (x \cdot y) = x$ and $x \cdot (x + y) = x$ for all x and y;
(c)   each element is an idempotent with respect to $+$ and $\cdot$;
(d)   if 0 and 1 are universal bounds, then $0' = 1$ and $1' = 0$;
(e)   for all x, if x' exists, then $(x')' = x$.

10.1.2   Prove that, if $\langle X, \geq, +, \cdot \rangle$ is a lattice, then $\langle X, \leq, \cdot, + \rangle$ is an isomorphic lattice.

10.1.3   Prove that every finite lattice has universal bounds.

10.1.4   A *sublattice* of a lattice X is a subset Y of X that is closed under $+$ and $\cdot$. Prove that the set of normal subgroups of a group $\langle G, \circ \rangle$ is a sublattice of the lattice of subgroups of G under the partial ordering "is a subgroup of." Prove that the least upper bound of two normal subgroups N_1 and N_2 of G is the product group

$$N_1 \circ N_2 = \{n_1 \circ n_2 | (n_1 \in N_1) \wedge (n_2 \in N_2)\}.$$

10.1.5 Let X be a lattice in which $+$ distributes over $\cdot$. Prove that $\cdot$ distributes over $+$.

10.1.6 Let X be a lattice with the partial ordering $\geq$. Prove that, for every x, y in X, we have $x \geq y$ if and only if $x + y = x$. What is the dual of this statement?

10.1.7 Let X be a complemented lattice. For all x and y in X, prove that, if x', y', and $(x + y)'$ exist, then $(x + y)' = x' \cdot y'$. This result is called De Morgan's Theorem.

10.1.8 Show the graph of the lattice of the subgroups of C_{24}, the cyclic group with 24 elements. Prove that this lattice is isomorphic to the lattice of the divisors of 24 under the partial ordering $x | y$.

10.1.9 Draw the graphs of the lattices of the subgroups of D_6 and D_8. What is the graph of the lattice D_{2n} when n is prime? What is the graph when $n = p_1 p_2$, where p_1 and p_2 are prime?

10.1.10 Prove that the partitions of a set form a lattice under the partial-ordering relation "refines." Recall that R_1 refines R_2 if and only if every pair of R_1-equivalent elements also is R_2-equivalent.

10.1.11 Prove that, in a distributive lattice, the subset of elements with complements forms a sublattice.

10.2 Boolean Algebras and Switching Functions

In this section we describe some of the major properties of boolean algebras, and we relate these results to the design of switching networks. For the algebraic structures studied in earlier chapters, we have paid a good deal of attention to the representation of the structures. Recall that groups, semigroups, and fields are representable by permutation groups, transformation semigroups, and polynomial algebras, respectively. In this section we discover that every finite boolean algebra is isomorphic to an algebra of sets under union and intersection. From this result it follows that every finite boolean algebra has 2^n elements for some n.

In deriving the representation of a boolean algebra of sets, we reveal another important property of boolean algebras. Every finite boolean algebra of order 2^n has n generators called atoms, and every element has a unique description as a sum of atoms. The sum-of-atoms description of an element is called its normal form.

The theory described here has had a substantial impact on the design of computers. Boolean algebra is the appropriate tool because the *n*-

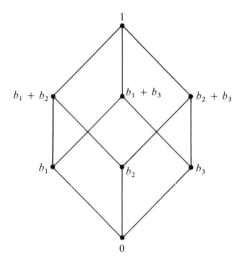

FIG. 10.2.1   An 8-Element Boolean Algebra

variable switching functions form a boolean algebra. The normal form of an *n*-variable switching function in this algebra corresponds in a one-to-one fashion to a network realization for the function, where the network building blocks are limited to three different types. Thus, with the aid of boolean algebra, we can translate a function specification directly into a modular switching network.

At the end of this section, we show that only one type of device is necessary to implement all switching functions. Indeed, combinational functions in computers normally are constructed entirely of such devices.

In the initial part of this section, our goal is to show that every boolean algebra is isomorphic to an algebra of sets. En route to our goal, we obtain the major result that every element in a boolean algebra has a unique description in terms of the generators of the algebra.

In seeking an isomorphism between a boolean algebra

$$\langle B, \geq, +, \cdot, ', 0, 1 \rangle$$

and an algebra of subsets of a set X, $\langle \mathscr{P}(X), \supseteq, \cup, \cap, ^-, \varnothing, X \rangle$, we proceed by identifying elements in B that correspond to the singleton sets $\{x_i\}$ for each x_i in X. We then find an isomorphism between the singleton sets and their correspondents in B that we can extend to an isomorphism of the entire structures $\mathscr{P}(X)$ and B.

The singleton sets in $\mathscr{P}(X)$ are generators of the algebraic structure because every element in the structure can be described completely in terms of these sets. The generators are called atoms, and the next definition establishes their essential properties.

Definition   An *atom* b in a boolean algebra B is a nonzero element such that either $b \cdot y = b$ or $b \cdot y = 0$ for every y in B.

Fig. 10.2.1 shows an 8-element boolean algebra with the atoms denoted as b_1, b_2, and b_3. Note that each of these elements is connected directly to the element 0. An atom cannot have an element y separating it from 0 in the graph, for then $b \cdot y = y \neq b$, contrary to the definition. Note that the singleton sets $\{x_i\}$ are atoms in the boolean algebra $\mathscr{P}(X)$ over $\cap$ and $\cup$.

We mention earlier that the atoms of an algebra are generators of the algebra. However, we can state a much stronger result. Not only is every element in a boolean algebra expressible in terms of the atoms, but the expression is unique for each element, to within the ordering of the terms of the expression. Later we show that this result enables us to construct networks for arbitrary switching functions using just three types of devices.

As a clue to the atomic description of an arbitrary element in the algebra on $\mathscr{P}(X)$, note that the set $\{x_{i_1}, x_{i_2}, \ldots, x_{i_m}\}$ is expressible as the union of atoms

$$\{x_{i_1}\} \cup \{x_{i_2}\} \cup \ldots \cup \{x_{i_m}\}.$$

The equivalent statement for the boolean algebra B is that every element in the algebra is representable as a sum of atoms of the form

$$b_{i_1} + b_{i_2} + \cdots + b_{i_m}.$$

The next two lemmas prepare the way for a proof of this result.

Lemma In a finite boolean algebra B, a sum of atoms $x = b_1 + b_2 + \cdots + b_m$ satisfies $x \geq y$ for an atom y if and only if $y = b_i$ for some i.

Proof: If $y = b_i$, then $x \geq b_i = y$, which proves the reverse implication of the lemma. To prove the forward implication, we assume that $x \geq y$, and we show that $y = b_i$ for some i. The product $y \cdot x$ is either 0 or y, because y is an atom. If the product is 0, then $0 = y \cdot x \geq y \cdot y = y$, and y must be 0. If the product is nonzero, then $y \cdot x = y$ because y is an atom. By distributivity, $y \cdot x = y \cdot b_1 + y \cdot b_2 + \cdots + y \cdot b_m$. Because each b_i is an atom and y is an atom, if $y \cdot b_i \neq 0$ for some b_i, then $y = y \cdot b_i = b_i$. Hence $y \cdot x$ is nonzero if and only if $y = b_i$ for some i. $\square$

Lemma Let x be an element of a finite boolean algebra B with atoms

$$b_1, b_2, \ldots, b_n.$$

Then $x \cdot b_i = 0$ for all i if and only if $x = 0$.

Proof: If $x = 0$, then $x \cdot b_i = 0$ for all i, which proves the reverse implication of the lemma. To prove the forward implication, let x be a nonzero element such that $x \cdot b_i = 0$ for all i Now find any atom y in the algebra with the property that $x \geq y$. Because the algebra is finite, we can always find such a y, where y may possibly be x itself. Because $x \geq y$, then $x \cdot y = y$, contrary to the hypothesis that $x \cdot b_i = 0$ for each atom b_i. This is a contradiction, which completes the proof. $\square$

Now we are ready to show that every element in a finite boolean algebra is uniquely representable as the sum of atoms.

Theorem   (*Normal-Form Theorem*). Let x be an element of a finite boolean algebra B with atoms $b_1, b_2, \ldots, b_n$. Then there is a sum of atoms $b_{i_1} + b_{i_2} + \cdots + b_{i_m}$ that is equal to x, and this sum is unique to within the ordering of the terms.

Proof: First we find the sum, and then we show that it is unique to within the ordering of the terms. Let $y = b_{i_1} + b_{i_2} + \cdots + b_{i_m}$, where an atom b_{i_j} appears in the sum if and only if $x \cdot b_{i_j} \neq 0$. We claim that $y = x$. Note that $x \geq y$ because x is an upper bound for each atom in the sum. Now we show that $y \geq x$.

To prove $y \geq x$, we prove the equivalent statement $x = x \cdot y$. Because $x = x \cdot 1 = x \cdot (y + y') = x \cdot y + x \cdot y'$, then $x = x \cdot y$ if $x \cdot y' = 0$. We prove that $x \cdot y' = 0$ with the aid of the preceding lemma by considering all products of the form $(x \cdot y') \cdot b_i$ for each atom b_i. The only atoms for which $x \cdot b_i$ is nonzero are those for which $y \geq b_i$, but for each of these we have $0 = y' \cdot y \geq y' \cdot b_i$, so that $(x \cdot y') \cdot b_i = 0$ for every atom b_i. Therefore, by the preceding lemma, $x \cdot y' = 0$. From this it follows that $x = y$, so x is indeed equal to the sum of atoms it bounds from above.

To prove that the sum of atoms for x is unique to within the ordering of the terms, suppose that there are two distinct sums for x:

$$x = b_{i_1} + b_{i_2} + \cdots + b_{i_m}$$

and

$$x = b_{j_1} + b_{j_2} + \cdots + b_{j_m}.$$

If the sums are different, then there is some atom, say b_{i_1}, that appears in one sum and not in the other. Then $x \cdot b_{i_1} = b_{i_1}$ when we use the first sum for x, and $x \cdot b_{i_1} = 0 \neq b_{i_1}$ when we use the second sum for x. This is a contradiction, so the sums must contain the same atoms. $\square$

The isomorphism that we seek is a direct consequence of the Normal-Form Theorem.

Theorem   A boolean algebra $\langle B, \geq, +, \cdot, ', 0, 1 \rangle$ with n atoms is isomorphic to every set algebra $\langle \mathscr{P}(X), \supseteq, \cup, \cap, {}^-, \varnothing, X \rangle$ for which $|X| = n$.

Proof: Let $f: B \to \mathscr{P}(X)$ be the mapping that carries each atom b_i into $f(b_i) = \{x_i\}$ in $\mathscr{P}(X)$. Let

$$f(b_{i_1} + b_{i_2} + \cdots + b_{i_m}) = \{x_{i_1}\} \cup \{x_{i_2}\} \cup \cdots \cup \{x_{i_m}\}.$$

By the Normal-Form Theorem, each element in B has a unique normal form, so that each element of B has a unique image in $\mathscr{P}(X)$, and f is one-to-one. Moreover, f preserves all of the properties of the boolean algebra. That is, for every y and z in B, $f(y + z) = f(y) \cup f(z)$, $f(y \cdot z) = f(y) \cap f(z)$, $f(y') = \overline{f(y)}$, and $y \geq z$ if and only if $f(y) \supseteq f(z)$, $f(0) = \varnothing$, and $f(1) = X$. Thus f is an isomorphism. $\square$

Corollary   Every finite boolean algebra has 2^n elements for some integer n.

   Proof: There are 2^n subsets in $\mathscr{P}(X)$ when $|X| = n$. $\square$

Corollary   All boolean algebras of order 2^n are isomorphic.

   Proof: Every boolean algebra of order 2^n is isomorphic to the set algebra on $\mathscr{P}(X)$ for $|X| = n$. $\square$

   Now we turn our attention to the application of these results to the design of computers. Recall that the signal lines in computers are binary-valued, so that the individual functional modules implement functions from $\{0, 1\}^n$ into $\{0, 1\}$. In the remainder of this section, we look into the problem of implementing combinatorial (or memoryless) n-variable switching functions. In Section 7.1 we show that it is possible to implement sequential functions from combinational networks and unit delays, so that the fundamental design problem is the implementation of combinational switching functions. We show presently that the n-variable switching functions form a boolean algebra, and then we use a modified version of the Normal-Form Theorem to show how to construct arbitrary switching functions.

   We deal with the boolean algebra of switching functions in terms of the 2-element boolean algebra on the set $\{0, 1\}$ with the compositions $+$ and $\cdot$ defined by

$$0 + 0 = 0, \quad 0 + 1 = 1 + 0 = 1 + 1 = 1,$$
$$1 \cdot 1 = 1, \text{ and } 0 \cdot 0 = 0 \cdot 1 = 1 \cdot 0 = 0.$$

Also, $0' = 1$ and $1' = 0$.

   Fig. 10.2.2 lists the set of functions from $\{0, 1\}^2$ into $\{0, 1\}$. Note that there are $2^{2^2} = 16$ functions in the set. Each function in the table is represented as a column of four elements drawn from the set $\{0, 1\}$. We label the functions so that the function whose column is $\langle i_0, i_1, i_2, i_3 \rangle^t$ is labeled f_i, where the binary representation of i is

$$i = \sum_{r=0}^{3} i_r 2^r.$$

Thus f_7 is the function whose column representation is $\langle 1, 1, 1, 0 \rangle^t$.

   In order to describe the functions in Fig. 10.2.2 by a boolean algebra, we must find operations on these functions that correspond to the operations $+$, $\cdot$, and $'$ in a boolean algebra. It is quite natural to define

x_1	x_2	f_0	f_1	f_2	f_3	f_4	f_5	f_6	f_7	f_8	f_9	f_{10}	f_{11}	f_{12}	f_{13}	f_{14}	f_{15}
0	0	0	1	0	1	0	1	0	1	0	1	0	1	0	1	0	1
0	1	0	0	1	1	0	0	1	1	0	0	1	1	0	0	1	1
1	0	0	0	0	0	1	1	1	1	0	0	0	0	1	1	1	1
1	1	0	0	0	0	0	0	0	0	1	1	1	1	1	1	1	1

FIG. 10.2.2   The Switching Functions of Two Variables

these operations in terms of the algebra on $\{0, 1\}$. For example, let $\odot$ be the operation on the two-variable switching functions that corresponds to $\cdot$ in a boolean algebra. Then for two functions f and g, we define $f \odot g$ as the function h such that $h(x_1, x_2) = f(x_1, x_2) \cdot g(x_1, x_2)$ for each pair $\langle x_1, x_2 \rangle$. Because both $f(x_1, x_2)$ and $g(x_1, x_2)$ lie in $\{0, 1\}$, the operation $f(x_1, x_2) \cdot g(x_1, x_2)$ is well defined. Because of the close relation between $\odot$ on the two-variable functions and $\cdot$ on $\{0, 1\}$ in this definition, we choose to denote both operations by the same notation $\cdot$, and we use the context of the expression to determine whether the operands of $\cdot$ are elements of $\{0, 1\}$ or are functions from $\{0, 1\}^2$ into $\{0, 1\}$. Following this line of reasoning, we define $f + g$ as the function $h(x_1, x_2) = f(x_1, x_2) + g(x_1, x_2)$, and we define f' to be the function $f'(x_1, x_2) = [f(x_1, x_2)]'$.

An example will clarify the details of these definitions. Consider the sum of the functions f_6 and f_{13} in Fig. 10.2.2. We compute $f_6 + f_{13}$ by computing

$$\langle 0, 1, 1, 0 \rangle^t + \langle 1, 0, 1, 1 \rangle^t = \langle (0 + 1), (1 + 0), (1 + 1), (0 + 1) \rangle^t$$
$$= \langle 1, 1, 1, 1 \rangle^t,$$

which is the column vector associated with f_{15}. Also observe that $f_3 + f_4 = f_7$, $f_3 + f_6 = f_7$, and $f_1 + f_7 = f_7$. As examples of the composition $\cdot$, we find that $f_6 \cdot f_1 = f_0$, $f_6 \cdot f_7 = f_6$, and $f_3 \cdot f_6 = f_2$. Similarly, $f_6' = f_9$, $f_0' = f_{15}$, and $f_7' = f_8$.

Theorem The set of n-variable switching functions is a boolean algebra of order 2^{2^n} under the operations $+$, $\cdot$, and $'$ on the column representations of the functions.

Proof: To prove the theorem, we show that the functions form a complemented distributive lattice with the partial ordering $\geq$, where $f \geq g$ if and only if $f \cdot g = g$. The details of the proof are tedious and uninstructive, so we only sketch the proof here. The structure is a lattice because the operations $+$ and $\cdot$ are defined uniquely for every pair of functions. The lattice is distributive because the underlying lattice on $\{0, 1\}$ is distributive. The universal bounds are f_0 and f_{2^n-1}. Finally, every function has a complement. $\square$

Because the n-variable switching functions form a boolean algebra of order 2^{2^n}, the algebra has 2^n atoms. What are these atoms? Inspection of Fig. 10.2.2 shows that the atoms there are the functions with precisely one 1 in their column representations. These are the functions f_{2^i} for $0 \leq i \leq n - 1$. Note that $f_{2^i} \cdot g = f_{2^i}$ or $f_{2^i} \cdot g = 0$ for each f_{2^i} and for each function g in the algebra of switching functions, so that each f_{2^i} satisfies the definition of atom. From the Normal-Form Theorem, we know that every function in the algebra has a unique expansion in terms of these functions. This observation leads to a result of far-reaching practical importance. Before we state this major result, we state a preparatory lemma.

Lemma Let f_j be an n-variable switching function, and let j have the binary expansion

$$j = \sum_{r=0}^{m-1} j_r 2^r,$$

where $m = 2^n$. Then f_j has the following expansion in terms of the atomic switching functions f_{2^i}:

$$f_j = \sum_{\substack{r=0, \\ j_r=1}}^{m-1} f_{2^r}.$$

In this case, the summation is a summation in the boolean algebra, and the terms in the summation are precisely those for which the coefficient j_r is unity.

Proof: For each r such that $0 \le r \le m - 1$, we have $f_j \cdot f_{2^r} = f_{2^r}$ if and only if $j_r = 1$. From the Normal-Form Theorem, for each atom b_r, we know that b_r is a term in normal form for f_j if and only if $f_j \cdot b_r = b_r$. Thus the given form for f_j is its normal form. $\square$

As an example of the lemma, we note from Fig. 10.2.2 that

$$f_7 = 1 \cdot f_1 + 1 \cdot f_2 + 1 \cdot f_4 + 0 \cdot f_8 = f_1 + f_2 + f_4.$$

This lemma establishes a general form for the implementation of switching-function networks. As a direct consequence of the lemma, we can state that it is possible to construct networks for any function f_j in the form shown in Fig. 10.2.3(a). The square modules in the network implement the atomic functions, and the outputs of these modules are summed in the adder. The sum is done in the boolean algebra on $\{0, 1\}$. Thus we can implement switching functions from modules that implement the atomic functions and a single boolean adder. Fig. 10.2.3(b) shows the implementation of f_7.

To achieve the central result of this discussion, we must show how to construct the networks for the atomic functions in Fig. 10.2.3. We shall discover that every switching function can be implemented by networks containing only three types of devices. To prepare for this result, we illustrate the three types of devices in Fig. 10.2.4. The OR gate produces a 1 output if any of its inputs is 1. Thus the OR gate computes the sum $x_1 + x_2 + \cdots + x_m$ of its m inputs. The OR gate takes its name from the logical connective $\vee$, because of the isomorphism between the two operations. The AND gate in Fig. 10.2.4 computes the boolean product of its n inputs. Thus it produces a 1 output only if all inputs are 1's. It takes its name from the logical connective $\wedge$. The NOT gate, which takes its name from the logical connective $\neg$, computes the complement of a signal. This type of gate sometimes is called an *inverter*. It has a single boolean-valued input.

For our discussion, we must make a detailed investigation of the atomic functions. We let $x_1, x_2, \ldots, x_n$ denote the n variables, and we denote the complement of a variable x_i as x_i'. A *minterm* is a product

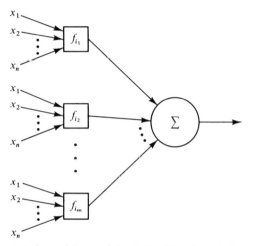

(a) General Form of the Network Implementation

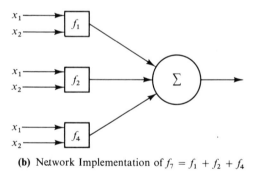

(b) Network Implementation of $f_7 = f_1 + f_2 + f_4$

FIG. 10.2.3   Network Implementations for Functions, Derived from the Normal Forms of the Functions

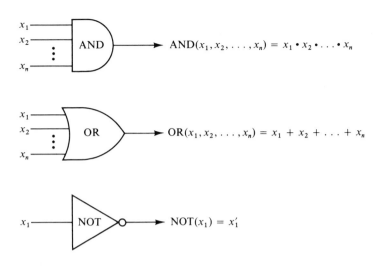

$$\text{AND}(x_1, x_2, \ldots, x_n) = x_1 \cdot x_2 \cdot \ldots \cdot x_n$$

$$\text{OR}(x_1, x_2, \ldots, x_n) = x_1 + x_2 + \ldots + x_n$$

$$\text{NOT}(x_1) = x_1'$$

FIG. 10.2.4   Three Types of Computer Devices Used in Switching-Function Networks

of the distinct variables or their complements. Thus $x_1 \cdot x_2 \cdot \ldots \cdot x_n$ and $x_1' \cdot x_2 \cdot x_3' \cdot x_4 \cdot x_5' \cdot \ldots \cdot x_n$ are minterms. Because there are n variables and each variable may be complemented or not, there are 2^n distinct minterms. Note that each of the minterms is an atom in the algebra of n-variable switching functions, because each minterm corresponds to a boolean function with exactly one 1 in its column representation.

The four atoms for the switching functions shown in Fig. 10.2.2 are the minterms $x_1' \cdot x_2' = f_1$, $x_1' \cdot x_2 = f_2$, $x_1 \cdot x_2' = f_4$, and $x_1 \cdot x_2 = f_8$. Now we are prepared to state a theorem that has major application to the construction of switching networks. It is called the *Disjunctive Normal-Form Theorem*.

Theorem   Every n-variable switching function has a unique representation as the boolean sum of a set of minterms. This representation is called the *disjunctive normal form*.

Proof: The proof follows immediately from the fact that each switching function in the algebra has a unique representation as a sum of atoms and, in this algebra, the atoms are the minterms.   □

As an example of this theorem, note that f_9 in Fig. 10.2.2 can be expressed as the sum $f_9 = x_1' \cdot x_2' + x_1 \cdot x_2$. A direct consequence of the theorem is that every switching function can be realized by a network of the form shown in Fig. 10.2.5(a). In the figure, each AND gate produces one minterm. The NOT gates at the inputs of the AND gates complement variables if they appear complemented in the corresponding minterm. The outputs of the AND gates are summed by the

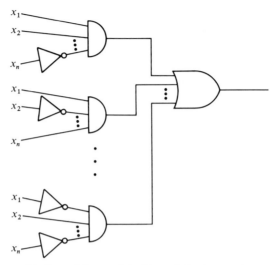

(a) General Form of the Network Implementation

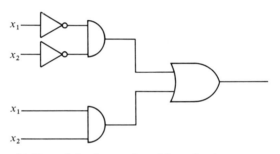

(b) Network Implementation of $f_9 = x_1' \cdot x_2' + x_1 \cdot x_2$

FIG. 10.2.5 Network Implementations for Functions, Derived from the Disjunctive Normal Forms of the Functions and Using the Three Types of Devices Shown in Fig. 10.2.4

OR gate. Thus the network implements a switching function by implementing it in its disjunctive normal form. In Fig. 10.2.5(b), we show the implementation of the function f_9.

It is clear from Fig. 10.2.5 that every function can be constructed solely from AND, OR, and NOT gates. This result is rather significant because it enables us to design and manufacture computers in a modular fashion. The three types of devices are our universal building blocks. Design and construction would be significantly more difficult if we needed special devices for the realization of each function in order to implement switching functions.

In practice, the implementation of switching functions requires even fewer distinct device types than we have indicated thus far. In fact, the

(a) The NOR Gate

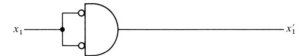

(b) A NOR-Gate Realization of a NOT Gate

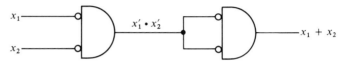

(c) A NOR-Gate Realization of an OR Gate

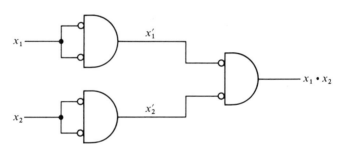

(d) A NOR-Gate Realization of an AND Gate

FIG. 10.2.6   The NOR Gate, and Realizations of NOT, OR, and AND Gates as Networks of NOR Gates

NOR gate, as shown in Fig. 10.2.6(a), is sufficient to implement every boolean function. The NOR gate produces the function $x_1' \cdot x_2' \cdot \ldots \cdot x_n'$. When $n = 2$, this is the function f_1 in Fig. 10.2.2.

Lemma   Every switching function can be implemented by a network of NOR gates.

   Proof: It is sufficient to show that we can construct AND, OR, and NOT gates with networks of NOR gates. This is shown in Fig. 10.2.6(b, c, d).   □

Summary   We show in this section that boolean algebra is isomorphic to set algebra, and that therefore every boolean algebra has 2^n elements. A fundamental result is the fact that every element in a boolean algebra is uniquely representable as a sum of atoms. This result in turn leads to techniques for implementing *n*-variable switching functions. Because the *n*-variable switching functions form a boolean algebra, they can be represented in normal form. The networks implement the functions in normal form.

It is unfair to leave the reader with the idea that switching functions are implemented in normal form. In fact, this is seldom the case. The existence of a normal form assures us that we can implement all functions with AND, OR, and NOT gates—or with NOR gates alone—but the normal form itself rarely is the best implementation of a function. It is usual design practice to use algebraic identities to change an algebraic expression from one form to an equivalent form, in order to construct a network with good or optimal properties. Among the useful identities are the following:

$$(x \cdot y)' = x' + y';$$
$$(x + y)' = x' \cdot y';$$
$$x \cdot y + x' \cdot z + y \cdot z = x \cdot y + x' \cdot z;$$
$$x \cdot (y + z) = x \cdot y + x \cdot z;$$
$$x + x' \cdot z = x + z;$$
$$x + x \cdot z = x.$$

Quine (1955) and McCluskey (1956) made one of the earliest contributions to the network-implementation problem. They investigated algorithms for finding the minimal number of AND gates required in a two-level network similar to those shown in Fig. 10.2.5. The Quine-McCluskey networks do not necessarily implement minterms at the outputs of the AND gates because the AND gates may have fewer than *n* variables. The Quine-McCluskey algorithm is quite suitable for functions of relatively few variables (approximately 10 or less), but it generally is computationally unattractive for functions of many variables.

After the work of Quine and McCluskey, much research centered on improvements to the minimization algorithm. Some researchers pursued the question of constructing optimal networks of NOR gates, with the work of Gimpel (1967) standing out as a notable contribution.

Device technology advanced considerably in the 1960s, to the extent that the criteria for optimal designs have changed drastically since the invention of the Quine-McCluskey algorithm. Thus design practices have changed rapidly, and today's designers dream up network implementations that undoubtedly would have horrified the designer of a decade ago. In the final analysis, logic design today is an art, and boolean algebra is an indispensable tool of the artist.

There are several excellent texts devoted to the subjects of switching theory and logic design. McCluskey (1965), Miller (1965), and Wood (1968), among others, cover thoroughly and lucidly the central points in switching theory. Harrison (1965) treats the theory in a more mathematical framework, and his text is recommended for the mathematically sophisticated reader. The material here on the representation of boolean algebras parallels his development of the topic. Booth (1971) and Bartee (1972) have written books oriented to the practice of computer design.

Exercises

10.2.1 Prove that the graph of a boolean algebra with 2^n elements is an n-dimensional cube.

10.2.2 A *symmetric switching function* of n variables is a switching function that is invariant under permutation of its n variables. For example, the 2-variable functions $x_1 + x_2$, $x_1 \cdot x_2$, and $x_1' \cdot x_2'$ are symmetric, but $x_1' + x_2$, $x_1 \cdot x_2'$, and $x_1' \cdot x_2$ are not. Prove that the symmetric functions are closed under $+$, $\cdot$, and complementation. Prove that the symmetric functions form a boolean algebra.

10.2.3 What is the order of the algebra found in Ex. 10.2.2? What are the atoms in this algebra?

10.2.4 The *dual* of an n-variable switching function $f(x_1, x_2, \ldots, x_n)$ is the function $f^D(x_1, x_2, \ldots, x_n)$ such that
$$f^D(x_1, x_2, \ldots, x_n) = \left[f(x_1', x_2', \ldots, x_n') \right]'.$$
A function f is said to be *self-dual* if $f^D = f$. Find all of the self-dual functions of two variables. Determine if these functions are closed under $+$, $\cdot$, and complementation.

10.2.5 Prove that the self-dual functions of n variables are closed under the operations $*$ and $\circ$ defined as follows, where x_i is one of the n variables:
$$f * g = x_1 \cdot f \cdot g + x_1 \cdot (f + g);$$
$$f \circ g = x_1' \cdot (f + g) + x_1 \cdot (f \cdot g).$$

10.2.6 Prove that the self-dual functions of n variables form a boolean algebra under the operations $*$ and $\circ$ (Ex. 10.2.5) and complementation. What is the partial ordering $\geq$ in this algebra? What are the atoms in this algebra? What is the order of the algebra?

10.2.7 Prove that every switching function can be implemented with only AND and NOT gates.

10.2.8 Use duality to prove that every switching function can be implemented using only gates that perform the function $f_7(x_1, x_2) = x_1' + x_2'$.

10.2.9 For the switching functions of three variables, write out the disjunctive normal forms of the functions f_{17}, f_{32}, f_{68}, and f_{134}.

10.2.10 Let f_i be a switching function of n variables. Prove that $f_i' = f_{2^n - 1 - i}$.

10.2.11 Draw the graph of the boolean algebra of 2-variable switching functions.

Bibliography

Abel, N. 1829. Mémoire sur une class particulière d'équations résolubles algé-
briquement. *Crelle* 4:131–156. Reprinted 1829. Also in *Oeuvres complètes de
Niels Henrik Abel*, ed. L. Sylow and S. Lie, vol. 1, pp. 478–507. Christiana:
Grøndahl and Son, 1881.

Arbib, M. A., ed. 1968. *Algebraic theory of machines, languages, and semigroups.*
New York: Academic Press.

_____. 1969. *Abstract theories of automata.* Englewood Cliffs, N.J.: Prentice-Hall.

Artin, E. 1959. *Galois theory.* Notre Dame Mathematical Lectures. Notre Dame,
Ind.: University of Notre Dame Press.

Avizienis, A. 1961. Signed digit number representations for fast parallel arithmetic. *IRE Trans. on Elec. Comp.* EC-10(3):389–400.

_____. 1964. Binary-compatible signed-digit arithmetic. *AFIPS, Proc. of the 1964
FJCC*, vol. 26, pp. 663–672. Baltimore, Md.: Spartan Books.

Bartee, T. C. 1972. *Digital computer fundamentals.* 3rd ed. New York: McGraw-
Hill.

Batcher, K. E. 1968. Sorting networks and their applications. *AFIPS, Proc. of
the 1968 SJCC*, vol. 32, pp. 307–314. Washington, D.C.: Thompson.

Berge, C. 1962. *The theory of graphs and its applications.* London: Methuen.

_____. 1971. *Principles of combinatorics.* New York: Academic Press.

Berlekamp, E. 1968. *Algebraic coding theory.* New York: McGraw-Hill.

Birkhoff, G., and S. MacLane. 1962. *A survey of modern algebra.* 2nd ed. New
York: Macmillan.

Booth, T. L. 1967. *Sequential machines and automata theory.* New York: John
Wiley and Sons.

_____. 1971. *Digital networks and computer systems.* New York: John Wiley
and Sons.

Borosh, I., and A. S. Fraenkel. 1966. Exact solutions of linear equations with
rational coefficients by congruence techniques. *Math. Comp.* 20:107–112.

Brown, D. T. 1960. Error detecting and error correcting codes for arithmetic
operations. *IRE Trans. on Elec. Comp.* EC-9(3):333–337.

Brzozowski, J. A. 1962. A survey of regular expressions and their applications.
IRE Trans. on Elec. Comp. EC-11(3):324–335.

Burnside, W. 1911. *Theory of groups of finite order.* Cambridge: Cambridge University Press. Reprinted 1955 by Dover, New York.

Carmichael, R. D. 1937. *Introduction to the theory of groups of finite order.* Boston:
Ginn. Reprinted 1956 by Dover, New York.

Cauchy, A. 1815. Sur le nombre des valeurs qu'une fonction peut acquérir.
J. École Polytechnique 17:1–37. Also in *Oeuvres complètes d'Augustin Cauchy*,
ser. 1, vol. 1, pp. 64–90. Paris: Gauthier-Villars, 1905.

_____. 1845. Papers on the theory of groups appearing in *Comptes Rendus* 21,
22 (1845–46), reprinted in *Oeuvres complètes d'Augustin Cauchy*, ser. 1, vols.
9–10, arts. 300 ff. Paris: Gauthier-Villars, 1906.

Cayley, A. 1854. On the theory of groups, as depending on the symbolic equation $\theta^n = 1$, parts I and II. *Phil. Mag.* 7:40–47, 408–409. Also in *The collected mathematical papers of Arthur Cayley*, vol. 2, pp. 123–132. Cambridge: Cambridge University Press, 1889.

———. 1878*a*. The theory of groups; graphical representations. *Am. J. of Math.* 1:174–176. Also in *The collected mathematical papers of Arthur Cayley*, vol. 10, pp. 403–405. Cambridge: Cambridge University Press, 1896.

———. 1878*b*. On the theory of groups. *Proc. London Math. Soc.* 9:126–133. Also in *The collected mathematical papers of Arthur Cayley*, vol. 10, pp. 324–332. Cambridge: Cambridge University Press, 1896.

———. 1889. On the theory of groups. *Am. J. Math.* 11:139–157. Also in *The collected mathematical papers of Arthur Cayley*, vol. 12, pp. 639–656. Cambridge: Cambridge University Press, 1897.

Chien, R. T. 1959. Orthogonal matrices, error-correcting codes, and load sharing matrix switches. *IRE Trans. on Elec. Comp.* EC-8(3):400.

Chien, R. T., and S. J. Hong. 1972. Error correction in high-speed arithmetic. *IEEE Trans. on Computers* C-21(5):433–438.

Copi, I., C. Elgot, and J. Wright. 1958. Realization of events by logical nets. *J. Assoc. Comp. Mach.* 5(2):181–196.

Cramér, H. 1945. *Mathematical methods of statistics*. Princeton, N.J.: Princeton University Press.

de Bruijn, N. G. 1959. Generalization of Pólya's fundamental theorem in enumerative combinatorial analysis. *Nederl. Akad. Wetensch. Proc. Ser. A, 62, Indag. Math.* 21:59–69.

———. 1964. Pólya's theory of counting. In *Applied combinatorial techniques*, ed. E. F. Beckenbach, pp. 144–184. New York: John Wiley and Sons.

DeRusso, P. M., R. J. Roy, and C. M. Close. 1965. *State variables for engineers*. New York: John Wiley and Sons.

Diamond, J. M. 1955. Checking codes for digital computers. *Proc. IRE* 43(4): 487–488.

Dickson, L. E. 1901. *Linear groups*. Leipzig: B. G. Teubner. Reprinted 1958 by Dover, New York.

Dyck, W. 1882. Gruppentheoretische studien. *Math. Ann.* 20:1–44.

Elspas, B. 1959. The theory of autonomous linear sequential networks. *IRE Trans. on Circuit Theory* CT-6(1):45–60. Also in Kautz 1965.

Frobenius, F. G. 1896. Über Gruppencharaktere. *Sitzungberichte der Akademie der Wissenschaften zu Berlin* 985–1021.

Gaal, L. 1971. *Classical Galois theory with examples*. Chicago: Markham.

Galois, É. 1830. Sur la théorie des nombres. *Bulletin des sciences mathématiques de Ferussac* 13:428–435. Also in *Écrits et memoires mathématiques d'Évariste Galois*, pp. 113–127. Paris: Gauthier-Villars, 1962.

———. 1832. Lettre de Galois à M. Auguste Chevalier. *Revue Encyclopédique* 55:568–576. Also in *Écrits et memoires mathématiques d'Évariste Galois*, pp. 173–185. Paris: Gauthier-Villars, 1962.

Garner, H. L. 1959. The residue number system. *IRE Trans. on Elec. Comp.* EC-8(2):140–147.

Gauss, K. F. 1876. *Werke*. Göttingen: Königliche Gesellschaft der Wissenschaften.

Gill, A. 1966. *Linear sequential circuits*. New York: McGraw-Hill.

Gimpel, J. F. 1967. The minimization of TANT networks. *IEEE Trans. on Elec. Comp.* EC-16(1):18–38.

Ginzburg, A. 1968. *Algebraic theory of automata*. New York: Academic Press.

Golomb, S. W. 1967. *Shift register sequences*. San Francisco: Holden-Day.

Golomb, S. W., et al. 1964. *Digital communications with space applications*. Englewood Cliffs, N.J.: Prentice-Hall.

Gries, D. 1971. *Compiler construction*. New York: John Wiley and Sons.

Grossman, I., and W. Magnus. 1964. *Groups and their graphs*. New York: Random House.

Hall, M. 1959. *Theory of groups*. New York: Macmillan.

Halmos, P. R. 1960. *Naive set theory*. Princeton, N.J.: Van Nostrand.

———. 1967. *Lectures on boolean algebras*. Princeton, N.J.: Van Nostrand.

Hamming, R. W. 1950. Error detecting and error correcting codes. *Bell Sys. Tech. J.* 29:147–160.

Harary, F. 1969. *Graph theory*. Reading, Mass.: Addison-Wesley.

Harrison, M. A. 1963. Combinatorial problems in boolean algebras and applications to the theory of switching. Unpublished doctoral thesis, University of Michigan, Ann Arbor, Mich.

———. 1965. *Introduction to switching and automata theory*. New York: McGraw-Hill.

———. 1969. *Lectures on linear sequential machines*. New York: Academic Press.

———. 1971. Counting theorems and their applications to switching functions. In *Recent developments in switching theory*, ed. A. Mukhopadhyay, pp. 85–120. New York: Academic Press.

Hartmanis, J., and R. E. Stearns. 1966. *Algebraic structure theory of sequential machines*. Englewood Cliffs, N.J.: Prentice-Hall.

Harvard 1951. *Synthesis of electronic computing and control circuits*, by the staff of the Harvard Computation Laboratory. Cambridge, Mass.: Harvard University Press.

Hellerman, L. 1963. A catalog of three-variable OR-invert and AND-invert logical circuits. *IEEE Trans. on Elec. Comp.* EC-12(3):198–223.

Hennie, F. C. 1968. *Finite-state models for logical machines*. New York: John Wiley and Sons.

Herstein, I. 1964. *Topics in algebra*. Waltham, Mass.: Xerox College Publishing.

Hohn, F. E. 1960. *Applied modern algebra*. New York: Macmillan.

Hopcroft, J. E. 1970. An $n \log n$ algorithm for minimizing states in a finite automaton. *Tech. Report CS-190*, Computer Science Dept., Stanford University, Stanford, Calif.

Hopcroft, J. E., and J. D. Ullman. 1969. *Formal languages and their relation to automata*. Reading, Mass.: Addison-Wesley.

Huffman, D. A. 1954. The synthesis of sequential switching circuits. *J. Franklin Inst.* 257:161–190, 275–303.

Jordan, C. 1870. *Traité des substitutions*. Paris: Gauthier-Villars. Reprinted 1957.

Kahn, D. 1967. *The codebreakers*. New York: Macmillan.

Kalman, R. E., P. L. Falb, and M. A. Arbib. 1969. *Topics in mathematical system theory*. New York: McGraw-Hill.

Kautz, W. H., ed. 1965. *Linear sequential switching circuits*. San Francisco: Holden-Day.

Kleene, S. C. 1956. Representation of events in nerve nets and finite automata. In *Automata studies*, ed. C. E. Shannon and J. McCarthy, pp. 3–42. Princeton, N.J.: Princeton University Press.

Knuth, D. E. 1968. *The art of computer programming*, vol. 1: *Fundamental algorithms*. Reading, Mass.: Addison-Wesley.

———. 1969. *The art of computer programming*, vol. 2: *Seminumerical algorithms*. Reading, Mass.: Addison-Wesley.

Krohn, K. B., and J. L. Rhodes. 1962. Algebraic theory of machines. *Proc. Symp. Math. Theory of Automata, Brooklyn, N.Y., 1962*, pp. 341–384. New York: John Wiley and Sons.

———. 1965. Results on finite semigroups derived from the algebraic theory of machines. *Proc. Nat. Acad. Sci. USA* 53:499–501.

Lagrange, J.-L. 1771. Réflexions sur la résolution algébrique des équations, III. *Mémoires de l'Académie des Sciences*. Also in *Oeuvres de Lagrange*, vol. 3, pp. 305–421. Paris: Gauthier-Villars, 1869.

Lechner, R. J. 1971. Harmonic analysis of switching functions. In *Recent developments in switching theory*, ed. A. Mukhopadhyay, pp. 122–228. New York: Academic Press.

Liu, C. L. 1968. *Introduction to combinatorial mathematics*. New York: McGraw-Hill.

McCluskey, E. J. 1956. Minimization of boolean functions. *Bell Sys. Tech. J.* 35(5):1417–1444.

———. 1965. *Introduction to the theory of switching circuits*. New York: McGraw-Hill.

MacLaren, M. D., and G. Marsaglia. 1965. Uniform random number generators. *J. Assoc. Comp. Mach.* 12:83–89.

Maisel, H., and G. Gnugnoli. 1972. *Simulation of discrete stochastic systems*. Chicago: Science Research Associates.

Massey, J. L. 1964. Survey of residue coding for arithmetic errors. *ICC Bull.* 3(4):195–209.

Massey, J. L., and O. N. Garcia. 1972. Error-correcting codes in computer arithmetic. *Advances in Information Systems Science* 4:273–326. New York: Plenum Press.

Mealy, G. H. 1955. A method for synthesizing sequential circuits. *Bell Sys. Tech. J.* 34:1045–1079.

Miller, R. E. 1965. *Switching theory*. New York: John Wiley and Sons.

Minnick, R. C., and J. L. Haynes. 1962. Magnetic core access switches. *IRE Trans. on Elec. Comp.* EC-11(3):352–368.

Moore, E. F. 1956. Gedanken experiments on sequential machines. In *Automata studies*, ed. C. E. Shannon and J. McCarthy, pp. 129–153. Princeton, N.J.: Princeton University Press.

Moore, E. H. 1893. A doubly-infinite system of simple groups. In *Mathematical papers, International Mathematical Congress of 1893*, pp. 208–242. New York: Macmillan. Abstracted in *Bull. Am. Math. Soc.* 3:73–78 (1893).

Myhill, J. 1957. Finite automata and the representation of events. *WADC Tech. Rpt. 57-624* (Wright-Patterson Air Force Base, Ohio).

Nelson, R. J. 1968. *Introduction to automata*. New York: John Wiley and Sons.

Nering, E. D. 1963. *Linear algebra and matrix theory*. New York: John Wiley and Sons.

Ninomiya, I. 1961. A study of the structures of boolean functions and its applications to the synthesis of switching circuits. *Mem. Faculty Eng. Nagoya Univ.* 13(2):149–363.

Ofman, U. 1962. On the algorithmic complexity of discrete functions. *Dokl. Akad. Nauk SSSR, Mathemat.* 145(1):48–51.

Oldham, I. B., R. T. Chien, and D. T. Tang. 1968. Error detection and correction in a photo-digital memory system. *IBM J. Res. and Dev.* 12(6):422–430.

Ott, G., and N. Feinstein. 1961. Design of sequential machines from their regular expressions. *J. Assoc. Comp. Mach.* 8(4):585–600.

Paull, M. C., and S. H. Unger. 1959. Minimizing the number of states in incompletely specified sequential switching functions. *IRE Trans. on Elec. Comp.* EC-8(3):356–367.

Pease, M. 1968. An adaptation of the fast Fourier transform for parallel processing. *J. Assoc. Comp. Mach.* 15(2):252–264.

Peterson, W. W. 1961. *Error-correcting codes*. Cambridge, Mass.: MIT Press.

Peterson, W. W., and E. J. Weldon, Jr. 1972. *Error-correcting codes*. 2nd ed. Cambridge, Mass.: MIT Press.

Pólya, G. 1937. Kombinatorische Anzahlbestimmungen für Gruppen, Graphen und chemische Verbindungen. *Acta Math.* 68:143–254.

Quine, W. V. 1955. A way to simplify truth functions. *Am. Math. Monthly* 62(9): 627–631.

Rabin, M. O., and D. Scott. 1959. Finite automata and their decision problems. *IBM J. Res. and Dev.* 3(2):114–125.

Rao, T. R. N., and O. N. Garcia. 1971. Cyclic and multiresidue codes for arithmetic operations. *IEEE Trans. on Info. Theory* IT-17(1):85–91.

Redfield, J. J. 1927. The theory of group reduced distributions. *Amer. J. Math.* 49:433–455.

Serret, J.-A. 1885. *Cours d'algèbre supérieure*. 5th ed. Vol. 2. Paris: Gauthier-Villars.

Shannon, C. E. 1938. A symbolic analysis of relay and switching circuits. *Trans. AIEE* 57:713–723.

_____. 1949. Communication theory of secrecy systems. *Bell Sys. Tech. J.* 28(4):656–715.

Shestakoff, V. I. 1941. Algébra dvuhpolúsnyh shém, postroénnyh iskúčitel' no iz dvuhpolúsnikov (Algébra Ashem). *Automatica i Telemekhanika* 2(6):15–24.

Slepian, D. 1953. On the number of symmetry types of boolean functions of n variables. *Can. J. Math.* 5(2):185–193.

_____. 1956. A class of binary signalling alphabets. *Bell Sys. Tech. J.* 35:203–234.

Spira, P. M. 1969. The time required for group multiplication. *J. Assoc. Comp. Mach.* 16(2):235–243.

Srinivasan, C. V. 1971. Codes for error correction in high-speed memory systems, I and II. *IEEE Trans. on Computers* C-20(8):882–888, C-20(12):1514–1520.

Stone, H. S. 1971. Parallel processing with the perfect shuffle. *IEEE Trans. on Computers* C-18(3):251–257.

_____. 1972. Dynamic memories with enhanced access. *IEEE Trans. on Computers* C-21(4):359–366.

Stone, H. S., and C. L. Jackson. 1969. Structures of the affine families of switching functions. *IEEE Trans. on Computers* C-18(3):251–257.

Suppes, P. 1957. *Introduction to logic*. Princeton, N.J.: Van Nostrand.

Svoboda, A., and M. Valach. 1955. Operátorové obvody. *Stroje Na Zpracování Informací* 3:247–295.

Szabo, N. S., and R. I. Tanaka. 1967. *Residue arithmetic and its application to computer technology*. New York: McGraw-Hill.

Tausworthe, R. C. 1964. Optimal ranging codes. *IEEE Trans. on Space Elec. and Telem.* SET-10:19–30.

_____. 1965. Random numbers generated by linear recurrence modulo two. *Math. Comp.* 19:201–209.

Titsworth, R. C. 1962. Correlation properties of cyclic sequences. Unpublished Ph.D. dissertation, California Institute of Technology, Pasadena, Calif.

Van der Waerden, B. L. 1931. *Modern algebra*. English trans. New York: Ungar.

Victor, W. K., R. Stevens, and S. W. Golomb. 1961. Radar exploration of Venus. *Jet Prop. Lab.* (Pasadena, Calif.) *Rept. No. 32-132*.

Warshall, S. 1962. A theorem on boolean matrices. *J. Assoc. Comp. Mach.* 9:11–12.

Watson, E. J. 1962. Primitive polynomials (mod 2). *Math. Comp.* 16:368–369.

Whitesitt, J. E. 1961. *Boolean algebra and its applications*. Reading, Mass.: Addison-Wesley.

Winograd, S. 1965. On the time required to perform addition. *J. Assoc. Comp. Mach.* 12(2):277–285.

_____. 1967. On the time required to perform multiplication. *J. Assoc. Comp. Mach.* 14(4):793–802.

Wood, P. E., Jr. 1968. *Switching theory*. New York: McGraw-Hill.

Zierler, N. 1955. Several binary-sequence generators. *MIT Lincoln Lab. Tech. Rept. No. 95*. Also in Kautz 1965.

_____. 1959. Linear recurring sequences. *J. SIAM* 7(1):31–48. Also in Kautz 1965.

INDEX

Abel, N. H., 54, 74, 93
Abelian group, 56, 97–99
Absurdity, 9
Access control mechanisms, 151–52
  for cyclic memory, 152
  for perfect shuffle memory, 157–61
Access time, 152–54, 160–61
Acyclic graph, 21
Addition
  checking of, 179–83
  modulo n, 44
  speed of, 138–47
Address register, 157
Adjacency matrix, 20
Algebra of polynomials, 269–75. *See also* Polynomials
Algebraic equations, solution of, 54, 103, 105–8
Algebraic expressions, 105–6
Algebraic structures, 45–52
  definition, 45
Alternating group, 101–2
A · n codes, 179–83
AND gate, 381
Antisymmetric relation, 24
Arbib, M. A., 103, 236
Arc, 20
A-register, 157
Arithmetic codes, 179–83
Arithmetic distance, 180
Arithmetic error model, 165–66
Arithmetic weight, 179
Artin, E., 288
Assembler, 204
Associated permutation, 327
Associative composition, 41
Associative composition of functions, 37–38
Associativity, group, 55
Atom, 376
Automorphism, 47
Autonomous machines, 296, 302–12
Avizienis, A., 146
Axioms, group, 55–64
Axioms, set, 11, 18

Bacon, Francis, 3
Bags, 12
Bartee, T. C., 387
Basis for vector space, 297
Batcher, K. E., 161
Berge, C., 21, 135
Berlekamp, E., 178, 360
Bijection, 32
Bilinearity, 265
Binary compositions. *See* Binary operations
Binary operations (compositions), 39–44
  associative, 41
  cancellability in, 42

Binary operations (compositions) (*cont'd*)
  commutative, 41
  definition, 39
  distributive, 41
Binary relations, 23. *See also* Relations
Binary representation, 138–39, 142–43
Birkhoff, G., 374
Block diagonal form, 310–11
Boolean algebra, 373–87
  definition, 373
  and switching functions, 375–87
Boolean lattice, 373. *See also* Boolean algebra
Booth, T. L., 255, 387
Borosh, I., 151
Bounded lattice, 372
Brown, D. T., 182, 183
Brzozowski, J., 255
Burnside, W., 73, 93, 116, 118
Burnside enumeration, 115–21. *See also* Pólya theory
  of enumeration
Burnside-Frobenius theorem, 118–21
Burst error model, 165–66

Cancellable, 42–43
  left, 195
Canonical feedback shift-register matrix. *See*
  Companion matrix
Canonical parity-check matrix, 175
Cardinality, 12
  of group, 56
Carmichael, R., 93
Carry propagation, 147–48
Cartesian product, 18, 60
Cascade decomposition, 228–34
Cauchy, A., 73, 75, 81, 93
Cayley, A., 73, 81, 93
Cayley diagram, 73
Cayley representation, 81
Ceiling function, 142
Center
  group, 94
  semigroup, 202
Characteristic
  of field, 274
  of integral domain, 267
Characteristic polynomial, 310
Check polynomial, 353
Check position, 175
Chien, R. T., 177, 178, 183
Chinese Remainder Theorem, 148–49
Circulating memories. *See* Dynamic memories
Circulating shift-register, 322
  and Pólya theory, 322–35
Cloak-and-dagger plots, 340
Clock, 206–7
Close, C. M., 296

Closure, group, 55
Closure, transitive, 29, 30
Code polynomial, 353
Code word, 166
Codes. *See* Error-detecting and -correcting codes;
    Group codes
Codomain, 30
Color groups, 73
Coloring problems. *See* Pólya theory of enumeration
Columbus, C., 3
Column major order, 161
Combinational network, 205
Communications systems, secure, 340–42
Commutative composition, 41
Commutative group, 56, 97–99
Commutative ring, 262
Commutative semigroup, 187
Commutator, 94
Companion matrix, 310–11
Comparison operation, 146, 151
Compilers, 192, 204–5, 238–39, 251–54
Complemented lattice, 372
Complement in lattice, 372
Complement, relative, 13
Complete graph, 20
Composition of functions, 35–39
    associative, 37–38
    definition, 35
    left, 35–36
    right, 35–36
Composition of permutations, 76
Compositions, 41. *See also* Binary operations
Composition table, 40
    of group, 57–59
    of permutation group, 76, 78–79
Conditionally transient state, 215
Congruence, 48–49
Congruence modulo n, 28, 49
Conjugacy class, 120
Conjugate, 94, 119–20
Conjugate sets, 279–80
Connectives, logical, 4
Constant function, 126
Contingency, 9
Contrapositive proofs, 6–7
Control register, 157
Copi, I., 254
Coset leader, 176
Coset partition, 83
Coset relation, 83–86
Cosets, 82–95
    definition, 82
    left, 82
    of normal subgroup, 85
    right, 82
Cover, 234, 238
Cramér, H., 167
C-register, 157
Cryptography, 340
Cycle-and-add property, 321, 346
Cycle (graph), 21
Cycle index polynomials, 126–30
    definition, 127
Cycle (permutation), 75
Cycle-set multiplication, 319–20
Cycle structure, 120–21
Cycle structure of matrix, 305–9
Cyclic groups, 95–97
Cyclic memories. *See* Dynamic memories

de Bruijn, N. G., 135
Decoder, Hamming code, 356–60
    algorithm for, 357
Decoding group codes, 176–78
Decoding rules, 168–69
Decomposition, machine
    cascade, 228–34
    parallel, 236
Defining relations, 70–73
    definition, 70
    use in graph construction, 71–73
Degree of permutation, 75
De Morgan's Theorem, 375
Dependent vectors, 297
DeRusso, P. M., 296
Diamond, J. M., 183
Dickson, L. E., 288
Dihedral groups, 99–101
Dimension of vector space, 297
Direct product, 60
    of cyclic groups, 97–99, 101
Disjunctive normal form, 383
    theorem, 383
Distance
    arithmetic, 170
    Hamming, 169–70
    minimum, 170
Distributive composition, 41
Distributive lattice, 371
Divisors of zero, 262
Domain, 30
Duality, 369–70
Dual lattice, 370
Dual of function, 387
Dyck, W., 73
Dynamic memories, 151–61
    access control mechanisms, 151–52
    access time, 152–54, 160–61
    interconnections for, 152–57
    perfect shuffle in, 156–61

Edge, 20
Eight queens problem, 111
Element, 11
Elementary symmetric function, 107
Elgot, C., 254
Elspas, B., 311, 320
Emerson, R. W., 9
Empty set, 14
Encoder, Hamming code, 350–56
    algorithm for, 354
Encoding group codes, 174–75
Endomorphism, 47
Enumeration
    Burnside method of, 115–21
    over conjugacy classes, 120
    Pólya theory of, 114–36
Enumerator, 121
Epimorphism, 47
Equality of sets, 11
Equivalence
    in Burnside enumeration, 115–16
    of functions, 124–25
    of generator relations, 70
    of states, 216
    of transition systems, 246–47
Equivalence classes, 27
    counting of (*see* Pólya theory of enumeration)

Equivalence partition, 27–28. *See also* Quotient structure
  refinement of, 217
Equivalence relation, 23, 27–28
Error-detecting and -correcting codes, 164–83. *See also* Group codes
  arithmetic, 179–83
  detection vs. correction in, 166–68
  error models for, 164–65
  Hamming, 176
    decoder and encoder for, 350–61
  parity, 166–78
  in radar ranging, 344–48
  simple (triplication), 166–69
Error n-tuple, 176
Error polynomial, 356
Errors, 164–66
  arithmetic, 165–66
  burst, 166
  independent, 165
Euclid, 55
Euclidean division theorem for polynomials, 269–70
Extension field, 271
Euler's totient function, 104–5
Even permutation, 101
Exchange-shuffle, 155–56. *See also* Perfect shuffle
Existential quantifier, 15–16
Exponent (group), 98

Factor group, 91
Failures. *See* Errors
Falb, P. L., 236
Faults. *See* Errors
Feedback circuit, 205–6
Feedback shift register. *See* Linear feedback
Feinstein, N., 255
Fields, 262–64
  finite (*see* Finite fields)
  skew, 355
Final state, 240
Finite fields, 267–76
  multiplicative group of, 277–78
  of order p, 267–68
  of order p^n, 271–94
  representation of, 290–94
  structure of, 276–89
Finite-state machines, 204–57
  definition, 208
  homomorphisms of, 223–28
  Mealy model of, 212
  Moore model of, 212
  and semigroups, 205–13
  semigroups of, 211
  as sequence recognizers, 238–57
  simulation of, 223–28
  state reduction in, 214–23 (*see also* Linear finite-state machines)
Floor function, 142
Fraenkel, A. S., 151
Free monoid, 193
Free semigroup, 191–93
Free variable, 6
French Academy of Sciences, 81
Frobenius, F. G., 116, 118, 135
Functions, 30–35
  composition of, 35–39
  constant, 126
  definition, 30
  equivalence of, 124–25

Functions (*cont'd*)
  inventories of, 121–26
  inverse, 31, 38
  set of all, 33
  weight of, 122
Fundamental theorem of group homomorphisms, 92

Gaal, L., 288
Galois, É., 54–55, 81, 93, 95, 288
Galois fields, 268. *See also* Finite fields
Garcia, O. N., 182, 183
Garner, H. L., 147
Gauss, K. F., 74, 93, 103
Generator relations, 64–74
  definition, 68
  properties, 68–70 (*see also* Defining relations)
Generators, 64–74
  definition, 65
  semigroup, 190–92
Gill, A., 301
Gimpel, J. F., 386
Ginzburg, A., 236
Gnugnoli, G., 349
Golumb, S., 349
Graphs, 20–23
  acyclic, 21
  complete, 20
  definition, 20
  group, 64–74, 86–89
  interconnection matrix of, 20
  of partial orderings, 25–26
  of relations, 23–24
  of semigroups, 191
  state, 209
  symmetries of, 108–10
  undirected, 21
Greatest lower bound, 30, 366
Gries, D., 255
Grossman, I., 73
Ground field, 271
Group codes, 171–83
  decoding, 176–78
  definition, 171
  encoding, 174–75 (*see also* Error-detecting and -correcting codes)
Group machine, 235
Groups, 54–112
  abelian, 56, 97–99
  alternating, 101–2
  axioms, 55–64
  basic properties of, 56–58
  commutative, 56, 97–99
  composition tables of, 57–59
  cyclic, 95–97
  definition, 55
  dihedral, 99–101
  graphs of, 64–74, 86–89
  homomorphisms of, 90–93
  order of, 56
  order of elements of, 79
  permutation, 74–81
  quaternion, 86–89
  regular representation of, 78
  simple, 102
  symmetric, 76, 101–3
  symmetry, 105–12 (*see also* Semigroups; Subgroups)

Hall, M., 93
Halmos, P. R., 18, 374

Hamming, R. W., 164, 169
Hamming code, 176
  decoder and encoder, 350–61
Hamming distance, 169–70
Hamming weight, 171
Harary, F., 21
Harrison, M. A., 135, 136, 255, 301, 333, 386
Hartmanis, J., 235
Harvard, 114, 135
Haynes, J. L., 178
Hellerman, L., 333
Hennie, F. C., 235, 255
Henry VIII, 4
Herstein, I., 93, 110, 288
Hohn, F. E., 374
Hollow validity, 10
Homomorphic image, 47
  of machine, 226
Homomorphisms, 46–52
  definition, 46
  group, 90–93
  machine, 223–28
  monoid, 198
  semigroup, 198–99
    composition of, 199–201
Hong, S. J., 183
Hopcroft, J. E., 192, 221
Huffman, D. A., 221

Ideal, 196
Idempotent, 42
Identical to within isomorphism, 47
Identity
  in composition, 41
  of group, 55
  of semigroup, 187
Image, 32
  homomorphic, 47
  of machine, 226
Implication, 4–7
Implication of generator relations, 69–70
Independent error model, 164
Independent vectors, 297
Index (group), 84
Induction, mathematical, 7–8
Infinite set, 14–15
Infix notation, 39
Information polynomial, 353
Information position, 175
Initial vertex, 20
Injection, 32
Integral domain, 262–63
Interconnection matrix, 20, 22–23
Interconnection of memories. See Dynamic memories
Intersection, 13
Into mapping, 32
Inventory of function, 121–26
  definition, 121
Inverse
  of composition, 42
  of function, 31, 38
  in group, 55
  left, 38
  right, 38
Inverter, 381
Irreducible polynomials, 271
  number of, 285–86
  table of, 273, 337
Isomorphism, 47

Jackson, C. L., 136
Jordan, C., 81, 93, 103, 288
Jordan-Holder theorem, 235

Kahn, D., 340, 349
Kalman, R. E., 236
Kautz, W. H., 301
k-equivalence, 216–20
Kernel, 92
Kleene, S. C., 242, 254
Kleene analysis theorem, 242–44
Kleene synthesis theorem, 245–51
Klein's four-group, 103
Knuth, D. E., 21, 151, 340, 349
Krohn, K. B., 212, 235
Krohn-Rhodes decomposition, 235–36

Labeling of graphs, 108–10. See also Pólya theory of enumeration
Lagrange, J., 74, 85, 93
Lagrange's theorem, 85, 93
Λ-transitions, 246
Lattices, 364–75
  boolean, 373
  bounded, 372
  complemented, 372
  definition, 367
  distributive, 371
  dual, 370
  universally bounded, 372 (see also Boolean algebra; Sublattice)
Least upper bound, 366
Lechner, R. J., 136
Left composition, 35–36
Left coset partition, 83
Left coset relation, 84
Left cosets, 82
Left ideal, 196
Left identity, 187
Left inverse, 38
Left regular representation, 78
Left regular transformational representation, 190
Left-simple, 197
Left zero, 188
Length of cycle, 75
Lexical analysis, 252–54
Linear associative algebra, 265
Linear combination, 297
Linear feedback shift-registers, 309–11
  canonical matrix for (see Companion matrix)
  circulating, 322
  cycle structure of, 313–21
  definition, 309
  as encoders and decoders, 350–61
  negacyclic, 334
  and Pólya theory, 322–35
  primitive, 335–50
Linear finite-state machines, 296–312
  autonomous, 296, 302–12
  definition, 297–98
  superposition in, 299–301 (see also Linear feedback shift-registers)
Linear function, 297
Linear transformation, 304
Liu, C. L., 135
Logical connectives, 4
Logical matrix multiplication, 22–23
Logic design, 377, 379–87
Logic devices, 39, 381, 385

Logic, elementary, 2–11
  use of, in proofs, 6–8
Logic functions. *See* Switching functions
Logic network, 16–17. *See also* Switching functions
Lower bound, 29–30, 366

McCluskey, E. J., 386
Machine decomposition, 223–38
  cascade, 228–34
  parallel, 236
Machine homomorphisms, 223–28
  definition, 226
Machine simulation, 223–28
  relation to machine homomorphisms, 227
MacLane, S., 374
MacLaren, M. D., 349
Magnitude comparison, 146, 151
  function, 147
Magnus, W., 73
Maisel, H., 349
Mapping, 31
Marsaglia, G., 349
Massey, J. L., 182, 183
Mathematical induction, 7–8
Matrix
  companion, 310–11
  cycle structure, 305–9
  nonsingular, 304
  similar, 306
Maximum-length sequence, 341
Maximum-likelihood estimation, 167, 170
Mealy, G. H., 212, 221
Mealy model, 212
Member, 11
Miller, R. E., 386
Minimum distance, 170
Minimum polynomial, 280
Minnick, R. C., 178
Minterm, 381–83
Modular representation, 148
Module, 264–65
  unitary, 265
Monic polynomial, 272
  period of, 313
Monoids, 187–95, 198
  definition, 187
  free, 193
  homomorphism, 198
Monomorphism, 47
Monte Carlo algorithms, 337
Moore, E. F., 212, 221
Moore, E. H., 288
Moore model, 212
Multiplication
  cycle-set, 319–20
  modulo n, 44
  speed of, 145
Multisets, 12
Myhill, J., 254

n-ary operations, 41
n-ary relation, 29
Negacyclic shift-register, 334
Nelson, R. J., 255
Nering, E. D., 296
Next-state function, 208
Nilpotent, 202
Ninomyia, I., 136
Node, 20

Nonsingular matrix, 304
NOR gate, 385
Normal form, 378
  theorem, 378 (*see also* Disjunctive normal form)
Normalizer, 94
Normal subgroups, 85–95
  definition, 85
NOT gate, 381
n queens problem, 111
n rooks problem, 112
n-tuple, 17–18
Null relation, 24
Null sequence, 239
Null set, 14
Null space, 173

Odd permutation, 101
Ofman, U., 146
Oldham, I. B., 177
One-to-one mapping, 32
One-to-one onto mapping, 32
Onto mapping, 32
Ordered pair, 18, 23–24
Ordering
  partial, 25–26, 365
  total, 26
  well, 30
Order of group, 56
Order of group element, 79
Order of permutation in group, 79
OR gate, 381
Ott, G., 255
Output function, 208
Overflow detection, 146, 151

Parallel decomposition, 236
Parity check, 173
  matrix, 173
    canonical, 174
Partially ordered set, 365
Partial ordering, 25–26
Partition, 27
  coset, 83
  equivalence, 27–28
Path, 20
Pattern, 125
  weight of, 125
Pattern inventory, 126
Paull, M. C., 221
Pease, M., 161
P-equivalence, 231
Perfect shuffle, 155–56
  in dynamic memories, 156–61
Period of monic polynomial, 313
Period of n-tuple, 323
Permutation groups, 74–81. *See also* Symmetric groups
  definition, 76
Permutations
  composition of, 76
  definition, 34, 57, 75
  degree of, 75
  even, 101
  identity, 75
  notation for, 75–76
  order of, in group, 79
  odd, 101 (*see also* Perfect shuffle)
Peterson, W. W., 177, 182, 360
Pigeonhole principle, 32–33
Pivot bit, 157

Pólya, G., 130, 134
Pólya theory of enumeration, 114–36. *See also*
    Burnside enumeration
    central theorem, 130–31
        use of, 131–34
    and circulating shift registers, 322–35
    historical background, 134–36
Polynomial equations, solution of, 54, 103, 105–8
Polynomials
    addition of, 269
    algebra of, 269–75
    characteristic, 310
    division of, 269–70
    irreducible, 271
    minimum, 280
    monic, 272
        period of, 313
    primitive, 294, 315
    reciprocal, 289
    roots of, 278–87
Power set, 13
    cardinality of, 14
Predicates, 6, 15–17
    evaluation of, 16–17
    quantification of, 15–16
Primitive element, 315
Primitive polynomials, 294, 315
Primitive shift registers, 335–37
    for radar ranging, 342–49
    for random number generation, 337–40
    for secure communications systems, 340–42
Products, 56
Product set, 94
Proof, 6–8
    by contradiction, 6–7
    contrapositive, 6–7
    inductive, 7–8
Proper ideal, 196
Proper subset, 12

Quantification, 15–16
Quaternion group, 86–89
Quine, W. V., 386
Quotient group, 91
Quotient structure, 85–95

Rabin, M. O., 254
Radar ranging systems, 342–49
Random-number generation, 337–40
Range, 30
Rao, T. R. N., 182
Reachable, 214
Read station, 151–53
Reciprocal polynomials, 289
Recognize, 239–40
Recognizers, 238–57
    in compilers, 251–54
Redfield, J. H., 130, 134
Redfield-Pólya theorem. *See* Pólya theory of
    enumeration, central theorem
Reduction of states. *See* State reduction
Refine, 217
Reflexive relation, 24
Regular expression, 240–41
Regular representation, 73
    and composition table, 73–74
    semigroup, 190
Regular set, 242–45
Relations, 23–30
    antisymmetric, 24

Relations (*cont'd*)
    binary, 23
    equivalence, 23, 27–28
    n-ary, 29
    reflexive, 24
    symmetric, 24
    transitive, 24 ·
    union and intersection of, 24
Relative complement, 13
Relatively prime integers, 64
Reset, 206, 211
Residue number system, 147–51
Residue representation, 148
Rhodes, J. L., 212, 235
Right composition, 35–36
Right coset partition, 83
Right coset relation, 84
Right cosets, 82
Right ideal, 196
Right identity, 187
Right inverse, 38
Right regular representation, 78
Right regular transformational representation, 190
Right-simple, 197
Right zero, 188
Rigid transformations, 100–101
Ring with identity, 262
Ring without divisors of zero, 262
Ring with unity, 265
Rings, 261–67
    commutative, 262
    definition, 261
Roots of polynomials, 278–87
Row major order, 161
Roy, R. J., 296

Scalars, 264
Scott, D., 254
Self-dual, 387
Self-inverse, 64
Self-loop, 21
Semigroups, 186–202
    commutative, 187
    definition, 187
    and finite-state machines, 205–13
    free, 191–93
    generators of, 190–92
    graphs of, 191
    homomorphisms of, 198–99
        composition of, 199–201
    of machines, 211
    simple, 197
Sequence, maximum length, 341
Sequence recognizers. *See* Recognizers
Sequence, well-defined, 10
Sequential machines. *See* Finite-state machines
Sequential network, 205
Serial correlation, 339
Serret, J. A., 93, 288
Set
    definition, 11
    difference, 13
    empty or null, 14
    infinite, 14–15
    power, 13
    theory, 11
Sets, 11–19
    equality of, 11
    from predicates, 15
    of sets, 12, 14–15

Sets (*cont'd*)
    well-known, 17
Shakespeare, William, 3
Shannon, C. E., 2, 349
Shestakoff, V. I., 2
Shift register. *See* Linear feedback shift-register
Sieve method, 272
Similar matrix, 306
Simple group, 102
Simple semigroup, 197
Simulation, machine, 223–28
Skew field, 262
Slepian, D., 135, 164
Solution of algebraic equations, 54, 103, 105–8
Spira, P. M., 145–46
S-register, 157
Srinivasan, C. V., 177
State, 208
    final, 240
    transient, 215
State cover, 234, 238
State equivalence, 214–23
    definition, 216
State graph, 209
Statement, 3
    compound, 3–5
State reduction, 214–23
    algorithm, 220–21
Stearns, R. E., 235
Stevens, R., 349
Stone, H. S., 136, 161
Strength of generator relations, 70
Strongly connected machine, 215
Structure-preserving maps, 45–52. *See also*
        Homomorphisms
Subgroups, 61–62
    definition, 61
    normal, 85–95
Sublattice, 374–75
Submonoids, 196–201
    definition, 196
Subsemigroups, 196–201
    definition, 196
Subset, 12
Substitution property, 230
Successor, 14
Suppes, P., 8, 18
Surjection, 32
Svoboda, A., 147
Switching functions, 114
    and boolean algebra, 375–87
    number of equivalence classes of, 114–16, 135,
        322–35
        algorithm for computing, 331–32
    symmetric, 387
Symmetric function, 107
Symmetric groups, 76, 101–3
Symmetric relation, 24
Symmetric switching function, 387
Symmetry groups, 105–12
Symmetry operations, 105–6
Syndrome, 177
Syntax analysis, 252–53
Szabo, N. S., 151

Tail machine, 229
Tanaka, R. I., 151
Tang, D. T., 177
Tausworthe, R. C., 349. *See also* Titsworth, R. C.
Tautology, 9

Terminal vertex, 20
Ternary operation, 41
Tic-tac-toe, three-dimensional, 111
Titsworth, R. C., 349. *See also* Tausworthe, R. C.
Total ordering, 26
Totient function, 104–5
Transformation, 189
Transformational representation, 190
Transient state, 215
Transition systems, 245–51
    definition, 245
Transitive closure, 29, 30
Transitive relation, 24
Transpositions, 101
Tree, 21

Ubiquitous element, 144
Ullman, J. D., 161, 192
Undirected graph, 21
Unger, S. H., 221
Union, 13
Unitary module, 265
Universal lower bound, 372
Universally bounded lattice, 372
Universal quantifier, 15–16
Universal upper bound, 372
Unordered pair, 13
Upper bound, 366

Valach, M., 147
Van der Waerden, B. L., 93, 110, 235, 288
Vectors, 264, 297
    dependent, 297
    independent, 297
Vector space, 265, 297
    basis for, 297
    dimension of, 297
Venn diagram, 13
Vertex, 20
Victor, W. K., 349

Warshall, S., 30
Watson, E. J., 337
Weight
    arithmetic, 179
    of code word (Hamming), 171
    of function, 122
    of pattern, 125
    of set of functions, 122
    of set of patterns, 126
Weldon, E. J., 177, 360
Well-defined operation, 49
Well-defined sequence, 10
Well ordering, 30
Whitesitt, J. E., 374
Winograd, S., 138, 144, 146, 147
Winograd's theorem, 138–46
Wood, P. E., Jr., 386
Wright, J., 254

Zermelo-Fraenkel axioms, 18
Zero
    for composition, 42
    divisors of, 262
    in ring, 261
    in semigroup, 188
0-input, 298
0-state, 298
Zierler, N., 311, 320

INDEX TO NOTATION

Notation	Explanation	Section
A	Matrix	9.2
A^{-1}	Inverse of A	9.2
$det(A)$	Determinant of A	9.2
A_n	Alternating group of order n	2.5.4
C	Complex numbers	1.2
C_n	Cyclic group of order n	2.5.1
$\delta_{g,s}$	Function which is 1 if $g(s) = s$	3.1
$\delta(s)$	Output function of machine	7.1
D_n	Dihedral group of order n	2.5.3
$deg[f(x)]$	Degree of $f(x)$	8.2
$D(x, y)$	Arithmetic distance	5.3
e	Identity element	1.7
ε	Error n-tuple	5.2.3
$\varepsilon(x)$	Error polynomial	9.6
$f(x)$	Function of x	1.5
f^{-1}	Inverse of f	1.5
f'	Complement of f	10.2
f^D	Dual of f	10.2
$f: X \rightarrow Y$	Function from X to Y	1.5
$f(X)$	Image of X under f	1.5
$\overline{f(x)}$	Reciprocal polynomial	8.3
G	Graph	1.3
G	Group	2.1
$|G|$	Order of group G	2.1
g	Element of G	2.1
g^{-1}	Inverse of g	2.1
$g \cdot H$	Left cosets of G relative to H	2.4
G/H	G modulo H	2.4
$[G:H]$	Index of H under G	2.4
gcd	Greatest common denominator	9.3
$GF(p)$	Galois field of order p	8.2
$H \cdot g$	Right cosets of G relative to H	2.4
H^t	Transpose of matrix H	5.2.2
H_s	Set of permutations that leave s fixed	3.1
$H(x, y)$	Hamming distance	5.2.2
I^*	Free monoid	7.4
I_g	Number of elements that are fixed by g	3.1
$I_p(n)$	Number of polynomials of degree n over $GF(p)$	8.3
I_r	$r \times r$ identity matrix	5.2.2
K_4	Klein's four-group	2.5.4
Λ	Null or empty sequence	6.1
lcm	Least common multiple	9.3
M	Interconnection matrix	1.3
$M^{(r)}$	Power of M by logical matrix multiplication	1.3
M	Matrix	4.3
$[M]_{i,j}$	i,jth element of M	4.3
M	Machine	7.1
$M_\alpha(x)$	Minimum polynomial of α	8.3

Notation	Explanation	Section
$\mathbf{N}$	Natural numbers	1.2
n^+	Successor of n	1.2
$N_p(n)$	Number of n-cycles	9.4
$\mathbf{P}$	Positive integers	1.2
$[P \mid I_r]$	Matrix concatenation	5.2.2
π	Permutation	3.2
$\phi_A(\lambda)$	Characteristic polynomial	9.2
$\phi_{(n)}$	Euler's totient function	2.5.4
$\varnothing$	Null set	1.2
$\mathscr{P}(S)$	Power set of S	1.2
$p(n)$	Predicate in n	1.1
$p(x)$	Predicate in x	1.2
$p(x)$	Polynomial	2.6
$\mathbf{Q}$	Rational numbers	1.2
$\mathbf{R}$	Real numbers	1.2
R	Relation	1.4
$[s]$	Set of elements equivalent to s	1.4
$[s]_P$	P-equivalence class containing s	7.3
$\lvert S \rvert$	Cardinality of S	1.2
$s/\delta(s)$	State/output	7.1
S_n	Symmetric group of degree n	2.3
$S_{n,r}$	Elementary symmetric function	2.6
V	Regular expression	7.4
$\lVert V \rVert$	Set of input sequences represented by V	7.4
V^*	Indefinite concatenation	7.4
$W(f)$	Weight of function f	3.2
$w(r)$	Weight of pattern r	3.2
$W(x)$	Weight of code word x	5.2.2
x^{-1}	Inverse of x	1.7
x'	Complement of x	10.2
$\lceil x \rceil$	Ceiling of x	4.1
$\lfloor x \rfloor$	Floor of x	4.1
X^+	Free semigroup	6.1
X^*	Free monoid	6.1
$\{x \mid p(x)\}$	Set definition using predicate	1.2
X/R	X modulo R	1.8
Y^X	Set of all functions from X to Y	1.5
z	Zero element	1.7
$\mathbf{Z}$	Integers	1.2
$\mathbf{Z}_n$	Integers modulo n	1.2
$\mathbf{Z}_n^*$	Direct-product decomposition of $\mathbf{Z}_n$	4.2
$Z_G(x_1, \ldots, x_m)$	Cycle index polynomial	3.3
$\wedge$	AND	1.1
$\vee$	OR	1.1
$\neg$	NOT	1.1
$\Rightarrow$	Implies	1.1
$\Leftrightarrow$	Equivalence	1.1
$\binom{n}{i}$	Binomial coefficient	1.1
$\{ \ldots \}$	Set brackets	1.2
$\in$	Member of set	1.2
$\subseteq$	Subset	1.2
$\subset$	Proper subset	1.2
$\cup$	Union	1.2
$\cap$	Intersection	1.2
$-$	Set difference	1.2
$\overline{T}$	Set complement	1.2
$\forall$	Universal quantifier	1.2
$\exists$	Existential quantifier	1.2

Notation	Explanation	Section
$\times$	Cartesian product	1.2
$\times$	Direct product	2.1
$\langle \ldots \rangle$	n-tuple	1.2
$\circ$	Composition of functions	1.6
$\circ$	Binary operation	1.7
$*$	Binary operation	1.7
$+_n$	Addition modulo n	1.7
$\cdot_n$	Multiplication modulo n	1.7
$\odot$	Binary operation	1.8
$\oplus$	Binary operation	1.8
$\oplus$	Addition	4.1
$\oplus$	Addition of residue representations	4.2
$\oplus$	Componentwise addition modulo-2	5.2.2
$\cdot$	Group composition	2.1
$\circ$	Group composition	2.1
$*$	Group composition	2.1
$\cong$	Isomorphism	2.5.1
$\begin{pmatrix} 1 & 2 & n \\ p(1) & p(2) \ldots & p(n) \end{pmatrix}$	Permutation	2.3
$(i\ p(i) \ldots p^{k-1}(i))$	Cycle	2.3
$(\ldots)(\ldots)\ldots(\ldots)$	Permutation	2.3

This book was designed by Beth Slye
set in 10/12 Times Roman with Univers display
by Holmes Typography of San Jose California
printed and bound by the Kingsport Press of Kingsport Tennessee
edited by Larry McCombs
sponsoring editor Stephen D. Mitchell
art by Bob Berns and John Foster
series designer Michael Rogondino
production editor George Oudyn